The University of Minnesota Press
gratefully acknowledges assistance provided
by the McKnight Foundation
for publication of this book.

Good Thinking

The Foundations of Probability and Its Applications

I. J. Good

UNIVERSITY OF MINNESOTA PRESS □ MINNEAPOLIS

Published by the University of Minnesota Press,
2037 University Avenue Southeast, Minneapolis, MN 55414
Printed in the United States of America.

Library of Congress Cataloging in Publication Data

Good, Irving John.
Good thinking.
Bibliography: p.
Includes index.
1. Probabilities. I. Title.
QA273.4.G66 519.2 81-24041
ISBN 0-8166-1141-6 AACR2
ISBN 0-8166-1142-4 (pbk.)

Contents

Acknowledgments

I am grateful to the original publishers, editors, and societies for their permission to reprint items from *Journal of the Royal Statistical Society*; *Uncertainty and Business Decisions* (Liverpool University Press); *Journal of the Institute of Actuaries*; *Science 129* (1959), 443-447; *British Journal for the Philosophy of Science*; *Logic, Methodology and Philosophy of Science* (Stanford University Press); *Journal of the American Statistical Association*; *Foundations of Statistical Inference* (Holt, Rinehart, and Winston of Canada, Ltd.); *The American Statistician* (American Statistical Association); (1) *PSA 1972,* (2) *Foundations of Probability Theory, Statistical Inference, and Statistical Theories of Science,* (3) *Philosophical Foundations of Economics,* (4) *Synthese* (respectively, © 1974, © 1976, © 1981, and © 1975, D. Reidel Publishing Company); *Machine Intelligence 8* (edited by Professor E. W. Elcock and Professor Donald Michie, published by Ellis Horwood Limited, Chichester); *Proceedings of the Royal Society*; *Trabajos de Estadística y de Investigación Operativa*; and *Journal of Statistical Computation and Simulation* (© Gordon and Breach); and *OMNI* for a photograph of the author. The diagram on the cover of the book shows all the possible moves on a chessboard; it appeared on the cover of *Theoria to Theory 6* (1972 July), in the *Scientific American* (June, 1979), and in *Personal Computing* (August, 1979). I am also beholden to Leslie Pendleton for her diligent typing, to Lindsay Waters and Marcia Bottoms at the University of Minnesota Press for their editorial perspicacity, to the National Institutes of Health for the partial financial support they gave me while I was writing some of the chapters, and to Donald Michie for suggesting the title *Good Thinking.*

Introduction

This is a book about applicable philosophy, and most of the articles contain all four of the ingredients philosophy, probability, statistics, and mathematics.

Some people believe that clear reasoning about many important practical and philosophical questions is impossible except in terms of probability. This belief has permeated my writings in many areas, such as rational decisions, statistics, randomness, operational research, induction, explanation, information, evidence, corroboration or weight of evidence, surprise, causality, measurement of knowledge, computation, mathematical discovery, artificial intelligence, chess, complexity, and the nature of probability itself. This book contains a selection of my more philosophical and less mathematical articles on most of these topics.

To produce a book of manageable size it has been necessary to exclude many of my babies and even to shrink a few of those included by replacing some of their parts with ellipses. (*Additions* are placed between pairs of brackets, and a few minor improvements in style have been made unobstrusively.) To compensate for the omissions, a few omitted articles will be mentioned in this introduction, and a bibliography of my publications has been included at the end of the book. (My work is cited by the numbers in this bibliography, for example, #**26**; a boldtype number is used when at least a part of an article is included in the present collection.)

About 85% of the book is based on invited lectures and represents my work on a variety of topics, but with a unified, rational approach. Because the approach is unified, there is inevitably some repetition. I have not deleted all of it because it seemed advisable to keep the articles fairly self-contained so that they can be read in any order. These articles, and some background information, will be briefly surveyed in this introduction, and other quick overviews can be obtained from the Contents and from the Index (which also lightly covers the bibliography of my publications).

Some readers interested in history might like to know what influences helped to form my views, so I will now mention some of my own background, leaving aside early childhood influences. My first introduction to probability, in high school, was from an enjoyable chapter in Hall and Knight's elementary textbook called *Higher Algebra.* I then found the writings on probability by J. M. Keynes and by F. P. Ramsey in the Hendon public library in North West London. I recall laboriously reading part of Keynes while in a queue for the Golders Green Hippodrome. My basic philosophy of probability is a compromise between the views of Keynes and Ramsey. In other words, I consider (i) that, even if physical probability exists, which I think is probable, it can be measured only with the aid of subjective (personal) probability and (ii) that it is not always possible to judge whether one subjective probability is greater than another, in other words, that subjective probabilities are only "partially ordered." This approach comes to the same as assuming "upper and lower" subjective probabilities, that is, assuming that a probability is interval valued. But it is often convenient to approximate the interval by a point.

I was later influenced by other writers, especially by Harold Jeffreys, though he regards sharp or numerical logical probabilities (credibilities) as primary. (See #1160, which is not included here, for more concerning Jeffreys's influence.) I did not know of the work of Bruno de Finetti until Jimmie Savage revealed it to speakers of English, which was after #13 was published. De Finetti's position is more radical than mine; in fact, he expresses his position paradoxically by saying that probability does not exist.

My conviction that rationality, rather than fashion, was a better approach to truth, and to many human affairs, was encouraged by some of the writings of Bertrand Russell and H. G. Wells. This conviction was unwittingly reinforced by Hitler, who at that time was dragging the world into war by making irrationality fashionable in Germany.

During Hitler's war, I was a cryptanalyst at the Government Code and Cypher School (GC & CS), also known as the Golf Club and Chess Society, in Bletchley, England, and I was the main statistical assistant in turn to Alan Turing (best known for the "Turing Machine"), Hugh Alexander (three times British chess champion), and Max Newman (later president of the London Mathematical Society). In this cryptanalysis the concepts of probability and weight of evidence, combined with electromagnetic and electronic machinery, helped greatly in work that eventually led to the destruction of Hitler.

Soon after the war I wrote *Probability and the Weighing of Evidence* (#13), though it was not published until January 1950. It discussed the theory of subjective or personal probability and the simple but powerful concept of Bayes factors and their logarithms or weights of evidence that I learned from Jeffreys and Turing. (For a definition, see weight of evidence in the Index of the present book.) I did not know until much later that the great philosopher of science C. S. Peirce had proposed the name "weight of evidence" in nearly the same technical sense in 1878, and I believe the discrepancy could be attributed to a mistake that Peirce made. (See #1382.) Turing, who called weight of evidence "decibannage," suggested in 1940 or 1941 that certain kinds of experiments could be evaluated by their expected weights of evidence (per observation) $\Sigma p_i \log(p_i/q_i)$, where p_i and q_i denote multinomial probabilities. Thus, he used weight of evidence as a quasiutility or epistemic utility. This was the first important practical use of expected weight of evidence in ordinary statistics, as far as I know, though Gibbs had used an algebraic expression of this form in statistical mechanics. (See equation 292 of his famous paper on the equilibrium of heterogeneous substances.) The extension to continuous distributions is obvious. Following Turing, I made good use of weight of evidence and its expectation during the war and have discussed it in about forty publications. The concept of weight of evidence completely captures that of the degree to which evidence corroborates a hypothesis. I think it is almost as much an intelligence amplifier as the concept of probability itself, and I hope it will soon be taught to all medical students, law students, and schoolchildren.

In his fundamental work on information theory in 1948, Shannon used the expression $-\Sigma p_i \log p_i$ and called it "entropy." His expression $\Sigma p_{ij} \log[p_{ij}/(p_{i.}p_{.j})]$ for mutual information is a special case of expected weight of evidence. Shannon's work made the words "entropy" and "information" especially fashionable, so that expected weight of evidence is now also called "discrimination information," or "the entropy of one distribution with respect to another," or "relative entropy," or "cross entropy," or "dinegentropy," or "dientropy," or even simply "entropy," though the last name is misleading. Among the many other statisticians who later emphasized this concept are S. Kullback, also an excryptanalyst, and Myron Tribus. The concept is used in several places in the following pages.

Statistical techniques and philosophical arguments that depend (explicitly) on subjective or logical probability—as, for example, in works of Laplace, Harold Jeffreys, Rudolf Carnap, Jimmy Savage, Bruno de Finetti, Richard Jeffrey, Roger Rosenkrantz, and myself in #13—are often called Bayesian or neo-Bayesian. The main controversy in the foundations of statistics is whether such techniques should be used. In 1950 the Bayesian approach was unpopular, largely because of Fisher's influence, and most statisticians were so far to my "right" that I was regarded as an extremist. Since then, a better description of my position would be "centrist" or "eclectic," or perhaps "left center" or advocating a Bayes/non-Bayes compromise. It seems to me that such a compromise is forced on anyone who regards nontautological probabilities as only interval valued because the degree of Bayesianity depends on the narrowness of the intervals. My basic position has not changed in the last thirty-five years, but the centroid of the statistical profession has moved closer to my position and many statisticians are now to my "left."

One of the themes of #13 was that the fundamental principles of probability, or of any scientific theory in almost finished form, can be categorized as "axioms," "rules," and "suggestions." The axioms are the basis of an abstract or mathematical theory; the rules show how the abstract theory can be applied; and suggestions, of which there can be an unlimited number, are devices for eliciting your own or others' judgments. Some examples of axioms, rules, and suggestions are also given in the present book.

#13 was less concerned with establishing the axiomatic systems for subjective probabilities by prior arguments than with describing the theory in the simplest possible terms, so that, while remaining realistic, it could be readily understood by any intelligent layperson. I wrote the book succinctly, perhaps overly so, in the hope that it would be read right through, although I knew that large books look more impressive.

With this somewhat perfunctory background let's rapidly survey the contents of the present book. It is divided into five closely related parts. (I had planned a sixth part consisting of nineteen of my book reviews, but the reviews are among the dropped babies. I shudder to think of readers spending hours in libraries hunting down these reviews, so, to save these keen readers some work, here are my gradings of these reviews (not of the books reviewed) on the scale 0 to 20. #761: 20/20; ##191, 294, 516, 697, 844, 956, 958, 1217, 1221, and 1235: 18/20; ##541A and 754: 17/20; #162: 16/20; ##115 and 875 with 948: 15/20; ##75 and 156: 14/20.)

The first part is about rationality, which can be briefly described as the maximization of the "mathematical expectation" of "utility" (value). #13 did not much emphasize utilities, but they were mentioned nine times. Article **#26** (Chapter 1 of Part I of this volume) was first delivered at a 1951 weekend conference of the Royal Statistical Society in Cambridge. It shows how very easy it is to extend the basic philosophy of #13 so as to include utilities and decisions. It also suggests a way of rewarding consultants who estimate probabilities—such as racing and stock-market tipsters, forecasters of the weather, guessers of Zener cards by ESP, and examinees in multiple-choice examinations—so as to encourage them to make accurate estimates. The idea has popular appeal and was reported in the Cambridge *Daily News* on September 25, 1951. Many years later I found that much the same idea had been proposed independently a year earlier by Brier, a meteorologist, but with a different (quadratic) fee. My logarithmic fee is related to weight of evidence and its expectation, and, when there are more than two alternatives, it is the only fee that does not depend on the probability estimates of events that do not later occur. (This fact apparently was first observed by Andrew Gleason.) The theory is improved somewhat in **#43** (Chapter 16). The topic was taken up by John McCarthy (1956) and by Jacob Marschak (1959), who said that the topic opened up an entirely new field of economics, the economics of information. There is now a literature of the order of a hundred articles on the topic, including my #690A, which was based on a lecture invited by Marschak for the second world conference on econometrics. #690A is somewhat technical and is not included here, but I think it deserves mention.

The logarithmic scoring system resembles the score in a time-guessing game that I invented at the age of thirteen when I was out walking with my father in Princes Risborough, Buckinghamshire. The score for an error of t minutes, rounded up to an exact multiple of a minute, is $-\log t$. When there are two players whose errors are t_1 and t_2, the "payment" is $\log(t_1/t_2)$. The logarithms of 1, 2, 3, . . . , 10, to base $10^{1/40}$, are conveniently close to whole numbers, which, when you think about it, is why there are twelve semitones in an octave. For probability estimators in our less sporting application, the "payment" is $\log(p/q)$, where p and q are the probability estimates of two consultants, rounded up to exact multiples of say 1/1000 so as not to punish too harshly people who carelessly estimate a nontautological probability as zero. If you hired only one consultant, you could define q as your *own* probability of the event that occurs. Then, the expected payoff is a "trientropy" of the form $\Sigma\pi_i\log(p_i/q_i)$, where the π_i is a "true" probability, and it pays the consultant in expectation to choose his or her p's equal to the π's. If the consultant agrees with your estimates, he or she earns nothing, except perhaps some constant fee, because $\log 1 = 0$.

One way to interpret the logarithmic payoff is that it rewards most highly the consultant who is the least surprised by the outcome, where surprise is also measured logarithmically. We return to this topic when reviewing Part IV.

Another subject discussed in **#26** is that of hierarchies of probabilities. For example, the probability of a statement containing probabilities of type 1 is itself of type 2. This hierarchical Bayesian approach is developed in several of my statistical articles, and I believe it is an essential "psychological" technique in the application of neo-Bayesian methods to problems containing many parameters, although Savage (1954, p. 58) summarily dismisses it. It is one example of the influence of philosophy on statistics. Some history of the hierarchical Bayesian technique is surveyed in **#1230** (Chapter 9 of this volume).

An article not included in this book is #290 because its ideas are mostly covered by later articles. In addition, it argues that the concept of Bayesian rationality should be understood by the managers of large firms and that this would greatly increase the value of operational research. Bayesian rationality is now widely taught in college courses in departments of economics, business, engineering, statistics, political science, and psychology, but this was not true twenty years ago. The article incidentally contains two natural measures of the precision of a probability judgment.

#679 (Chapter 2) is an attempt to collect together and to codify the basic foundations of rationality, as I understand it, into twenty-seven brief "priggish principles." Axiomatic systems for probability and rationality are esteemed by mathematicians because of their succinctness and sharpness, and I thought it useful to attempt a codification of my own views that goes beyond a formal axiomatic system. Note that "Type II rationality," in principle number 6, again leads naturally to a Bayes/non-Bayes compromise, or synthesis, as well as to other compromises, such as between subjectivism and credibilism. **#765** (Chapter 3) is another attempt at classification, this time of the varieties of philosophies that have been called Bayesian. The number of categories turns out to exceed the number of members of the American Statistical Association.

#838 (Chapter 4) describes some of the Bayesian influence in statistics and also some of the tricks used in non-Bayesian techniques to cover up subjectivity. This article includes an elaboration of the codification **#679**.

Part II deals mainly with probability. **#85A** (Chapter 5), half of which is omitted, asks whether probability or statistics is historically or logically primary and comes up with an eggs-and-hens answer. It contains some early criticisms of some conventional statistical concepts. **#182** (Chapter 6) discusses kinds of probability, again with some historical background. **#230** (Chapter 7), originally a 1960 lecture, describes the theory of #13 as a "black box" theory and shows how it can be used to generate an axiomatic system for upper and lower subjective or logical probabilities. The resulting axiomatic system is somewhat related to work by B. O. Koopman (1940a, b). It also turned out to be related to C. A. B. Smith (1961), and this was not surprising since Smith's aim was to produce a prior justification for the philosophy of partially ordered subjective probabilities.

#815 (Chapter 8) is a discussion of randomness that was delivered in a symposium on randomness dedicated to the memory of Jimmie Savage. The main topic of the article is the place of randomization in statistics. An omitted page gives a clear intuitive reason, in terms of "generalized decimals," for believing in the mathematical "existence" of infinite random sequences, though no such sequence can be explicitly defined.

I have already mentioned **#1230**, which is included as Chapter 9.

In #13 it had been mentioned that, if we wish to talk about the probabilities of mathematical theorems, we need to make a small adjustment to the usual axioms of probability. The same point occurs when we say that the output of a computer gives information. The most standard notation, among mathematicians and statisticians, for the probability of A given B is $P(A|B)$, and a familiar beginner's error is to forget about B when discussing the probability of A. But even with A and B both mentioned, one's subjective probability $P(A|B)$ can vary in time just by thinking or by computing. This is especially clear in a game of "perfect information" such as chess. I have called such probabilities "evolving," "sliding," or "dynamic." Dynamic probability is the really fundamental variety of subjective probability. This may be true even in the limiting case of ordinary logic, a case briefly considered by Henri Poincaré. The statement "all statements by Cretans are false," itself asserted by a Cretan, oscillates rapidly between truth and falsehood as one considers its implications until one conjectures that the word "other" needs to be inserted into the statement. The concept of dynamic or evolving probability *must* be invoked, as in #599, to resolve an important paradox in the philosophy of science that had been brought to my attention by Joseph Agassi. I have reluctantly omitted #599, which is one of my favorites, because it is largely superseded by **#1000**, Chapter 23 in Part V, and because **#938**, Chapter 10 in Part II, deals in detail with dynamic probability.

For further surveys of degrees of belief and of axiomatic systems for probability, see my encyclopedia articles ##1300 and 1313, neither of which is included in this book. I wish there had been room for them.

Part III deals mainly with corroboration, hypothesis testing, and simplicity. ##**518** and **600** (Chapter 11 and 12) resolve "Hempel's paradox of confirmation," in which "confirmation" means "corroboration" and can be best interpreted as weight of evidence, although this fact is not essential to the statement or to the resolution of the paradox. The somewhat surprising conclusion reached in these two short articles is that seeing a black crow does not necessarily confirm the hypothesis that all crows are black. Also an explanation is given for why this seems surprising. It is incidentally unfortunate that many philosophers have been confusing themselves by sometimes using "confirmation" to mean "probability." Bad choices of terminology often have bad effects, even though logically they shouldn't.

#**603B** (Chapter 13) was written in an attempt to decide whether the Titius-Bode "law" of planetary distances could be due to chance, an issue that many astronomers had decided by means of snap judgments but not always in the same direction. The Bayesian evaluation of scientific theories, in more than purely qualitative terms, has not yet been well developed, and I thought it would be interesting to begin with a fairly "numerological" law. The Titius-Bode law is not entirely numerological, and it receives some indirect support especially from a similar law for the moons of Saturn, which I called "Dabbler's law." I pointed out that Dabbler's law had some predictive value regarding the satellite Phoebe, but it "predicts" another moon (or ring?) at a distance of 4 million miles from Saturn, and this has not been observed.

The interesting question about both the Titius-Bode law and Dabbler's law is not whether they are "true," for obviously they are not, but whether they deserve an explanation. Most of #**603B** is omitted because it is long and technical and because it would, in fairness, be necessary to discuss later literature on the topic. (See Bode in the Index and Efron [1971], and Nieto [1972].) The editor had invited me to include a summary of my views on probability and induction and that is almost the only part of the article reprinted here. It is somewhat repetitive of material in other articles, but it is extremely succinct. The reader who likes some fun could refer to the original for the heated discussion.

#**1234** (Chapter 14) discusses hypothesis testing using various possible tools: tail-area probabilities, Bayes factors or weights of evidence, surprise indexes, and a Bayes/non-Bayes compromise. Of course, the probability of an observation, described in complete detail, given a single hypothesis, is usually exceedingly small and *by itself* does not help us to evaluate the hypothesis, and that is why tail-area probabilities were invented.

The third part ends with #**846** (Chapter 15), which surveys the topic of corroboration. This article includes a few pages on my philosophy of probability and rationality as well as sections on complexity and on checkability. A section on "explicativity" is mostly omitted here because it is covered by #**1000**. Another omitted article that is relevant to Part III is #1330, which deals with the philosophy of the analysis of data. The reason qualitative Bayesian thinking is basic to this topic is that the analysis of data necessarily involves the implicit or explicit formulation of hypotheses, and these hypotheses can have various degrees of "kinkosity."

Part IV begins with some parts of #**43** (Chapter 16), much of what is omitted, having been covered by #**26**. Here Warren Weaver's surprise index is generalized and related to Shannon's entropy. The topic was continued in #82, which is not included. As mentioned in #82, "Perhaps the main biological function of surprise is to jar us into reconsidering the validity of some hypothesis that we had previously accepted." That article also points out the relationship between surprise and simplicity. (See also #755.) ##**43** and 82 anticipated measures of information often attributed to the eminent probabilist Rényi.

G. L. S. Shackle had proposed that business decisions were usually based on the concept of potential surprise and not on subjective probability. I argued that, since surprise can be measured in terms of subjective probability, feelings of potential surprise can help you to make probability judgments and vice versa, so that the two forms of judgment can be used to enrich one another. I did not persuade Shackle of this. I think that anyone who disagrees strongly with my view had his mind in shackles.

#77 deals with the terminology and notation of information theory, but it is not included here. In it I proposed, simultaneously with Lindley, that information in Shannon's sense could be used as a quasi-utility in the design of experiments. The idea had been suggested in 1953 by Cronbach in a technical report, but the idea of using expected weight of evidence for this purpose had been long anticipated by Turing, as I mentioned before, and comes to essentially the same thing. Fisher had even earlier used his own measure of information for the same purpose, though he never thought of it as a substitute for utility. For an account of Turing's statistical work during World War II, see #1201 which is not included in the present book.

Perhaps the term "quasiutility" should be reserved for a quantity whose expectation it is reasonable to maximize in the design of a statistical experiment. A necessary condition for a quasiutility is additivity for independent experiments. Examples of "quasiutilities" are (i) Fisher's measure of information concerning the value of a parameter, (ii) Shannon's mutual information, (iii) more generally weight of evidence, (iv) logarithmic surprise indexes, and (v) explicativity (see #**1000**).

There is a theorem which owes its existence to Abraham Wald, that a "least favorable" prior is minimax. (Readers interested in less technical subjects could skip the rest of this paragraph.) He intended "least favorable" to be interpreted in terms of utilities, but it was pointed out in ##618 and 622, which are omitted and are therefore denoted by timid type, that we can apply the theorem with the utilities replaced by quasiutilities and that this shows that the "principle of maximum entropy" and the "principle of minimum discriminability" for formulating hypotheses (Kullback, 1959; and very explicitly p. 913 of #322) and Jeffreys's "invariant prior," are all minimax procedures. This fact sheds light on both the advantages and the disadvantages of these techniques. (Maximum entropy was originally suggested by E. T. Jaynes for the generation of credibilistic priors.) A principle of minimum Fisherian information could also be tried for hypothesis formulation in the absence of a sample: it can be seen to lead, for example, to the hypothesis of independence for multivariate normal distributions, as does also the principle of maximum entropy. Similarly, the maximization of entropy leads to the hypothesis of independence for contingency tables and to that of the vanishing of interactions of various orders for multidimensional contingency tables (see #322). But, when the prior probabilities of the hypotheses are already known, or assumed, the minimax method is meaningless, so there is no principle of minimum explicativity unless the probabilities of the hypotheses are interval valued instead of being sharp.

#508 (Chapter 17) explains in detail, in terms of rationality, why it pays, at any rate in one's own judgment, to acquire new evidence when it is free, a question that was raised by A. J. Ayer at a conference of distinguished philosophers of science, none of whom had come up with the answer. But **#855** (Chapter 18) shows that, from the point of view of another person who knows more than you do, it can sometimes in expectation be to your disadvantage to acquire a small amount of new evidence.

#659 (Chapter 19) is a quick survey of my work on the topics of information, evidence, surprise, causality, explanation, and utility. Its appendix was **#679**, which constitues Chapter 2.

#814 (Chapter 20) points out that, contrary to an opinion expressed by Eddington, we should not be surprised that our galaxy is unusually large.

Part V deals with probabilistic causality and explanation. Most philosophical writings on causality interpret causality and explanation in a strictly deterministic sense, but probabilistic causality is of more practical importance (for example, when assigning blame and credit), and anyway strict causality is a special case. The only serious philosophical work on probabilistic causality that I know of, preceding my work, was by Hans Reichenbach and by Norbert Wiener. **#223B** (Chapter 21), which uses a desideratum-explicatum approach, is more ambitious than Reichenbach's work in the sense of suggesting a quantitative explicatum (in terms of weight of evidence) for the degree to which one event tends to cause another one. Wiener's work dealt with stochastic processes instead of with events. The writing of **#223B** kept me off the streets for a year of evenings and weekends. A relationship between Wiener's work and mine might well emerge from the application of my explication to regression theory, as in #1317, but this mathematical article is not included. **#1336** (Chapter 22) simplifies some of the argument of **#223B**. Although **#223B** is somewhat mathematical, I thought its omission would damage too much the picture of my work on foundations. Recently, there has been increasing interest in the topic of probabilistic causality. See, for example, Suppes (1970), Salmon (1980), and Humphreys (1980), reviewed or answered, respectively, in ##754, 1263, 1331, and 1333. See also Mayr (1961), Simon (1978), Sayre (1977), and Rosen (1978), and ##928 and 1157.

The final article is #**1000** (Chapter 23), which deals with "explicativity," a topic already mentioned in this introduction. Some of the ideas had already appeared in #599. Explicativity is a measure of the explanatory strength of a hypothesis or theory in relation to given observations, and its maximization can be regarded as a sharpening of the razor of Duns and Ockham. The explication proposed can be regarded as a compromise, or synthesis, between Bayesian and Popperian views. The article contains a brief discussion of "predictivity." A feature of #**1000** of interest in applied philosophy is that the explication for explicativity can be used in statistical problems of both estimation and significance testing. The explicatum, which again uses logarithms of probabilities, is thus supported not just by the strong prior desiderata but by the deduction of sensible results resembling those in classical statistical theory. Nevertheless, these statistical applications are somewhat technical and have been omitted from this book. The complete article was reprinted, as #1161, in the volume of essays in honor of Harold Jeffreys. That volume also contained #1160, which dealt with the contributions of Jeffreys to Bayesian statistics. Although the statistical part of #**1000** has been omitted, this article makes an appropriate conclusion to the present book because of its intimate blend of philosophy and statistics.

Part I. Bayesian Rationality

CHAPTER 1

Rational Decisions (#26)

SUMMARY

This paper deals with the relationship between the theory of probability and the theory of rational behavior. A method is then suggested for encouraging people to make accurate probability estimates, a connection with the theory of information being mentioned. Finally Wald's theory of statistical decision functions is summarised and generalized and its relation to the theory of rational behavior is discussed.

1. INTRODUCTION

I am going to discuss the following problem. Given various circumstances, to decide what to do. What universal rule or rules can be laid down for making rational decisions? My main contention is that our methods of making rational decisions should not depend on whether we are statisticians. This contention is a consequence of a belief that consistency is important. A few people think there is a danger that overemphasis of consistency may retard the progress of science. Personally I do not think this danger is serious. The resolution of inconsistencies will always be an essential method in science and in cross-examinations. There may be occasions when it is best to behave irrationally, but whether there are should be decided rationally.

It is worth looking for unity in the methods of statistics, science and rational thought and behavior; first in order to encourage a scientific approach to non-scientific matters, second to suggest new statistical ideas by analogy with ordinary ideas, and third because the unity is aesthetically pleasing.

Clearly I am sticking my neck out in discussing this subject. In most subjects people usually try to understand what other people mean, but in philosophy and near-philosophy they do not usually try so hard.

2. SCIENTIFIC THEORIES

In my opinion no scientific theory is really satisfactory until it has the following form:

(i) There should be a very precise set of axioms from which a purely abstract theory can be rigorously deduced. In this abstract theory some of the words or symbols may remain undefined. For example, in projective geometry it is not necessary to know what points, lines and planes are in order to check the correctness of the theorems in terms of the axioms.

(ii) There should be precise rules of application of the abstract theory which give meaning to the undefined words and symbols.

(iii) There should be suggestions for using the theory, these suggestions belonging to the technique rather than to the theory. The suggestions will not usually be as precisely formulated as the axioms and rules.

The adequacy of the abstract theory cannot be judged until the rules of application have been formulated. These rules contain indications of what the undefined words and symbols of the abstract theory are all about, but the indications will not be complete. It is the theory as a whole, i.e., the axioms and rules combined, which gives meaning to the undefined words and symbols. It is mainly for this reason that a beginner finds difficulty in understanding a scientific theory.

It follows from this account that a scientific theory represents a decision and a recommendation to use language and symbolism in a particular way, and possibly also to think and act in a particular way. Consider, for example, the principle of conservation of energy, or energy and matter. Apparent exceptions to the principle have been patched up by extending the idea of energy, to potential energy, for example. Nevertheless the principle is not entirely tautological.

Some theoreticians formulate theories without specifying the rules of application, so that the theories cannot be understood at all without a lot of experience. Such formulations are philosophically unsatisfactory.

Ordinary elementary logic can be regarded as a scientific theory. The recommendations of elementary logic are so widely accepted and familiar, and have had so much influence on the educated use of language, that logic is often regarded as self-evident and independent of experience. In the empirical sciences the selection of the theories depends much more on experience. The theory of probability occupies an intermediate position between logic and empirical sciences. Some people regard any typical theory of probability as self-evident, and others say it depends on experience. The fact is that, as in many philosophical disputes, it is a question of degree: the theory of probability does depend on experience, but does not require much more experience than does ordinary logic. There are a number of different methods of making the theory seem nearly tautological by more or less *a priori* arguments. The two main methods are those of "equally probable cases" and of limiting frequencies. [It is questionable whether these are still the main methods.] Both methods depend on idealizations, but it would be extremely surprising if either method could be proved to lead to inconsistencies. When actually estimating probabilities, most of us use both methods. It may be possible in principle to trace back all probability estimates to individual experiences of frequencies, but this has not yet been done. Two examples in which beliefs do not depend in an obvious way on frequencies are (i) the estimation of the probability that a particular card will be drawn from a well-suffled pack of 117 cards; (ii) the belief which newly-born piglings appear to have that it is a good thing to walk round the mother-pig's leg in order to arrive at the nipples. (This example is given for the benefit of those who interpret a belief as a tendency to act.)

3. DEGREES OF BELIEF

I shall now make twelve remarks about degrees of belief.

(i) I *define* the theory of probability as the logic of degrees of belief. Therefore degrees of belief, either subjective or objective, must be introduced. Degrees of belief are assumed, following Keynes, to be partially ordered only, i.e., some pairs of beliefs may not be comparable.

(ii) F. Y. Edgeworth, Bertrand Russell, and others use the word "credibilities" to mean objective rational degrees of belief. A credibility has a definite but possibly unknown value. It may be regarded as existing independently of human beings.

(iii) A subjective theory of probability can be developed without assuming that there is necessarily a credibility of E given F for every E and F, where E and F are propositions. This subjective theory can be applied whether credibilities exist or not. It is therefore more general and economical not to assume the existence of credibilities as an axiom.

(iv) Suppose Jeffreys is right that there is a credibility of E given F, for every E and F. Then either the theory will tell us what this credibility is, and we must adjust our degree of belief to be equal to the credibility. Or on the other hand the theory will not tell us what the credibility is, and then not much is gained, except perhaps a healthier frame of mind, by supposing that the credibility exists.

(v) A [simple] statistical hypothesis H is an idealized proposition such that for some E, $P(E|H)$ is a credibility with a specified value. Such credibilities may be called "tautological probabilities."

(vi) There is an argument for postulating the existence of credibilities other than tautological probabilities, namely that probability judgments by different people have some tendency to agree.

(vii) The only way to assess the cogency of this argument, if it can be assessed at all, is by the methods of experimental science whose justification is by means of a subjective theory.

(viii) My own view is that it is often quite convenient to accept the postulate that credibilities exist, but this should be regarded as a suggestion rather than an axiom of probability theory.

(ix) This postulate is useful in that it enables other people to do some of our thinking for us. We pay more attention to some people's judgment than to others'.

(x) If a man holds unique beliefs it is possible that everybody else is wrong. If we want him to abandon some of his beliefs we may use promises, threats, hypnotism and suggestion, or we may prefer the following more rational method: By asking questions we may obtain information about his beliefs. Some of the questions may be very complicated ones, of the form, "I put it to you that the following set of opinions is cogent: . . . ." We may then show, by applying a subjective theory of probability, that the beliefs to which the man has paid lip-service are not self-consistent.

(xi) Some of you may be thinking of the slogan "science deals only with what is objective." If the slogan were true there would be no point for scientific purposes in introducing subjective judgments. But actually the slogan is false. For example, intuition (which is subjective) is the main instrument of original scientific research, according to Einstein. The obsession with objectivity arises largely from the desire to be convincing in published work. There are, however, several activities in which it is less important to convince other people than to find out the truth for oneself. There is another reason for wanting an objective theory, namely that there is a tendency to wishful thinking in subjective judgments. But objectivity is precisely what a subjective theory of probability is for: its function is to introduce extra rationality, and therefore objectivity, into your degrees of belief.

(xii) Once we have decided to objectify a rational degree of belief into a credibility it begins to make sense to talk about a degree of belief concerning the numerical value of a credibility. It is possible to use probability type-chains (to coin a phrase) with more than two links, such as a degree of belief equal to ½ that the credibility of H is 1/3 where H is a statistical hypothesis such that $P(E|H) = \frac{1}{4}$. It is tempting to talk about reasonable degrees of belief of higher and higher types, but it is convenient to think of all these degrees of belief as being of the same kind (usually as belonging to the same body of beliefs in the sense of section 4) by introducing *propositions* of different kinds. In the above example the proposition which asserts that the credibility of H is 1/3 may itself be regarded as a statistical hypothesis "of type 2." Our type-chains can always be brought back ultimately to a subjective degree of belief. All links but the first will usually be credibilities, tautological or otherwise.

4. UTILITIES

The question whether utilities should be regarded as belonging to the theory of probability is very largely linguistic. It therefore seems appropriate to begin with a few rough definitions.

Theory of reasoning: A theory of logic plus a theory of probability.

Body of beliefs: A set of comparisons between degrees of belief of the form that one belief is held more firmly than another one, or if you like a set of judgments that one probability is greater than (or equal to) another one.

Reasonable body of beliefs: A body of beliefs which does not give rise to a contradiction when combined with a theory of reasoning.

A reasonable degree of belief is one which occurs in a reasonable body of beliefs. A *probability* is an expression of the form $P(E|F)$ where E and F are propositions. It is either a reasonable degree of belief "in E given F," or else it is something introduced for formal convenience. Degrees of belief may be called "probability estimates."

Principle of rational behavior: The recommendation always to behave so as to maximize the expected utility per time unit.

Theory of rational behavior: Theory of reasoning plus the principle of rational behavior.

Body of decisions: A set of judgments that one decision is better than another. Hypothetical circumstances may be considered as well as real ones (just as for a body of beliefs).

Reasonable body of decisions: A body of decisions which does not give rise to a contradiction when combined with a theory of rational behavior.

A reasonable decision is one which occurs in a reasonable body of decisions.

We see that a theory of reasoning is a recommendation to think in a particular way while a theory of rational behavior is a recommendation to act in a particular way.

Utility judgments may also be called "value judgments." The notion of utility is not restricted to financial matters, and even in financial matters utility is not strictly proportional to financial gain. Utilities are supposed to include all human values such as, for example, scientific interest. Part of the definition of utility is provided by the theory of rational action itself.

It was shown by F. P. Ramsey (1931) how one could build up the theory of probability by starting from the principle of maximizing expected utilities. L. J. Savage has recently adopted a similar approach in much more detail in some unpublished notes. The main argument for developing the subject in the Ramsey-Savage manner is that degrees of belief are only in the mind or expressed verbally, and are therefore not immediately significant operationally in the way that behavior is. Actions speak louder than words. I shall answer this argument in four steps:

(i) It is convenient to classify knowledge into subjects which are given names and are discussed without very much reference to the rest of knowledge. It is possible, and quite usual, to discuss probability with little reference to utilities. If utilities are introduced from the start, the axioms are more complicated and it is debatable whether they are more "convincing." The plan which appeals to me is to develop the theory of probability without much reference to utilities, and then to adjoin the principle of rational behavior in order to obtain a theory of rational behavior. The above list of definitions indicates how easily the transition can be made from a theory of probability to a theory of rational behavior.

(ii) People's value judgments are, I think, liable to disagree more than their probability judgments. Values can be judged with a fair amount of agreement when the commodity is money, but not when deciding between, say, universal education and universal rowing, or between your own life and the life of some other person.

(iii) The principle of maximizing the expected utility can be made to look fairly reasonable in terms of the law of large numbers, provided that none of the utilities are very large. It is therefore convenient to postpone the introduction of the principle until after the law of large numbers has been proved.

(iv) It is not quite clear that infinite utilities cannot occur in questions of salvation and damnation (as suggested, I think, by Pascal), and expressions like $\infty - \infty$ would then occur when deciding between two alternative religions. To have to argue about such matters as a necessary preliminary to laying down any of the axioms of probability would weaken the foundations of that subject.

5. AXIOMS AND RULES

The theory of probability which I accept and recommend is based on six axioms, of which typical ones are—

A1. $P(\mathrm{E}|\mathrm{F})$ is a non-negative number (E and F being propositions).

A4. If E is logically equivalent to F then $P(\mathrm{E}|\mathrm{G}) = P(\mathrm{F}|\mathrm{G})$, $P(\mathrm{G}|\mathrm{E}) = P(\mathrm{G}|\mathrm{F})$.

There is also the possible modification—

A4′. If you have *proved* that E is logically equivalent to F then $P(\mathrm{E}|\mathrm{G}) = P(\mathrm{F}|\mathrm{G})$, etc. The adoption of A4′ amounts to a weakening of the emphasis on consistency and enables you to talk about the probability of purely mathematical propositions.

The main rule of application is as follows: Let $P'(\mathrm{E}|\mathrm{F}) > P'(\mathrm{G}|\mathrm{H})$ mean that you judge that your degree of belief in E given F (i.e., if F were assumed) would exceed that of G given H. Then in the abstract theory you may write $P(\mathrm{E}|\mathrm{F}) > P(\mathrm{G}|\mathrm{H})$ and conversely.

Axiom A1 may appear to contradict the assumption of section 3 that degrees of belief are only partially ordered. But when the axioms are combined with the above rule of application it becomes clear that we cannot necessarily effect the comparison between any pair of beliefs. The axioms are therefore stronger than they need be for the applications. Unfortunately if they are weakened they become much more complicated. (See Koopman, 1940a.)

6. EXAMPLES OF SUGGESTIONS

(i) Numerical probabilities can be introduced by imagining perfect packs of cards perfectly shuffled, or infinite sequences of trials under essentially similar conditions. Both methods are idealizations, and there is very little to choose between them. It is a matter of taste: that is why there is so much argument about it.

(ii) Any theorem of probability theory and anybody's methods of statistical inference may be used in order to help you to make probability judgments.

(iii) If a body of beliefs is found to be unreasonable after applying the abstract theory, then a good method of patching it up is by being honest (using unemotional judgment). This suggestion is more difficult to apply to utility judgments because it is more difficult to be unemotional about them.

7. RATIONAL BEHAVIOR

I think that once the theory of probability is taken for granted, the principle of maximizing the expected utility per unit time [or rather its integral over the future, with a discounting factor decreasing with time, depending on life expectancy tables] is the only fundamental principle of rational behavior. It teaches us, for example, that the older we become the more important it is to use what we already know rather than to learn more.

In the applications of the principle of rational behavior some complications arise, such as—

(i) We must weigh up the expected time for doing the mathematical and statistical calculations against the expected utility of these calculations. Apparently less good methods may therefore sometimes be preferred. For example, in an emergency, a quick random decision is [usualiy] better than no decision. But of course theorizing has a value apart from any particular application.

(ii) We must allow for the necessity of convincing other people in some circumstances. So if other people use theoretically inferior methods we may be encouraged to follow suit. It was for this reason that Newton translated his calculus arguments into a geometrical form in the *Principia*. Fashions in modern statistics occur partly for the same reason.

(iii) We may seem to defy the principle of rational action when we insure articles of fairly small value against postal loss. It is possible to justify such insurances on the grounds that we are buying peace of mind, knowing that we are liable to lapse into an irrational state of worry.

(iv) Similarly we may take on bets of negative expected financial utility because the act of gambling has a utility of its own.

(v) Because of a lack of precision in our judgment of probabilities, utilities, expected utilities and "weights of evidence" we may often find that there is nothing to choose between alternative courses of action, i.e., we may not be able to say which of them has the larger expected utility. Both courses of action may be reasonable and a decision may then be arrived at by the operation known as "making up one's mind." Decisions reached in this way are not usually reversed, owing to the negative utility of vacillation. People who attach too large a value to the negative utility of vacillation are known as "obstinate."

(vi) Public and private utilities do not always coincide. This leads to ethical problems.

Example.—An invention is submitted to a scientific adviser of a firm. The adviser makes the following judgments:

(1) The probability that the invention will work is p.

(2) The value to the firm if the invention is adopted and works is V.

(3) The loss to the firm if the invention is adopted and fails to work is L.

(4) The value to the adviser personally if he advises the adoption of the invention and it works is v.

(5) The loss to the adviser if he advises the adoption of the invention and it fails to work is ℓ.

(6) The losses to the firm and to the adviser if he recommends the rejection of the invention are both negligible, because neither the firm nor the adviser have rivals.

Then the firm's expected gain if the invention is accepted is $pV - (1-p)L$ and the adviser's expected gain in the same circumstances is $pv - (1-p)\ell$. The firm has positive expected gain if $p/(1-p) > L/V$, and the adviser has positive expected gain if $p/(1-p) > \ell v$. If $\ell v > p/(1-p) > L/V$, the adviser will be faced with an ethical problem, i.e., he will be tempted to act against the interests of the firm. Of course real life is more complicated than this, but the difficulty obviously arises. In an ideal society public and private expected utility gains would always be of the same sign.

What can the firm do to prevent this sort of temptation from arising? In my opinion the firm should ask the adviser for his estimates of p, V and L, and should take the onus of the actual decision on its own shoulders. In other words, leaders of industry should become more probability-conscious.

If leaders of industry did become probability-conscious there would be quite a reaction on statisticians. For they would have to specify probabilities of hypotheses instead of merely giving advice. At present a statistician of the Neyman-Pearson school is not permitted to talk about the probability of a statistical hypothesis.

8. FAIR FEES

The above example raises the question of how a firm can encourage its experts to give fair estimates of probabilities. In general this is a complicated problem, and I shall consider only a simple case and offer only a tentative solution. Suppose that the expert is asked to estimate the probability of an event E in circumstances where it will fairly soon be known whether E is true or false, e.g., in weather forecasts.

It is convenient at first to imagine that there are two experts A and B whose estimates of the probability of E are $p_1 = P_1(\mathrm{E})$ and $p_2 = P_2(\mathrm{E})$. The suffixes refer to the two bodies of belief, and the "given" propositions are taken for granted and omitted from the notation. We imagine also that there are objective probabilities, or credibilities, denoted by P. We introduce hypotheses H_1 and H_2 where H_1 (or H_2) is the hypothesis that A (or B) has objective judgment. Then

$$p_1 = P(\mathrm{E}|\mathrm{H}_1), p_2 = P(\mathrm{E}|\mathrm{H}_2).$$

Therefore, taking "H_1 or H_2" for granted, the factor in favour of H_1 (i.e., the ratio of its final to its initial odds) if E happens is p_1/p_2. Such factors are multiplicative if a series of independent experiments are performed. By taking logs we obtain an additive measure of the difference in the merits of A and B, namely $\log p_1 - \log p_2$ if E occurs or $\log(1-p_1) - \log(1-p_2)$ if E does not occur. By itself $\log p_1$ (or $\log(1-p_1)$) is a measure of the merit of a probability estimate, when it is theoretically possible to make a correct prediction with certainty. It is never positive, and represents the amount of information lost through not knowing with certainty what will happen.

A reasonable fee to pay an expert who has estimated a probability as p_1 is $k \log(2p_1)$ if the event occurs and $k \log(2-2p_1)$ if the event does not occur. If $p_1 > \frac{1}{2}$ the latter payment is really a fine. (k is independent of p_1 but may depend on the utilities. It is assumed to be positive.) This fee can easily be seen to have the desirable property that its expectation is maximized if $p_1 = p$, the true probability, so that it is in the expert's own interest to give an objective estimate. It is also in his interest to collect as much evidence as possible. Note that no fee is paid if $p_1 = \frac{1}{2}$. The justification of this is that if a larger fee were paid the expert would have a positive expected gain by saying that $p_1 = \frac{1}{2}$, without looking at the evidence at all. If the class of problems put to the expert have the property that the average value of p is x, then the factor 2 in the above formula for the fee should be replaced by $x^{-x}(1-x)^{-(1-x)} = b$, say. (For more than two alternatives the corresponding formula for b is $\log b = -\Sigma x_i \log x_i$, the initial "entropy." [But see p. 176.]) Another modification of the formula should be made in order to allow for the diminishing utility of money (as a function of the amount, rather than as a function of time). In fact if Daniel Bernoulli's logarithmic formula for the utility of money is assumed, the expression for the fee ceases to contain a logarithm and becomes $c\{(bp_1)^k - 1\}$ or $-c\{1 - (b - bp_1)^k\}$, where c is the initial capital of the expert.

This method could be used for introducing piece-work into the Meteorological Office. The weather-forecaster would lose money whenever he made an incorrect forecast.

When making a probability estimate it may help to *imagine* that you are to be paid in accordance with the above scheme. It is best to tabulate the amount to be paid as a function of p_1.

9. LEGAL AND STATISTICAL PROCEDURES COMPARED

In legal proceedings there are two men A and B known as lawyers and there is a hypothesis H. A is paid to pretend that he regards the probability of H as 1 and B is paid to pretend that he regards the probability of H as 0. Experiments are performed which consist in asking witnesses questions. A sequential procedure is adopted in which previous answers influence what further questions are asked and what further witnesses are called. (Sequential procedures are very common indeed in ordinary life.) But the jury, which has to decide whether to accept or to reject H (or to remain undecided), does not control the experiments. The two lawyers correspond to two rival scientists with vested interests and the jury corresponds to the general scientific public. The decision of the jury depends on their estimates of the final probability of H. It also depends on their judgments of the utilities. They may therefore demand a higher threshold for the probability required in a murder case than in a case of petty theft. The law has never bothered to specify the thresholds numerically. In America a jury may be satisfied with a lower threshold for condemning a black man for the rape of a white woman than vice versa (*News-Chronicle* 32815 [1951], 5). Such behavior is unreasonable when combined with democratic bodies of decisions.

The importance of the jury's coming to a definite decision, even a wrong one, was recognized in law at the time of Edward III (c. 1350). At that time it was regarded as disgraceful for a jury not to be unanimous, and according to some reports such juries could be placed in a cart and upset in a ditch (*Enc. Brit.*, 11th ed., 15, 590). This can hardly be regarded as evidence that they believed in credibilities in those days. I say this because it was not officially recognized that juries could come to wrong decisions except through their stupidity or corruption.

10. MINIMAX SOLUTIONS

For completeness it would be desirable now to expound Wald's theory of statistical decision functions as far as his definition of Bayes solutions and minimax solutions. He gets as far as these definitions in the first 18 pages of *Statistical Decision Functions*, but not without introducing over 30 essential symbols and 20 verbal definitions. Fortunately it is possible to generalize and simplify the definitions of Bayes and minimax solutions with very little loss of rigour.

A number of mutually exclusive statistical hypotheses are specified (one of them being true). A number of possible decisions are also specified as allowable. An example of a decision is that a particular hypothesis, or perhaps a disjunction of hypotheses, is to be acted upon without further experiments. Such a decision is called a "terminal decision." Sequential decisions are also allowed. A sequential decision is a decision to perform further particular experiments. I do not think that it counts as an allowable decision to specify the final probabilities of the hypotheses, or their expected utilities. (My use of the word "allowable" here has nothing to do with Wald's use of the word "admissible.") The terminal and sequential decisions may be called "non-randomized decisions." A "randomized decision" is a decision to draw lots in a specified way in order to decide what non-randomized decision to make.

Notice how close all this is to being a classification of the decisions made in ordinary life, i.e., you often choose between (i) making up your mind, (ii) getting further evidence, (iii) deliberately making a mental or physical toss-up between alternatives. I cannot think of any other type of decision. But if you are giving advice you can specify the relative merits or expected utilities of taking various decisions, and you can make probability estimates.

A non-randomized decision function is a (single-valued) function of the observed results, the values of the function being allowable non-randomized decisions. A randomized decision function, δ, is a function of the observed results, the values of the function being randomized decisions.

Minus the expected utility [strictly, a constant should be added to the utility to make its maximum value zero, for any given F and variable decisions] for a given statistical hypothesis F and a given decision function δ is called the *risk* associated with F and δ, $r(F, \delta)$. (This is intended to allow for utilities including the cost of experimentation. Wald does not allow for the cost of theorizing.) If a distribution ξ of initial probabilities of the statistical hypotheses is assumed, the expected value of $r(F, \delta)$ is called $r^*(\xi, \delta)$, and a decision function δ which minimizes $r^*(\xi, \delta)$ is called a *Bayes solution* relative to ξ.

A decision function δ is said to be a *minimax solution* if it minimizes $\max_F r(F, \delta)$.

An initial distribution ξ is said to be *least favorable* if it maximizes $\min_\delta r^*(\xi, \delta)$. Wald shows under weak conditions that a minimax solution is a Bayes solution relative to a least favorable initial distribution. Minimax solutions seem to assume that we are living in the worst of all possible worlds. Mr. R. B. Braithwaite suggests calling them "prudent" rather than "rational."

Wald does not in his book explicitly recommend the adoption of minimax solutions, but he considered their theory worth developing because of its importance in the theory of statistical decision functions as a whole. In fact the book is more concerned with the theory than with recommendations as to how to apply the theory. There is, however, the apparently obvious negative recommendation that Bayes solutions cannot be applied when the initial distribution ξ is unknown. The word "unknown" is rather misleading here. In order to see this we consider the case of only two hypotheses H and H′. Then ξ can be replaced by the probability, p, of H. I assert that in most practical applications we regard p as bounded by inequalities something like $\cdot 1 < p < \cdot 8$. For if we did not think that $\cdot 1 < p$ we would not be prepared to accept H on a small amount of evidence. Is ξ unknown if $\cdot 1 < p < \cdot 8$? Is it unknown if $\cdot 4 < p < \cdot 5$; if $\cdot 499999 < p < \cdot 500001$? In each of these circumstances it would be reasonable to use the Bayes solution corresponding to a value of p selected arbitrarily within its allowable range.

In what circumstances is a minimax solution reasonable? I suggest that it is reasonable if and only if the least favorable initial distribution is reasonable according to your body of beliefs. In particular a minimax solution is always reasonable provided that only reasonable ξ's are entertained. But then the minimax solution is only one of a number of reasonable solutions, namely all the Bayes solutions corresponding to the various ξ's.

It is possible to generalize Wald's theory by introducing a distribution ζ of the ξ's themselves. We would then be using a probability type-chain of type 3. (See section 3.) The expected value of $r^*(\xi, \delta)$ could be called $r^{**}(\zeta, \delta)$, and a decision function δ which minimizes $r^{**}(\zeta, \delta)$ could be called a "Bayes solution of type 3" relative to ζ. For consistency we could then call Wald's Bayes solutions "Bayes solutions of type 2." When there is only one available statistical hypothesis F we may define a "Bayes solution of type 1" relative to F as one which maximizes $r(F, \delta)$. The use of Bayes solutions of any type is an application of the principle of rational behavior.

One purpose in introducing Bayes solutions of the third type is in order to overcome feelings of uneasiness in connection with examples like the one mentioned above, where $\cdot 1 < p < \cdot 8$. One feels that if $p = \cdot 09$ has been completely ruled out, then $p = \cdot 11$ should be *nearly* ruled out, and this can only be done by using probability type-chains of the third type. It may be objected that the higher the type the woollier the probabilities. It will be found, however, that the higher the type the less the woolliness matters, provided the calculations do not become too complicated. Naturally any lack of definition in ζ is reflected in ambiguity of the Bayes solution of type 3. This ambiguity can be avoided by introducing a *minimax solution of type 2*, i.e., a decision function which minimizes $\max_\xi r^*(\xi, \delta)$. [A more accurate definition is given in #80.]

By the time we had gone as far as type-chains of type 3 I do not think we would be inclined to objectify the degrees of belief any further. It would therefore probably not be necessary to introduce Bayes solutions of type 4 and minimax solutions of type 3.

Minimax solutions of type 1 were in effect originated by von Neumann in the theory of games, and it is in this theory that they are most justifiable. But even here in practice you would prefer to maximize your expected gain. You would probably use minimax solutions when you had a fair degree of belief that your opponent was a good player. Even when you use the minimax solution you may be maximizing your expected gain, since you may already have worked out the details of the minimax solution, and you would probably not have time to work out anything better once a game had started. To attempt to use a method other than the minimax method would lead to too large a probability of a large loss, especially in a game like poker. (As a matter of fact I do not think the minimax solution has been worked out for poker.)

I am much indebted to the referee for his critical comments.

CHAPTER 2

Twenty-seven Principles of Rationality (#679)

In the body of my paper for this symposium I originally decided not to argue the case for the use of subjective probability since I have expressed my philosophy of probability, statistics, and (generally) rationality on so many occasions in the past. But after reading the other papers I see that *enlightenment* is still required. So, in this appendix I give a succinct list of 27 priggish principles. I have said and stressed nearly all of them before, many in my 1950 book, but have not brought so many of them together in one short list. As Laplace might have said, they are *au fond le bon sens*, but they cannot be entirely reduced to a calculus. [The writers who influenced me are mentioned in the Introduction.]

1. Physical probabilities probably *exist* (I differ from de Finetti and L. J. Savage here) but they can be *measured* only with the help of subjective probabilities. There are several kinds of probability. (##**182**, 522. The latter paper contains a dendroidal categorization.)

2. A familiar set of axioms of subjective probability are to be used. Kolmogorov's axiom (complete additivity) is convenient rather than essential. The axioms should be combined with rules of application and less formal suggestions for aiding the judgment. Some of the suggestions depend on theorems such as the laws of large numbers which make a frequency definition of probability unnecessary. (It is unnecessary and insufficient.)

3. In principle these axioms should be used in conjunction with inequality judgments and therefore they often lead only to inequality discernments. The axioms can themselves be formulated as inequalities but it is easier to incorporate the inequalities in the rules of application. In other words most subjective probabilities are regarded as belonging only to some interval of values the end points of which may be called the lower and upper probabilities. (Keynes, 1921; Koopman, 1940a; ##13, **230**; C. A. B. Smith, 1961.)

4. The principle of rationality is the recommendation to maximize expected utility.

5. The input and output to the abstract theories of probability and rationality are judgments of inequalities of probabilities, odds, Bayesian factors (ratios of final to initial odds), log-factors or weights of evidence, information, surprise indices, utilities, *and any other functions of probabilities and utilities.* (For example, Good, 1954.) It is often convenient to forget about the inequalities for the sake of simplicity and to use precise estimates (see Principle 6).

6. When the expected time and effort taken to think and do calculations are allowed for in the costs, then one is using the principle of *rationality of type II.* This is more important than the ordinary principle of rationality, but is seldom mentioned because it contains a veiled threat to conventional logic by incorporating a time element. It often justifies *ad hoc* [and non-Bayesian] procedures such as confidence methods and this helps to decrease controversy.

7. The purposes of the theories of probability and rationality are to enlarge bodies of beliefs and to check them for consistency, and thus to improve the objectivity of subjective judgments. This process can never be [the words "known to be" should be inserted here to cope with Dr. Barnard's comment that followed this paper] completed even in principle, in virtue of Gödel's theorem concerning consistency. Hence the type II principle of rationality is a logical necessity.

8. For clarity in your own thinking, and especially for purposes of communication, it is important to state what judgments you have used and which parts of your argument depend on which judgments. The advantage of likelihood is its mathematical independence of initial distributions (priors), and similarly the advantage of weight of evidence is its mathematical independence of the initial odds of the null hypothesis. The subjectivist states his judgments whereas the objectivist sweeps them under the carpet by calling assumptions *knowledge*, and he basks in the glorious objectivity of science.

9. The vagueness of a probability judgment is defined either as the difference between the upper and lower probabilities or else as the difference between the upper and lower log-odds (#290). I conjecture that the vagueness of a judgment is strongly correlated with its variation from one judge to another. This could be tested.

10. The distinction between type I and type II rationality is very similar to the distinction between the standard form of subjective probabilities and what I call *evolving* or *sliding* [or *dynamic*] *probabilities.* These are probabilities that are currently judged and they can change in the light of thinking only, without change of empirical evidence. The fact that probabilities change when *empirical* evidence changes is almost too elementary a point to be worth mentioning in this distinguished assembly, although it was overlooked by . . . R. A. Fisher in his fiducial argument. More precisely he talked about the probabilities of certain events or propositions without using the ordinary notation of the vertical stroke or any corresponding notation, and thus fell into a fallacy. . . . Evolving probabilities are essential for the refutation of Popper's views on simplicity (see #599 [or p. 223]). [Great men are not divine.]

11. My theories of probability and rationality are theories of consistency

only, that is, consistency between judgments and the basic axioms and rules of application of the axioms. Of course, these are usually judgments about the objective world. In particular it is incorrect to suppose that it is necessary to inject an initial probability distribution from which you are to infer a final probability distribution. It is just as legitimate logically to assume a final distribution and to infer from it an initial distribution. (For example, pp. 35 and 81 of #13.) To modify an aphorism quoted by Dr. Geisser elsewhere, "Ye priors shall be known by their posteriors."

12. This brings me to the *device of imaginary results* (#13) which is the recommendation that you can derive information about an initial distribution by an imaginary (*Gedanken*) experiment. Then you can make discernments about the final distribution after a real experiment. . . .

13. *The Bayes/non-Bayes compromise* (p. 863 of #127, and also many more references in #398 under the index entry "Compromises"). Briefly: use Bayesian methods to produce statistics, then look at their tail-area probabilities and try to relate these to Bayes factors. A good example of both the device of imaginary results and of the Bayes/non-Bayes compromise was given in #547. I there found that Bayesian significance tests in multiparameter situations seem to be much more sensitive to the assumed initial distribution than Bayesian estimation is.

14. The weakness of Bayesian methods for significance testing is also a strength, since by trying out your assumed initial distribution on problems of significance testing, you can derive much better initial distributions and these can then be used for problems of estimation. This improves the Bayesian methods of estimation!

15. Compromises between subjective probabilities and credibilities are also desirable because standard priors might be more general-purpose than non-standard ones. In fact it is mentally healthy to think of your subjective probabilities as estimates of credibilities (p. 5). Credibilities are an ideal that we cannot reach.

16. The need to compromise between simplicity of hypotheses and the degree to which they explain the facts was discussed in some detail in #599, and the name I gave for the appropriate and formally precise compromise was "a Sharpened Razor." Ockham (actually his eminent predecessor John Duns Scotus) in effect emphasized simplicity alone, without reference to *degrees* of explaining the facts. (See also #**1000**.)

17. The relative probabilities of two hypotheses are more relevant to science than the probabilities of hypotheses *tout court* (pp. 60, 83-84 of #13).

18. The objectivist or his customer reaches precise results by throwing away evidence; for example (a) when he keeps his eyes averted from the precise choice of random numbers by using a Statistician's Stooge; (b) when his customer uses confidence intervals for betting purposes, which is legitimate *provided that he regards the confidence statement as the entire summary of the evidence.*

19. If the objectivist is prepared to bet, then we can work backwards to infer

constraints on his implicit prior beliefs. These constraints are of course usually vague, but we might use precise values in accordance with type II rationality.

20. When you don't trust your estimate of the initial probability of a hypothesis you can still use the Bayes factor or a tail-area probability to help you decide whether to do more experimenting (p. 70 of #13).

21. Many statistical techniques are legitimate and useful but we should not knowingly be inconsistent. The Bayesian flavor vanishes when a probability is judged merely to lie in the interval (0,1), but this hardly ever happens.

22. A hierarchy of probability distributions, corresponding in a physical model to populations, superpopulations, etc., can be helpful to the judgment even when these superpopulations are not physical. I call these *distributions of types I, II, III, . . .*" (partly in order to be noncommittal about whether they are physical) but it is seldom necessary to go beyond the third type (##**26**, 398, 547). [When we are prepared to be committal, the current names for types II and III are "priors" and "hyperpriors."]

23. Many compromises are possible, for example, one might use the generalizations of the likelihood ratio mentioned on p. 80 of #198.

24. *Quasi- or pseudoutilities.* When your judgments of utilities are otherwise too wide it can be useful [in the planning of an experiment] to try to maximize the expectation of something else that is of value, known as a quasiutility or pseudoutility. Examples are (a) weight of evidence, when trying to discriminate between two hypotheses; (b) information in Fisher's sense when estimating parameters; (c) information in Shannon's sense when searching among a set of hypotheses; (d) strong explanatory power [explicativity] when explanation is the main aim: this includes example (c); (e) and (ea) tendency to cause (or a measure of its error) if effectiveness of treatment (or its measurement) is the aim; (f) f(error) in estimation problems, where $f(x)$ depends on the application and might, for example, reasonably be $1 - e^{-\lambda x^2}$ (or might be taken as x^2 for simplicity) [the sign should be changed]; (g) financial profit when other aims are too intangible. In any case the costs in money and effort have to be allowed for. [*After* an experiment is designed, a minimax inference minimizes the quasiutility. Minimax procedures cannot be fully justified but they lead to interesting proposals. See #622 or p. 198 of #618.]

25. The time to make a decision is largely determined by urgency and by the current rate of acquisition of information, evolving or otherwise. For example, consider chess timed by a clock.

26. In logic, the probability of a hypothesis does not depend on whether it was typed accidentally by a monkey, or whether an experimenter pretends he has a train to catch when he stops a sequential experiment. But in practice we do allow for the degree of respect we have for the ability and knowledge of the person who propounds a hypothesis.

27. All scientific hypotheses are numerological but some are more numerologi-

cal than others. Hence a subjectivistic analysis of numerological laws is relevant to the philosophy of induction (**#603B**).

I have not gone systematically through my writings to make sure that the above list is complete. In fact there are, for example, a few more principles listed in ##**85A**, 293. But I believe the present list is a useful summary of my position.

CHAPTER 3

46656 Varieties of Bayesians (#765)

Some attacks and defenses of the Bayesian position assume that it is unique so it should be helpful to point out that there are at least 46656 different interpretations. This is shown by the following classification based on eleven facets. The count would be larger if I had not artificially made some of the facets discrete and my heading would have been "On the Infinite Variety of Bayesians."

All Bayesians, as I understand the term, believe that it is usually meaningful to talk about the probability of a hypothesis and they make some attempt to be consistent in their judgments. Thus von Mises (1942) would not count as a Bayesian, on this definition. For he considered that Bayes's theorem is applicable only when the prior is itself a physical probability distribution based on a large sample from a superpopulation. If he is counted as a Bayesian, then there are at least 46657 varieties, which happens to rhyme with the number of Heinz varieties. But no doubt both numbers will increase on a recount.

Here are the eleven facets:

1. *Type II rationality.* (a) Consciously recognized; (b) not. Here Type II rationality is defined as the recommendation to maximize expected utility allowing for the cost of theorizing (#290). It involves the recognition that judgments can be revised, leading at best to consistency of *mature* judgments.

2. *Kinds of judgments.* (a) Restricted to a specific class or classes, such as preferences between actions; (b) all kinds permitted, such as of probabilities and utilities, and any functions of them such as expected utilities, weights of evidence, likelihoods, and surprise indices (#82; Good, 1954). This facet could of course be broken up into a large number.

3. *Precision of judgments.* (a) Sharp; (b) based on inequalities, i.e. partially ordered, but sharp judgments often assumed for the sake of simplicity (in accordance with 1[a]).

4. *Extremeness.* (a) Formal Bayesian procedure recommended for all applications; (b) non-Bayesian methods used provided that some set of axioms of intuitive probability are not seen to be contradicted (the Bayes/non-Bayes compromise: Hegel and Marx would call it a synthesis); (c) non-Bayesian methods used only after they have been given a rough Bayesian justification.

5. *Utilities.* (a) Brought in from the start; (b) avoided, as by H. Jeffreys; (c) utilities introduced separately from intuitive probabilities.

6. *Quasiutilities.* (a) Only one kind of utility recognized; (b) explicit recognition that "quasiutilities" (##690A, 755) are worth using, such as amounts of information or "weights of evidence" (Peirce, 1978 [but see #1382]; #13); (c) using quasiutilities without noticing that they are substitutes for utilities. The *use* of quasiutilities is as old as the words "information" and "evidence," but I think the name "quasiutility" serves a useful purpose in focussing the issue.

7. *Physical probabilities.* (a) Assumed to exist; (b) denied; (c) used as if they exist but without philosophical commitment (#617).

8. *Intuitive probability.* (a) Subjective probabilities regarded as primary; (b) credibilities (logical probabilities) primary; (c) regarding it as mentally healthy to think of subjective probabilities as estimates of credibilities, without being sure that credibilities really exist; (d) credibilities in principle definable by an international body. . . .

9. *Device of imaginary results.* (a) Explicit use; (b) not. The device involves imaginary experimental results used for judging final or posterior probabilities from which are inferred discernments about the initial probabilities. For examples see ##13, 547.

10. *Axioms.* (a) As simple as possible; (b) incorporating Kolmogorov's axiom (complete additivity); (c) using Kolmogorov's axiom when mathematically convenient but regarding it as barely relevant to the *philosophy* of applied statistics.

11. *Probability "types."* (a) Considering that priors can have parameters with "Type III" distributions, as a convenient technique for making judgments; (b) not. Here (a) leads, by a compromise with non-Bayesian statistics, to such techniques as Type II maximum likelihood and Type II likelihood-ratio tests (#547).

Thus there are at least $2^4 \cdot 3^6 \cdot 4 = 46656$ categories. This is more than the number of professional statisticians so some of the categories must be empty. Thomas Bayes hardly wrote enough to be properly categorized; a partial attempt is b--aaa?-b--. My own category is abcbcbccaca. What's yours?

CHAPTER 4

The Bayesian Influence, or How to Sweep Subjectivism under the Carpet (#838)

ABSTRACT

On several previous occasions I have argued the need for a Bayes/non-Bayes compromise which I regard as an application of the "Type II" principle of rationality. By this is meant the maximization of expected utility when the labour and costs of the calculations are taken into account. Building on this theme, the present work indicates how some apparently objective statistical techniques emerge logically from subjective soil, and can be further improved if their subjective logical origins (if not always histörical origins) are not ignored. There should in my opinion be a constant interplay between the subjective and objective points of view, and not a polarization separating them.

Among the topics discussed are, two types of rationality, 27 "Priggish Principles," 46656 varieties of Bayesians, the Black Box theory, consistency, the unobviousness of the obvious, probabilities of events that have never occurred (namely all events), the Device of Imaginary Results, graphing the likelihoods, the hierarchy of types of probability, Type II maximum likelihood and likelihood ratio, the statistician's utilities versus the client's, the experimenter's intentions, quasiutilities, tail-area probabilities, what is "more extreme"?, "deciding in advance," the harmonic mean rule of thumb for significance tests in parallel, density estimation and roughness penalities, evolving probability and pseudorandom numbers and a connection with statistical mechanics.

1. PREFACE

. . . There is one respect in which the title of this paper is deliberately ambiguous: it is not clear whether it refers to the *historical* or to the *logical* influence of "Bayesian" arguments. In fact it refers to both, but with more emphasis on the logical influence. Logical aspects are more fundamental to a science or philosophy than are the historical ones, although they each shed light on the other. The logical development is a candidate for being the historical development on *another* planet.

I have taken the expression the "Bayesian influence" from a series of lectures in mimeographed form (#750). In a way I am fighting a battle that has already been won to a large extent. For example, the excellent statisticians L. J. Savage, D. V. Lindley, G. E. P. Box (R. A. Fisher's son-in-law) and J. Cornfield were converted to the Bayesian fold years ago. For some years after World War II, I stood almost alone at meetings of the Royal Statistical Society in crusading for a Bayesian point of view. Many of the discussions are reported in the *Journal, series B*, but the most detailed and sometimes heated ones were held privately after the formal meetings in dinners at Berterolli's restaurant and elsewhere, especially with Anscombe, Barnard, Bartlett, Daniels, and Lindley. [Lindley was a non-Bayesian until 1954.] These protracted discussions were historically important but have never been mentioned in print before as far as I know. There is an unjustifiable convention in the writing of the history of science that science communication occurs only through the printed word. . . .

II. INTRODUCTION

On many previous occasions, and especially at the Waterloo conference of 1970, I have argued the desirability of a Bayes/non-Bayes compromise which, from one Bayesian point of view, can be regarded as the use of a "Type II" principle of rationality. By this is meant the maximization of expected utility when the labour and costs of calculations and thinking are taken into account. Building on this theme, the present paper will indicate how some apparently objective statistical techniques emerge logically from subjective soil, and can be further improved by taking into account their logical, if not always historical, subjective origins. There should be in my opinion a constant interplay between the subjective and objective points of view and not a polarization separating them.

Sometimes an orthodox statistician will say of one of his techniques that it has "intuitive appeal." This is I believe always a guarded way of saying that it has an informal approximate Bayesian justification.

Partly as a matter of faith, I believe that *all* sensible statistical procedures can be derived as approximations to Bayesian procedures. As I have said on previous occasions, "To the Bayesian all things are Bayesian."

Cookbook statisticians, taught by non-Bayesians, sometimes give the impression to *their* students that cookbooks are enough for all practical purposes. Any one who has been concerned with complex data analysis knows that they are wrong: that subjective judgment of probabilities cannot usually be avoided, even if this judgment can later be used for constructing apparently non-Bayesian procedures in the approved sweeping-under-the-carpet manner.

(a) What Is Swept under the Carpet?

I shall refer to "sweeping under the carpet" several times, so I shall use the abbreviations UTC and SUTC. One part of this paper deals with what is swept under the carpet, and another part contains some examples of the SUTC process. (The past tense, etc., will be covered by the same abbreviation.)

Let us then consider what it is that is swept under the carpet. Maurice Bartlett once remarked, in a discussion at a Research Section meeting of the Royal Statistical Society, that the word "Bayesian" is ambiguous, that there are many varieties of Bayesians, and he mentioned for example, "Savage Bayesians and Good Bayesians," and in a letter in the *American Statistician* I classified 46656 varieties (#765). There are perhaps not that number of practicing Bayesian statisticians, but the number comes to 46656 when your cross-classify the Bayesians in a specific manner by eleven facets. Some of the categories are perhaps logically empty but the point I was making was that there is a large variety of possible interpretations and some of the arguments that one hears against *the* Bayesian position are valid only against *some* Bayesian positions. As so often in controversies "it depends what you mean." The particular form of Bayesian position that I adopt might be called non-Bayesian by some people and naturally it is my own views that I would like most to discuss. I speak for some of the Bayesians all the time and for all the Bayesians some of the time. In the spoken version of this paper I named my position after "the Tibetan Lama K. Caj Doog," and I called my position "Doogian." Although the joke wears thin, it is convenient to have a name for this viewpoint, but "Bayesian" is misleading, and "Goodian" or "Good" is absurd, so I shall continue with the joke even in print. (See also Smith, 1961, p. 18, line minus 15, word minus 2.)

Doogianism is mainly a mixture of the views of a few of my eminent pre-1940 predecessors. Many parts of it are therefore not original, but, taken as a whole I think it has some originality; and at any rate it is convenient here to have a name for it. It is intended to be a general philosophy for reasoning and for rationality in action and not just for statistics. It is a philosophy that applies to all activity, to statistics, to economics, to the practice and philosophy of science, to ordinary behavior, and, for example, to chess-playing. Of course each of these fields of study or activity has its own specialized problems, but, just as the theories of each of them should be consistent with ordinary logic, they should in my opinion be consistent also with the theory of rationality as presented here and in my previous publications, a theory that is a useful and practically necessary extension of ordinary logic. . . .

At the Waterloo conference (#679), I listed 27 Priggish Principles that summarize the Doogian philosophy, and perhaps the reader will consult the Proceedings and some of its bibliography for a more complete picture, and for historical information. Here it would take too long to work systematically through all 27 principles and instead I shall concentrate on the eleven facets of the Bayesian Varieties in the hope that this will give a fairly clear picture. I do not claim that any of these principles were "discovered last week" (to quote Oscar Kempthorne's off-the-cuff contribution to the spoken discussion), in fact I have developed, acquired or published them over a period of decades, and most of them were used by others before 1940, in one form or another, and with various degrees of bakedness or emphasis. The main merit that I claim for the Doogian philosophy is that it codifies and exemplifies an adequately complete and simple theory of rationality, complete in the sense that it is I believe not subject to the criticisms that are usually directed at other forms of Bayesianism, and simple in the sense that it attains realism with the minimum of machinery. To pun somewhat, it is "minimal sufficient."

(b) Rationality, Probability, and the Black Box Theory

In some philosophies of rationality, a rational man is defined as one whose judgments of probabilities, utilities, and of functions of these, are all both consistent and sharp or precise. Rational men do not exist, but the concept is useful in the same way as the concept of a reasonable man in legal theory. A rational man can be regarded as an ideal to hold in mind when we ourselves wish to be rational. It is sometimes objected that rationality as defined here depends on betting behavior, and people sometimes claim they do not bet. But since their every decision is a bet I regard this objection as unsound: besides they could in principle be forced to bet in the usual monetary sense. It seems absurd to me to suppose that the *rational* judgment of probabilities would normally depend on whether you were forced to bet rather than betting by free choice.

There are of course people who argue (rationally?) against rationality, but presumably they would agree that it is sometimes desirable. For example, they would usually prefer that their doctor should make rational decisions, and, when they were fighting a legal case in which they were sure that the evidence "proved" their case, they would presumably want the judge to be rational. I believe that the dislike of rationality is often merely a dishonest way of maintaining an indefensible position. Irrationality is intellectual violence against which the pacifism of rationality may or may not be an adequate weapon.

In practice one's judgments are not sharp, so that to use the most familiar axioms it is necessary to work with judgments of inequalities. For example, these might be judgments of inequalities between probabilities, between utilities, expected utilities, weights of evidence (in a sense to be defined . . .), or any other convenient function of probabilities and utilities. We thus arrive at a theory that can be regarded as a combination of the theories espoused by F. P. Ramsey (1926/31/50/64), who produced a theory of precise subjective probability and utility, and of J. M. Keynes (1921), who emphasized the importance of inequalities (partial ordering) but regarded logical probability or credibility as the fundamental concept, at least until he wrote his obituary on Ramsey (Keynes, 1933).

To summarize then, the theory I have adopted since about 1938 is a theory of subjective (personal) probability and utility in which the judgments take the form of inequalities (but see Section III [iii] below). This theory can be formulated as the following "black box" theory. . . .[See pp. 75-76.]

To extend this theory to rationality, we need merely to allow judgments of preferences also, and to append the "principle of rationality," the recommendation to maximize expected utility. (##13, **26, 230.**)

The axioms, which are expressed in conditional form, have a familiar appearance (but see the reference to "evolving probability" below), and I shall not state them here.

There is emphasis on human judgment in this theory, based on a respect for the human brain. Even infrahuman brains are remarkable instruments that cannot yet be replaced by machines, and it seems unlikely to me that decision-making in general, and statistics in particular, can become independent of the human brain until the first ultraintelligent machine is built. Harold Jeffreys once remarked that the brain may not be a perfect reasoning machine but is the only one available. That is still true, though a little less so than when Jeffreys first said it. It [the brain] has been operating for millions of years in billions of individuals and it has developed a certain amount of magic. On the other hand I believe that some formalizing is useful and that the ultraintelligent machine will also use a subjectivistic theory as an aid to its reasoning.

So there is this respect for judgment in the theory. But there is also respect for logic. Judgments and logic must be combined, because, although the human brain is clever at perceiving facts, it is also cunning in the rationalization of falsity for the sake of its equilibrium. You *can* make bad judgments so you need a black box to check your subjectivism and to make it more objective. That then is the purpose of a subjective theory; to increase the objectivity of your judgments, to check them for consistency, to detect the inconsistencies and to remove them. Those who want their subjective judgments to be free and untrammeled by axioms regard themselves as objectivists: paradoxically, it is the subjectivists who are prepared to discipline their own judgments!

For a long time I have regarded this theory as almost intuitively obvious, partly perhaps because I have used it in many applications without inconsistencies arising, and I know of no other theory that I could personally adopt. It is the one and only True Religion. My main interest has been in developing and applying the theory, rather than finding *a priori* justification for it. But such a justification has been found by C. A. B. Smith (1961), based on a justification by Ramsey (1926/31/50/64), de Finetti (1937/64), and L. J. Savage (1954) of a slightly different theory (in which sharp values of the probabilities and utilities are assumed). These justifications show that, on the assumption of certain compelling desiderata, a rational person will hold beliefs, and preferences, *as if* he had a system of subjective probabilities and utilities satisfying a familiar set of axioms. He might as well be explicit about it: after all it is doubtful whether any of our knowledge is better than of the "as if" variety.

Another class of justifications, in which utilities are not mentioned, is exemplified by Bernstein (1921/22), Koopman (1940a, b), and R. T. Cox (1946, 1961). (See also pp. 105-106 of #13; also Good, 1962d.) A less convincing, but simpler justification is that the product and addition axioms are forced (up to a monotonic transformation of probabilities) when considering ideal games of chance, and it would be *surprising* if the same axioms did not apply more generally. Even to deny this seems to me to show poor or biased judgment.

Since the degrees of belief, concerning events over which he has no control, of a person with ideally good judgment, should surely not depend on whether he intends to use his beliefs in any specific manner, it seems desirable to have justifications that do not mention preferences or utilities. But utilities necessarily come in whenever the beliefs are to be used in a practical problem involving action.

(c) Consistency and the Unobviousness of the Obvious

Everybody knows, all scientists know, all mathematicians know, all players of chess know, that from a small number of sharp axioms you can develop a very rich theory in which the results are by no means obvious, even though they are, in a technical sense, tautological. This is an exciting feature of mathematics, especially since it suggests that it might be a feature of the universe itself. Thus the completion of the basic axioms of probability by Fermat and Pascal led to many interesting results, and the further near completion of the axioms for the *mathematical* theory of probability by Kolmogorov led to an even greater expansion of mathematical theory.

Mathematicians are I think often somewhat less aware that a system of rules and suggestions of an axiomatic system can also stimulate useful technical advances. The effect can be not necessarily nor even primarily to produce theorems of great logical depth, but sometimes more important to produce or to emphasize attitudes and techniques for reasoning and for statistics that seem obvious enough after they are suggested, but continue to be overlooked even then.

One reason why many theoretical statisticians prefer to prove mathematical theorems rather than to emphasize logical issues is that a theorem has a better chance of being indisputably novel. The person who proves the theorem can claim all the credit. But with general principles, however important, it is usually possible to find something in the past that to some extent foreshadows it. There is usually something old under the sun. Natural Selection was stated by Aristotle, but Darwin is not denied credit, for most scientists were still overlooking this almost obvious principle.

Let me mention a personal example of how the obvious can be overlooked.

In 1953, I was interested in estimating the physical probabilities for large contingency tables (using essentially a log-linear model) when the entries were very small, including zero. In the first draft of the write-up I wrote that I was concerned with the estimation of *probabilities of events that had never occurred before* (#83). Apparently this *concept* was itself an example of such an event, as far as the referee was concerned because he felt it was too provocative, and I deleted it in deference to him. Yet this apparently "pioneering" remark is obvious: every event in life is unique, and every real-life probability that we estimate in practice is that of an event that has never occurred before, provided that we do enough cross-classification. Yet there are many "frequentists" who still sweep this fact UTC.

A statistical problem where this point arises all the time is in the estimation of physical probabilities corresponding to the cells of multidimensional contingency tables. Many cells will be empty for say a 2^{20} table. A Bayesian proposal for this problem was made in Good (p. #75 of #398), and I am hoping to get a student to look into it; and to compare it with the use of log-linear models which have been applied to this problem during the last few years. One example of the use of a log-linear model is, after taking logarithms of the relative frequencies, to apply a method of smoothing mentioned in #146 in relation to factorial experiments: namely to treat non-significant interactions as zero (or of course they could be "flattened" Bayesianwise instead for slightly greater accuracy).

Yet another problem where the probabilities of events that have never occurred before are of interest is the species sampling problem. One of its aspects is the estimation of the probability that the next animal or word sampled will be one that has not previously occurred. The answer turns out to be approximately equal to n_1/N, where n_1 is the number of species that have so far occurred just once, and N is the total sample size: see ##38 & 86; this work was originated with an idea of Turing's (1940) which anticipated the empirical Bayes method in a special case. (See also Robbins, 1968.) The method can be regarded as non-Bayesian but with a Bayesian influence underlying it. More generally, the probability that the next animal will be one that has so far been represented r times is approximately $(r + 1)n_{r+1}/N$, where n_r is the "frequency of the frequency r," that is, the number of species each of which has already been represented r times. (In practice it is necessary to smooth the n_r's when applying this formula, to get adequate results, when $r>1$.) I shall here give a new proof of this result. Denote the event of obtaining such an animal by E_r. Since the order in which the N animals were sampled is assumed to be irrelevant (a Bayesian-type assumption of permutability), the required probability can be estimated by the probability that E_r would have occurred on the last occasion an animal was sampled if a random permutation were applied to the order in which the N animals were sampled. But E_r would have occurred if the last animal had belonged to a species represented $r + 1$ times *altogether*. This gives the result, except that for greater accuracy we should remember that we are talking about the $(N+1)$st trial, so that a more accurate result is $(r + 1)\mathcal{E}_{N+1}(n_{r+1})/(N + 1)$. Hence the expected physical probability q_r corresponding to those n_r species that have so far occured r times is

$$\mathcal{E}(q_r) = \frac{r+1}{N+1} \cdot \frac{\mathcal{E}_{N+1}(n_{r+1})}{\mathcal{E}_N(n_r)}.$$

This is formula (15) of #38 which was obtained by a more Bayesian argument. The "variance" of q_r was also derived in that paper, and a "frequency" proof of it would be more difficult. There is an interplay here between Bayesian and frequency ideas.

One aspect of Doogianism which dates back at least to F. P. Ramsey (1926/31/50/64) is the emphasis on *consistency*: for example, the axioms of probability can provide only *relationships* between probabilities and cannot manufacture a probability out of nothing. Therefore there must be a hidden assumption in

Fisher's fiducial argument. (This assumption is pinpointed in **#659** on its p. 139 omitted herein. The *reason* Fisher overlooked this is also explained there.)

The idea of consistency seems weak enough, but it has the following immediate consequence which *is* often overlooked.

Owing to the adjectives "initial" and "final" or "prior" and "posterior," it is usually assumed that initial probabilities must be assumed before final ones can be calculated. But there is nothing in the theory to prevent the implication being in the reverse direction: we can make judgments of initial probabilities and infer final ones, or we can equally make judgments of final ones and infer initial ones by *Bayes's theorem in reverse*. Moreover this can be done corresponding to entirely *imaginary* observations. This is what I mean by the Device of Imaginary Results for the judging of initial probabilities. (See, for example, Index of #13). I found this device extremely useful in connection with the choice of a prior for multinomial estimation and significance problems (#547) and I believe the device will be found to be of the utmost value in future Bayesian statistics. Hypothetical experiments have been familiar for a long time in physics, and in the arguments that led Ramsey to the axioms of subjective probability, but the use of Bayes's theorem in reverse is less familiar. "Ye priors shall be known by their posteriors" (p. 17). Even the slightly more obvious technique of imaginary bets is still disdained by many decision makers who like to say "That possibility is purely hypothetical." Anyone who disdains the hypothetical is a philistine.

III. THE ELEVENFOLD PATH OF DOOGIANISM

As I said before, I should now like to take up the 46656 varieties of Bayesians, in other words the eleven facets for their categorization. I would have discussed the 27-fold path of Doogianism if there had been space enough.

(i) Rationality of Types I and II

I have already referred to the first facet. Rationality of Type I is the recommendation to maximize expected utility, and Type II is the same except that it allows for the cost of theorizing. It means that in any practical situation you have to decide when to stop thinking. You can't allow the current to go on circulating round and round the black box or the cranium forever. You would like to reach a sufficient maturity of judgments, but you have eventually to reach some conclusion or to make some decision and so you must be prepared to sacrifice strict logical consistency. At best you can achieve consistency as far as you have seen to date (p. 49 of #13). There is a time element, as in chess, and this is realistic of most practice. It might not appeal to some of you who love ordinary logic, but it is a mirror of the true situation.

It may help to convince some readers if I recall a remark of Poincaré's that some antinomies in ordinary (non-probabilistic) logic can be resolved by bringing in a time element. ["Temporal," "evolving" or "dynamic" logic?]

The notion of Type II rationality, which I believe resolves a great many of the controversies between the orthodox and Bayesian points of view, also involves a shifting of your probabilities. The subjective probabilities shift as a consequence of thinking. . . . [See p. 107.] The conscious recognition of Type II rationality, or not, constitutes the two aspects of the first facet.

Another name for the principle of Type II rationality might be the *Principle of Non-dogmatism.*

(ii) Kinds of Judgment

Inequalities between probabilities and between expected utilities are perhaps the most standard type of judgment, but other kinds are possible. Because of my respect for the human mind, I believe that one should allow any kind of judgments that are relevant. One kind that I believe will ultimately be regarded as vying in importance with the two just mentioned is a judgment of "weights of evidence" (defined later) a term introduced by Charles Sanders Peirce (1878) although I did not know this when I wrote my 1950 book. . . .

It will encourage a revival of reasoning if statisticians adopt this appealing terminology . . . . [But Peirce blew it. See #1382.]

One implication of the "suggestion" that all types of judgments can be used is to encourage you to compare your "overall" judgments with your detailed ones; for example, a judgment by a doctor that it is better to operate than to apply medical treatment, on the grounds perhaps that this would be standard practice in the given circumstances, can be "played off" against separate judgments of the probabilities and utilities of the outcomes of the various treatments.

(iii) Precision of Judgments

Most theories of subjective probability deal with numerically precise probabilities. These would be entirely appropriate if you could always state the lowest odds that you would be prepared to accept in a gamble, but in practice there is usually a degree of vagueness. Hence I assume that subjective probabilities are only partially ordered. In this I follow Keynes and Koopman, for example, except that Keynes dealt primarily with logical probabilities, and Koopman with "intuitive" ones (which means either logical or subjective). F. P. Ramsey (1926/31/50/64) dealt with subjective probabilities, but "sharp" ones, as mentioned before.

A theory of "partial ordering" (inequality judgments) for probabilities is a compromise between Bayesian and non-Bayesian ideas. For if a probability is judged merely to lie between 0 and 1, this is equivalent to making no judgment about it at all. The vaguer the probabilities the closer is this Bayesian viewpoint to a non-Bayesian one.

Often, in the interests of simplicity, I assume sharp probabilities, as an approximation, in accordance with Type II rationality.

(iv) Eclecticism

Many Bayesians take the extreme point of view that Bayesian methods should always be used in statistics. My view is that non-Bayesian methods are acceptable *provided that they are not seen to contradict your honest judgments, when combined with the axioms of rationality.* This facet number (iv) is an application of Type II rationality. I believe it is sometimes, but not by any means always, easier to use "orthodox" (non-Bayesian) methods, and that they are often *good enough.* It is always an application of Type II rationality to say that a method is good enough.

(v) Should Utilities Be Brought in from the Start in the Development of the Theory?

I have already stated my preference for trying to build up the theory of subjective probability without reference to utilities and to bring in utilities later. The way the axioms are introduced is not of great practical importance, provided that the same axioms are reached in the end, but it is of philosophical interest. Also there is practical interest in seeing how far one can go without making use of utilities, because one might wish to be an "armchair philosopher" or "fun scientist" who is more concerned with discovering facts about Nature than in applying them. ("Fun scientist" is not intended to be a derogatory expression.) Thus, for example, R. A. Fisher and Harold Jeffreys never used ordinary utilities in their statistical work as far as I know (and when Jeffreys chaired the meeting in Cambridge when I presented my paper **#26** he stated that he had never been concerned with economic problems in his work on probability). See also the following remarks concerned with quasiutilities.

(vi) Quasiutilities

Just as some schools of Bayesians regard subjective probabilities as having sharp (precise) values, some assume that utilities are also sharp. The Doogian believes that this is often not so. It is not merely that utility inequality judgments of course vary from one person to another, but that utilities for individuals can also often be judged by them only to lie in wide intervals. It consequently becomes useful and convenient to make use of substitutes for utility which may be called *quasiutilities* or *pseudoutilities.* Examples and applications of quasiutilities will be considered later in this paper. The conscious recognition or otherwise of quasiutilities constitutes the sixth facet.

(vii) Physical Probability

Different Bayesians have different attitudes to the question of physical probability. de Finetti regards it as a concept that can be defined in terms of subjective probability, and does not attribute any other "real existence" to it. My view, or that of my alter ego, is that it seems reasonable to suppose that physical probabilities do exist, but that they can be measured only be means of a theory of subjective probability. For a fuller discussion of this point see de Finetti (1968/70) and #617. The question of the real existence of physical probabilities relates to the problem of determinism versus indeterminism and I shall have something more to say on this.

When I refer to physical probability I do not assume the long-run frequency definition: physical probability can be applied just as well to unique circumstances. Popper suggested the word "propensity" for it, which I think is a good term, although I think the suggestion of a word cannot by itself be regarded as the propounding of a "theory." [See also p. 405 of Feibleman, 1969.] As I have indicated before, I think good terminology is important in crystallizing out ideas. Language can easily mislead, but part of the philosopher's job is to find out where it can *lead*. Curiously enough Popper has also stated that the words you use do not matter much: what is important is what they mean in your context. Fair enough, but it can lead to Humpty-Dumpty-ism, such as Popper's interpretation of simplicity [or Carnap's usage of "confirmation" which has misled philosophers for decades].

(viii) Which is Primary, Logical Probability (Credibility) or Subjective Probability?

It seems to me that subjective probabilities are primary because they are the ones you have to use whether you like it or not. But I think it is mentally healthy to think of your subjective probabilities as estimates of credibilities, whether these really "exist" or not. Harold Jeffreys said that the credibilities should be laid down by an international body. He would undoubtedly be the chairman. As Henry Daniels once said (c. 1952) when I was arguing for subjectivism, "all statisticians would like their models to be adopted," meaning that in some sense everybody is a subjectivist.

(ix) Imaginary Results

This matter has already been discussed but I am mentioning it again because it distinguishes between some Bayesians in practice, and so forms part of the categorization under discussion. I shall give an example of it now because this will help to shed light on the tenth facet.

It is necessary to introduce some notation. Let us suppose that we throw a sample of N things into t pigeon holes, with statistically independent physical probabilities $p_1, p_2, \ldots, p_t$, these being unknown, and that you obtain frequencies $n_1, n_2, \ldots, n_t$ in the t categories or cells. This is a situation that has much interested philosophers of induction, but for some reason, presumably lack of familiarity, they do not usually call it multinomial sampling. In common with many people in the past, I was interested (##398, 547) in estimating the physical probabilities $p_1, p_2, \ldots, p_t$. . . . [See pp. 100-103.]

That then is an example of a philosophical attitude leading to a practical solution of a statistical problem. As a matter of fact, it wasn't just the estimation of the p's that emerged from that work, but, more important, a significance test for whether the p's were all equal. The method has the pragmatic advantage that it can be used for all sample sizes, whereas the ordinary chi-squared test breaks down when the cell averages are less then 1. Once you have decided on a prior (the initial relative probabilities of the components of the non-null hypothesis), you can calculate the weight of evidence against the null hypothesis without using asymptotic theory. (This would be true for any prior that is a linear combination of Dirichlet distributions, even if they were not symmetric, because in this case the calculations involve only one-dimensional integrations.) That then was an example of the device of imaginary results, for the selection of a prior, worked out in detail.

The successful use of the device of imaginary results for this problem *makes it obvious that it can and will also be used effectively for many other statistical problems. I believe it will revolutionize multivariate Bayesian statistics.*

(x) Hierarchies of Probabilities

When you make a judgment about probabilities you might sit back and say "Is that judgment probable." This is how the mind works—it is natural to think that way, and this leads to a hierarchy of types of probabilities (#**26**) which in the example just mentioned, I found useful, as well as on other occasions. Now an objection immediately arises: There is nothing in principle to stop you integrating out the higher types of probability. But it remains a useful suggestion to help the mind in making judgments. It was used in #547 and has now been adopted by other Bayesians, using different terminology, such as priors of the second "order" (instead of "type" or "two-stage Bayesian models." A convenient term for a parameter in a prior is "hyperparameter." [See also #**1230**.]

New techniques arose out of the hierarchical suggestion, again apparently first in connection with the multinomial distribution (in the same paper), namely the concept of Type II maximum likelihood (maximization of the Bayes factor against the null hypothesis by allowing the hyperparameters to vary), and that of a Type II likelihood ratio for significance tests. I shall discuss these two concepts when discussing likelihood in general.

(xi) The Choice of Axioms

One distinction between different kinds of Bayesians is merely a mathematical one, whether the axioms should be taken as simple as possible, or whether, for example, they should include Kolmogorov's axiom, the axiom of complete additivity. I prefer the former course because I would want people to use the axioms even if they do not know what "enumerable" means, but I am prepared to use Kolmogorov's axiom whenever it seems to be sufficiently mathematically convenient. Its interest is mathematical rather than philosophical, except perhaps for the philosophy of mathematics. This last facet by the way is related to an excellent lecture by Jimmie Savage of about 1970, called "What kind of probability do you want?".

So much for the eleven facets. Numbers (i) to (vii) and number (ix) all involve a compromise with non-Bayesian methods; and number (xiii) a compromise with the "credibilists."

IV. EXAMPLES OF THE BAYESIAN INFLUENCE AND OF SUTC

(a) The Logical and Historical Origins of Likelihood

One aspect of utility is communicating with other people. There are many situations where you are interested in making a decision without communicating. But there are also many situations, especially in much statistical and scientific practice where you do wish to communicate. One suggestion, "obvious," and often overlooked as usual, is that you should make your assumptions clear and you should try to separate out the part that is disputable from the part that is less so. One immediate consequence of this suggestion is an emphasis on likelihood, because, as you all know, in Bayes's theorem you have the initial probabilities, and then you have the likelihoods which are the probabilities of the event, given the various hypotheses, and then you multiply the likelihoods by the probabilities and that gives you results proportional to the final probabilities. That is Bayes's theorem expressed neatly, the way Harold Jeffreys (1939/61) expressed it. Now the initial probability of the null hypothesis is often highly disputable. One person might judge it to be between 10^{-3} and 10^{-1} whereas another might judge it to be between 0.9 and 0.99. There is much less dispute about likelihoods. There is no dispute about the numerical values of likelihoods if your basic parametric model is accepted. Of course you usually have to use subjective judgment in laying down your parametric model. Now the *hidebound* objectivist tends to hide that fact; he will not volunteer the information that he uses judgment at all, but if pressed he will say "I do, in fact, have good judgment." So there are good and bad subjectivists, the bad subjectivists are the people with bad or dishonest judgment and also the people who do not make their assumptions clear when communicating with other people. But, on the other hand, there are no good 100% (hidebound) objectivists; they are all bad because they sweep their judgments UTC.

> *Aside*: In the spoken discussion the following beautiful interchanges took place. *Kempthorne* (who also made some complimentary comments): Now, on the likelihood business, the Bayesians discovered likelihood Goddamit! Fisher knew all this stuff. Now look Jack, you are an educated guy. Now please don't pull this stuff. This really drives me up the wall! *Lindley*: If Fisher understood the likelihood principle why did he violate it? *Kempthorne*: I'm not saying he understood it and I'm not saying you do or you—nobody understands it. But likelihood ideas, so to speak, have some relevance to the data. That's a completely non-Bayesian argument. *Good*: It dates back to the 18th century. *Kempthorne*: Oh it dates back; but there are a lot of things being (?) Doogian, you know. They started with this guy

Doog. Who is this bugger? Doog is the guy who spells everything backwards.

In reply to this entertaining harangue, which was provoked by a misunderstanding that was perhaps my fault, although I did refer to Fisherian information, I mention the following points. Bayes's theorem (Bayes, 1763/65, 1940/58; Laplace, 1774) cannot be stated without introducing likelihoods; *therefore likelihood dates back at least to 1774.* Again, *maximum* likelihood was used by Daniel Bernoulli (1774/78/1961); see, for example, Todhunter (1865, p. 236) or Eisenhart (1964, p. 29). Fisher introduced the name *likelihood* and emphasized the method of maximum likelihood. Such emphasis is important and of course merits recognition. The fact that he was anticipated in its use does not deprive him of the major part of the credit or of the blame especially as the notion of defining [his kind of] amount of information in terms of likelihood was his brilliant idea and it led to the Aitken-Silverstone information inequality (the minimum-variance bound). [Perhaps *not* due to Aitken and Silverstone.]

Gauss (1798/1809/57/1963) according to Eisenhart, used inverse probability combined with a Bayes *postulate* (uniform initial distribution) and an assumption of normal error, to give one of the interpretations of the method of least squares. He could have used maximum likelihood in this context but apparently did not, so perhaps Daniel Bernoulli's use of maximum likelihood had failed to convince him or to be noticed by him. Further historical research might be required to settle this last question if it is possible to settle it at all.

So likelihood is important as all statisticians agree now-a-days, and it *takes sharper values* than initial probabilities. But some people have gone to extremes and say that initial probabilities don't mean anything. Now I think one reason for their saying so is trade unionism of a certain kind. It is very nice for a statistician to be able to give his customer absolutely clear-cut results. It is unfortunate that he can't do it so he is tempted to cover up, to pretend he has not had to use any judgment. Those Bayesians who *insist* on sharp initial probabilities are I think also guilty of "trade unionism," unless they are careful to point out that these are intended only as crude approximations, for I do not believe that sharp initial probabilities usually correspond to their honest introspection. If, on the other hand, they agree that they are using only approximations we might need more information about the degree of the approximations, and then they would be forced to use inequality judgments, thus bringing them closer to the True Religion. (I believe Dr. Kyburg's dislike of the Bayesian position, as expressed by him later in this conference, depended on his interpreting a Bayesian as one who uses sharp initial probabilities.) The use of "vague" initial probabilities (inequality judgments) does not prevent Bayes's theorem from establishing the likelihood principle. For Dr. Kempthorne's benefit, and perhaps for some others, I mention that to me the likelihood principle means that the likelihood function exhausts all the information about the parameters that can be obtained from an experiment or observation, provided of course that there is an undisputed set of exhaustive simple statistical hypotheses such as is provided, for example, by a parametric model. (In practice, such assumptions are often undisputed but are never indisputable. This is the main reason why significance tests, such as the chi-squared test, robust to changes in the model, are of value. Even here there is a Doogian interpretation that can be based on beliefs about the distribution of the test statistic when it is assumed that the null hypothesis is false. I leave this point on one side for the moment.) Given the likelihood, the inferences that can be drawn from the observations would, for example, be unaffected if the statistician arbitrarily and falsely calimed that he had a train to catch, although he really had decided to stop sampling because his favorite hypothesis was ahead of the game. (This might cause you to distrust the statistician, but if you believe his observations, this distrust would be immaterial.) On the other hand, the "Fisherian" tail-area method for significance testing violates the likelihood principle because the statistician who is prepared to pretend he has a train to catch (optional stopping of sampling) can reach arbitrarily high significance levels, given enough time, even when the null hypothesis is true. For example, see Good (1956).

(b) Weight of Evidence

Closely related to the concept of likelihood is that of weight of evidence, which I mentioned before and promised to define.

Let us suppose that we have only two hypotheses under consideration, which might be because we have decided to consider hypotheses two at a time. Denote them by H and $\bar{\text{H}}$, where the bar over the second H denotes negation. (These need not be "simple statistical hypotheses," as defined in a moment.) Suppose further that we have an event, experimental result, or observation denoted by E. The conditional probability of E is either $P(\text{E}|\text{H})$ or $P(\text{E}|\bar{\text{H}})$, depending on whether H or $\bar{\text{H}}$ is assumed. If H and $\bar{\text{H}}$ are "simple statistical hypotheses," then these two probabilities have sharp uncontroversial values given tautologically by the meanings of H and $\bar{\text{H}}$. Even if they are *composite* hypothesis, not "simple" ones, the Bayesian will still be prepared to talk about these two probabilities. In either case we can see, by four applications of the product axiom, or by two applications of Bayes's theorem, that

$$\frac{P(\text{E}|\text{H})}{P(\text{E}|\bar{\text{H}})} = \frac{O(\text{H}|\text{E})}{O(\text{H})}$$

where O denotes *odds*. (The odds corresponding to a probability p are defined as $p/(1-p)$.) Turing (1941) called the right side of this equation *the factor in favor of the hypothesis H provided by the evidence E,* for obvious reasons. Its logarithm is the *weight of evidence* in favor of H, as defined independently by Peirce (1878), #13, and Minsky and Selfridge (1961). [But see #1382.] It was much used by Harold Jeffreys (1939/61), except that in that book he identified it with the final log-odds because his initial probabilities were taken as 1/2. He had previously (1936) used the general form of weight of evidence and had called it "support." The non-Bayesian uses the left side of the equation, and calls it the probability ratio, provided that H and $\bar{\text{H}}$ are simple statistical hypotheses. He SUTC the right side, because he does not talk about the probability of a hypothesis. The Bayesian, the doctor, the judge and the jury can appreciate the importance of the right side even with only the vaguest estimates of the initial odds of H. For example, the Bayesian (or at least the Doogian) can logically argue in the following manner (p. 70 of #13): If we assume that it was sensible to start a sampling experiment in the first place, and if it has provided appreciable weight of evidence in favor of some hypothesis, and it is felt that the hypothesis is not yet convincing enough, then it is sensible to enlarge the sample since we know that the final odds of the hypothesis have increased whatever they are. Such conclusions can be reached even though judgments of the relevant initial probability and of the utilities have never been announced. Thus, even when the initial probability is extremely vague, the axioms of subjective probability (and weight of evidence) can be applied.

When one or both of H and $\bar{H}$ are composite, the Bayesian has to assume relative initial probabilities for the simple components of the composite hypothesis. Although these are subjective, they typically seem to be less subjective than the initial probability of H itself. To put the matter more quantitatively, although this is not easy in so general a context, I should say that the judgment of the factor in favor of a hypothesis might typically differ from one person to another by up to about 5, while the initial odds of H might differ by a factor of 10 or 100 or 1000. Thus the separation of the estimation of the weight of evidence from the initial or final probability of H serves a useful purpose, especially for communication with other people, just as it is often advisable to separate the judgments of initial probabilities and likelihoods.

It often happens that the weight of evidence is so great that a hypothesis seems convincing almost irrespective of the initial probability. For example, in quantum mechanics, it seems convincing that the Schrödinger equation is approximately true (subject to some limitations), given the rest of some standard formulation of quantum mechanics, because of great quantities of evidence from a variety of experiments,such as the measurements of the frequencies of spectral lines to several places of decimals. The large weight of evidence makes it seem, to people who do not stop to think, that the initial probability of the equation, conditional on the rest of the theory, is irrelevant; but really there has to be an implicit judgment that the initial probability is not too low; for example, not less than 10^{-50}. (In a fuller discussion I would prefer to talk of the relative odds of *two* equations in competition.) How we judge such inequalities, whether explicitly or implicitly, is not clear: if we knew how we made judgments we would not call them judgments (#183). It must be something to do with the length of the equation (just as the total length of [the "meaningful" nonredundant parts of the] chromosomes in a cell could be used as a measure of complexity of an organism) and with its analogy with the classical wave equation and heat equation. (The latter has even suggested to some people, for example, Weizel [1953], that there is some underlying random motion that will be found to "explain" the equation.) At any rate the large weight of evidence permits the initial probability to be SUTC and it leads to an apparent objectivism (the reliance on the likelihoods alone) that is really multisubjectivism. The same happens in many affairs of ordinary life, in perception (p. 68 of #13), in the law, and in medical diagnosis (for example, #755).

On a point of terminology, the factor in favor of a hypothesis is equal to the likelihood ratio, in the sense of Neyman, Pearson, and Wilks, only when both H and $\bar{H}$ are simple statistical hypotheses. This is another justification for using Turing's and Peirce's expressions, apart from their almost self-explanatory nature, which provides their potential for improving the reasoning powers of all people. Certainly the expression "weight of evidence" captures one of the meanings that was intended by ordinary language. It is not surprising that it was an outstanding philosopher who first noticed this: for one of the functions of philosophy is to make such captures. [It is a pity that Peirce's discussion contained an error.]

George Barnard, who is one of the Likelihood Brethren, has rightly emphasized the merits of graphing the likelihood function. A Bayesian should support this technique because the initial probability density can be combined with the likelihood afterwards. If the Bayesian is a subjectivist he will know that the initial probability density varies from person to person and so he will see the value of graphing of the likelihood function for communication. A Doogian will consider that even his own initial probability density is not unique so he should approve even more. Difficulties arise in general if the parameter space has more than two dimensions, both in picturing the likelihood hypersurface or the posterior density hypersurface. The problem is less acute when the hypersurfaces are quadratic in the neighborhood of the maximum. In any case the Bayesian can in addition reduce the data by using such quantities as expected utilities. Thus he has all the advantages claimed by the likelihood brotherhood, but has additional flexibility. [See also #862, p. 711 and #1444.]

(c) Maximum Likelihood, Invariance, Quasiutilities, and Quasilosses

Let us now consider the relationship between Bayesian methods and *maximum* likelihood.

In a "full-dress" Bayesian estimation of parameters, allowing for utilities, you compute their final distribution and use it, combined with a loss function, to find a single recommended value, if a point estimate is wanted. When the loss function is quadratic this implies that the point estimate should be the final expectation of the parameter (even for vector parameters if the quadratic is non-singular). The final expectation is also appropriate if the parameter is a physical probability because the subjective expectation of a physical probability of an event is equal to the current subjective probability of that event.

If you do not wish to appear to assume a loss function, you can adopt the argument of Jeffreys (1939/61, Section 4.0). He points out that for a sample of size n (n observations), the final probability density is concentrated in a range of order $n^{-1/2}$, and that the difference between the maximum-likelihood value of the parameter and the mode of the final probability density is of the order $1/n$. (I call this last method, the choice of this mode, a Bayesian method "in mufti.") "Hence if the number of observations is large, the error committed by taking the maximum likelihood solution as the estimate is less than the uncertainty inevitable in any case. . . . The above argument shows that in the great bulk of cases its results are indistinguishable from those given by the principle of inverse probability, which supplies a justification for it." It also will not usually make much difference if the parameter is assumed to have a uniform initial distribution. (Jeffreys, 1939/61, p. 145; p. 55 of #13. L. J. Savage, 1959/62, p. 23, named estimation that depends on this last point "stable estimation.")

By a slight extension of Jeffreys's argument, we can see that a point estimate based on a loss function, whether it is the expectation of the parameter or some other value (which will be a kind of average) induced by the loss function, will also be approximated by using the Bayes method in mufti, and by the maximum-likelihood estimate, when the number of observations is large. Thus the large-sample properties of the maximum-likelihood method cannot be used for distinguishing it from a wide class of Bayesian methods, whether full-dress or in mufti. This is true whether we are dealing with point estimates or interval estimates. Interval estimates and posterior distributions are generally more useful, but point estimates are easier to talk about and we shall concentrate on them for the sake of simplicity.

One may also regard the matter from a more geometrical point of view. If the graph of the likelihood function is sharply peaked, then the final density will also usually be sharply peaked at nearly the same place. This again makes it clear that there is often not much difference between Bayesian estimation and maximum-likelihood estimation, provided that the sample is large. This argument applies provided that the number of parameters is itself not large.

All this is on the assumption that the Bayesian assumptions are not dogmatic in the sense of ascribing zero initial probability to some range of values of the parameter; though "provisional dogmatism" is often justifiable to save time, where you hold at the back of your mind that it might be necessary to make an adjustment in the light of the evidence. Thus I do not agree with the often-given dogmatic advice that significance tests *must* be chosen before looking at the results of an experiment, although of course I appreciate the point of the advice. It is appropriate advice for people of bad judgment.

It is perhaps significant that Daniel Bernoulli introduced the method of maximum likelihood, in a special case, at almost the same time as the papers by Bayes and Laplace on inverse probability were published. But, as I said before, it is the logical rather than the historical connections that I wish to emphasize most. I merely state my belief that the influence of informal Bayesian thinking on apparently non-Bayesian methods has been considerable at both a conscious and a less conscious level, ever since 1763, and even from 1925 to 1950 when non-Bayesian methods were at their zenith relative to Bayesian ones.

Let us consider loss functions in more detail. In practice, many statisticians who do not think of themselves as Bayesians make use of "squared-error loss," and regard it as Gauss-given, without usually being fully aware that a loss is a negative utility and smacks of Bayes. The method of least squares is not always regarded as an example of minimizing squared loss (see Eisenhart, 1964), but it *can* be thought of that way. It measures the value of a putative regression line for fitting given observations. Since statisticians might not always be happy to concur with this interpretation, perhaps a better term for "loss" when used in this conventional way would be "quasiloss" or "pseudoloss." We might use it, *faute de mieux*, when we are not sure what the utility is, although posterior distributions for the parameters of the regression line would preserve more of the information.

If the loss is an analytic function it has to be quadratic in the neighborhood of the correct answer, but it would be more realistic in most applications to assume the loss to be asymptotic to some value when the estimate is far away from the correct value. Thus a curve or surface having the shape of an upside-down normal or multinormal density would be theoretically better than "squared loss" (a parabola or paraboloid). But when the samples are large enough the "tails of the loss function" perhaps do not usually affect the estimates much, except when there are outliers.

Once the minimization of the sum of squared residuals is regarded as an attempt to maximize a utility, it leads us to ask what other substitutes for utility might be used, *quasiutilities* if you like. This question dates back over a quarter of a millennium in estimation problems. Like quasilosses, which are merely quasiutilities with a change of sign, they are introduced because it is often difficult to decide what the real utilities are. This difficulty especially occurs when the applications of your work are not all known in advance, as in much pure science (the "knowledge business" to use Kempthorne's term). A quasiutility might be somewhat *ad hoc*, used partly for its mathematical convenience in accordance with Type II rationality. It is fairly clear that this was an important reason historically for the adoption of the method of least squares on a wide scale.

Fisher once expressed scorn for economic applications of statistics, but he introduced his ingenious concept of amount of information in connection with the estimation of parameters, and it can be regarded as another quasiutility. It measures the expected value of an experiment for estimating a parameter. Then again Turing made use of expected weight of evidence for a particular application in 1941. It measures the expected value of an experiment for discriminating between two hypotheses. The idea of using the closely related Shannon information in the design of statistical experiments has been proposed a number of times (Cronbach, 1953; Lindley, 1956; #77), and is especially pertinent for problems of search such as in dendroidal medical diagnosis (for example, #592). It measures the expected value of an experiment for distinguishing between several hypotheses. *In this medical example the doctor should switch to more "real" utilities for his decisions when he comes close enough to the end of the search to be able to "backtrack."* A number of other possibile quasiutilities are suggested in ##592 & 755, some of which are invariant with respect to transformations of the parameter space.

In all these cases, it seems to me that the various concepts are introduced essentially because of the difficulty of making use of utility in its more standard economic sense. I believe the term "quasiutility" might be useful in helping to crystallize this fact, and thus help to clarify and unify the logic of statistical inference. The quasiutilities mentioned so far are all defined in terms of the probability model alone, but I do not regard this feature as part of the definition of a quasiutility.

Even in the law, the concept of weight of evidence (in its ordinary linguistic sense, which I think is usually the same as its technical sense though not formalized) helps to put the emphasis on the search for the truth, leaving utilities to be taken into account later. One might even conjecture that the expressions "amount of information" and "weight of evidence" entered the *English language* because utilities cannot always be sharply and uncontroversially estimated. Both these expressions can be given useful quantitative meanings defined in terms of probabilities alone, and so are relevant to the "knowledge business."

These various forms of quasiutility were not all suggested with the conscious idea of replacing utility by something else, but it is illuminating to think of them in this light, and, if the *word* "quasiutility" had been as old as quasiutilities themselves, the connection could not have been overlooked. It shows how influential Bayesian ideas can be in the logic if not always in the history of statistics. The history is difficult to trace because of the tendency of many writers (i) to cover up their tracks, (ii) to forget the origins of their ideas, deliberately or otherwise, and (iii) not to be much concerned with "kudology," the fair attributions of credit, other than the credit of those they happen to like such as themselves.

The word "quasiutility" provokes one to consider whether there are other features of utility theory that might be interesting to apply to quasiutilities, apart from the maximization of their expectations for decision purposes. One such feature is the use of minimax procedures, that is, cautious decision procedures that minimize the maximum expected loss (or quasiloss here). Although minimax procedures are controversial, they have something to be said for them. They can be used when all priors are regarded as possible, or more generally when there is a class of possible priors (Hurwicz, 1951; #**26**, where this generalized minimax procedure was independently proposed: "Type II minimax"), so that there is no unique Bayesian decision: then the minimax procedure corresponds to the selection of the "least favorable" prior, in accordance with a theorem of Wald (1950). When the quasiutility is invariant with respect to transformations of the parameter space, then so is the corresponding minimax procedure and it therefore has the merit of decreasing arbitrariness. When the quasiutility is Shannon (or Szilard) information, the minimax procedure involves choosing the prior of maximum entropy (##618, 622), a suggestion made for other reasons by Jaynes (1957). The maximum-entropy method was reinterpreted as a method for *selecting null hypotheses* in #322. I especially would like to emphasize this interpretation because the formulation of hypotheses is often said to lie outside the statistician's domain of formal activity, *qua* statistician. It has been pointed out that Jeffreys's invariant prior (Jeffreys, 1946) can be regarded as a minimax choice when quasiutility is measured by weight of evidence (##618, 622). Thus other invariant priors could be obtained from other invariant quasiutilities (of which there is a one-parameter family mentioned later).

Jeffreys's invariant prior is equal to the square root of the determinant of Fisher's information matrix, although Jeffreys (1946) did not express it this way explicitly. Thus there can be a logical influence from non-Bayesian to Bayesian methods, and of course many other examples of influence in this direction could be listed.

Let us return to the discussion of Maximum Likelihood (ML) estimation. Since nearly all methods lead to ROME (Roughly Optimal Mantic Estimation) when samples are large, the real justification for choosing one method rather than another one must be based on samples that are not large.

One interesting feature of ML estimation, a partial justification for it, is its invariance property. That is, if the ML estimate of a parameter θ is denoted by $\hat{\theta}$, then the ML estimate of $f(\theta)$, for any monotonic function f, even a discontinuous one, is simply $f(\hat{\theta})$. Certainly invariant procedures have the attraction of decreasing arbitrariness to some extent, and it is a desideratum for an *ideal* procedure. But there are other invariant procedures of a more Bayesian tone to which I shall soon return: of course a completely Bayesian method would be invariant if the prior probabilities and utilities were indisputable. Invariance, like patriotism, is not enough. An example of a very bad invariant method is to choose as the estimate the least upper bound of all possible values of the parameter if it is a scalar. This method is invariant under all increasing monotonic transformations of the parameter!

Let us consider what happens to ML estimation for the physical probabilities of a multinomial distribution, which has been used as a proving ground for many philosophical ideas.

In the notation used earlier, let the frequencies in the cells be $n_1, n_2, \ldots, n_t$, with total sample size N. Then the ML estimates of the physical probabilities are n_i/N, $i = 1, 2, \ldots, t$. Now I suppose many people would say that a sample size of $n = 1{,}000$ is large, but even with this size it could easily happen that one of the n_i's is zero, for example, the letter Z could well be absent in a sample of 1,000 letters of English text. Thus a sample might be large in one sense but effectively small in another (##38, 83, 398). If one of the letters is absent ($n_i = 0$), then the maximum-likelihood estimate of p_i is zero. This is an appallingly bad estimate if it is used in a gamble, because if you believed it (which you wouldn't) it would cause you to give arbitrarily large odds against that letter occurring on the next trial, or perhaps ever. Surely even the Laplace-Lidstone estimate $(n_i + 1)/(N + t)$ would be better, although it is not optimal. The estimate of Jeffreys (1946), $(n_i + 1/2)/(N + t/2)$, which is based on his "invariant prior," is also better (in the same sense) than the ML estimate. Still better methods are available which are connected with reasonable "Bayesian significance tests" for multinomial distributions (##398, 547).

Utility and quasiutility functions are often invariant in some sense, although "squared loss" is invariant only under *linear* transformations. For example, if the utility in estimating a vector parameter θ as ϕ is $u(\theta,\phi)$, and if the parameter space undergoes some one-one transformation $\theta^* = \psi(\theta)$ we must have, for consistency, $\phi^* = \psi(\phi)$ and $u^*(\theta^*,\phi) = u(\theta,\phi)$, where u^* denotes the utility function in the transformed parameter space.

The principle of selecting the least favorable prior when it exists, in accordance with the minimax strategy, may be called *the principle of least utility*, or, when appropriate, *the principle of least quasiutility*. Since the minimax procedure must be invariant with respect to transformations of the problem into other equivalent languages, it follows that the principle of least utility leads to an invariant prior. This point was made in ##618, 622. It was also pointed out there (see also ##699, 701, 810 and App. C of **#815**) that there is a class of invariant quasiutilities *for distributions*. Namely, the quasiutility of assuming a distribution of density $g(x)$, when the true distribution of x if $F(x)$, was taken as

$$\int \log\left\{g(\mathrm{x})\ [\det \Delta\ (\mathrm{x})]^{-1/2}\right\} dF(\mathrm{x})$$

where

$$\Delta(\theta) = \left\{ - \left. \frac{\partial^2 u(\theta,\phi)}{\partial\phi_i \partial\phi_j} \right|_{\phi = 0} \right\} \quad i, j = 1,2, \ldots .$$

From this it follows further that

$$[\det\Delta(\mathrm{x})]^{1/2}$$

is an invariant prior, though it might be "improper" (have an infinite integral). In practice improper priors can always be "shaded off" or truncated to give them propriety (p. 56 of #13).

If θ is the vector parameter in a distribution function $F(\mathbf{x}|\theta)$ of a random variable $\mathbf{x}$, and θ is not to be used for any other purpose, then in logic we must identify $u(\theta,\phi)$ with the utility of taking the distribution to be $F(\mathbf{x}|\phi)$ instead of $F(\mathbf{x}|\theta)$. One splendid example of an invariant utility is expected weight of evidence per observation for discriminating between θ and ϕ or "dinegentropy,"

$$u_0^0(\theta,\phi) = \int \log \frac{dF(\mathbf{x}|\theta)}{dF(\mathbf{x}|\phi)} dF(\mathbf{x}|\theta),$$

which is invariant under non-singular transformations both of the *random* variable and of the parameter space. (Its use in statistical mechanics dates back to Gibbs.) Moreover it is additive for entirely independent problems, as a utility function should be. With this quasituility, $\Delta(\theta)$ reduces to Fisher's information matrix, and the square root of the determinant of $\Delta(\theta)$ reduces to Jeffreys's invariant prior. The dinegentropy was used by Jeffreys (1946) as a measure of distance between two distributions. The distance of a distribution from a correct one *can* be regarded as a kind of loss function. Another additive invariant quasiutility is (#82; Rényi, 1961; p. 180 of #755) the "generalized dinegentropy,"

$$u_c(\theta,\phi) = \frac{1}{c} \log \left[\frac{dF(\mathrm{x}|\theta)}{dF(\mathrm{x}|\phi)}\right]^c dF(\mathrm{x}|\theta) \quad (c > 0),$$

the limit of which as $c \to 0$ is the expected weight of evidence, $u_0(\theta,\phi)$, somewhat surprising at first sight. The square root of the determinant of the absolute value of the Hessian of this utility at $\phi = \theta$ is then an invariant prior indexed by the non-negative number c. Thus there is a continuum of additive invariant priors of which Jeffreys's is an extreme case. For example, for the mean of a univariate normal distribution the invariant prior is uniform, mathematically independent of c. The invariant prior for the variance ϕ is $\sigma^{-1}\sqrt{\{2(1+c)\}}$, which is proportional to σ^{-1} and so is again mathematically independent of c.

In more general situations the invariant prior will depend on c and will therefore not be unique. In principle it might be worth while to assume a ("type III") distribution for c, to obtain an average of the various additive invariant priors. It might be best to give extra weight to the value $c = 0$ since weight of evidence seems to be the best general-purpose measure of corroboration (##211, 599).

It is interesting that Jeffreys's invariant prior, and its generalizations, and also the principles of maximum entropy and of minimum discriminaability (Kullback, 1959) can all be regarded as applications of the principle of least quasiutility. This principle thus unifies more methods than has commonly been recognized. The existing criticisms of minimax procedures thus apply to these special cases.

The term "invariance" can be misleading if the class of transformations under which invariance holds is forgotten. For the invariant priors, although this class of transformations is large, it does not include transformations to a different *application* of the parameters. For example, if θ has a physical meaning, such as height of a person, it might occur as a parameter in the distribution of her waist measurement or her bust measurement, and the invariance will not apply between these two applications. This in my opinion (and L. J. Savage's, July 1959) is a logical objection to the use of invariant priors when the parameters have clear physical meaning. To overcome this objection completely it would perhaps be necessary to consider the joint distribution of all the random variables of potential interest. In the example this would mean that the joint distribution of at least the "vital statistics," given θ, should be used in constructing the invariant prior.

There is another argument that gives a partial justification for the use of the invariant priors in spite of Savage's objection just mentioned. It is based on the notion of "marginalism" in the sense defined by Good (pp. 808-809 of #174; p. 61 of #603B; p. 15 of #732). I quote from the last named. "It is only in marginal cases that the choice of the prior makes much difference (when it is chosen to give the non-null hypothesis a reasonable chance of winning on the size of sample we have available). Hence the name marginalism. It is a trick that does not give accurate final probabilities, but it protects you from missing what the data is trying to say owing to a careless choice of prior distribution." In accordance with this principle one might argue, as do Box and Tiao (1973, p. 44) that a prior should, at least on some occasions, be uninformative relative to the experiment being performed. From this idea they derive the Jeffreys invariant prior.

It is sometimes said that the aim in estimation is not necessarily to minimize loss but merely to obtain estimates close to the truth. But there is an implicit assumption here that it is better to be closer than further away, and this is equivalent to the assumption that the loss function is monotonic and has a minimum (which can be taken as zero) when the estimate is equal to the true value. This assumption of monotonicity is not enough to determine a unique estimate nor a unique interval estimate having an assigned probability of covering the true value (where the probability might be based on information before or after the observations are taken). But for large enough samples (*effectively* large, for the purpose in hand), as I said, all reasonable methods of estimation lead to Rome, if Rome is not too small.

(d) A Bayes/Non-Bayes Compromise for Probability Density Estimation

Up to a few years ago, the only nonparametric methods for estimating probability densities, from observations $x_1, x_2, \ldots, x_N$, were non-Bayesian. These, methods, on which perhaps a hundred papers have been written, are known as *window methods*. The basic idea, for estimating the density at a point x, was to see how many of the N observations lie in some interval or region around x, where the number ν of such observations tends to infinity while $\nu/N \to 0$ when $N \to \infty$. Also less weight is given to observations far from x than to those close to x, this weighting being determined by the shape of the window.

Although the window methods have some intuituve appeal it is not clear in what way they relate to the likelihood principle. On the other hand, if the method of ML is used it leads to an unsatisfactory estimate of the density function, namely a collection of fractions $1/N$ of Dirac delta functions, one at each of the observations. (A discussant: Go all the way to infinity if they are Dirac functions. Don't be lazy! IJG: Well I drew them a little wide so they are less high to make up for it.) There is more than one objection to this estimate; partly it states that the next observation will certainly take a value that it almost certainly will not, and partly it is not smooth enough to satisfy your subjective judgment of what a density function should look like. It occurred to me that it should make sense to apply a "muftian" Bayesian method, which in this application means finding some formula giving a posterior density in the function space of all density functions for the random variable X, and then maximizing this posterior density so as to obtain a single density function (single "point in function space") as the "best" estimate of the whole density function for X. But this means that from the log-likelihood $\Sigma \log f(x_i)$ we should subtract a "roughness penalty" before maximizing. (##733, 699, 701, 810, 1200.) There is some arbitrariness in the selection of this roughness penalty (which is a functional of the required density function f), which was reduced to the estimation of a single hyperparameter, but I omit the details. The point I would like to make here is that the method can be interpreted in a non-Bayesian manner, although it was suggested for Bayesian reasons. Moreover, in the present state of the art, only the Bayesian interpretation allows us to make a comparison between two hypothetical density functions. The weight of evidence by itself is not an adequate guide for this problem. Then again the non-Bayesian could examine the operational characteristics of the Bayesian interpretation. The Doogian should do this because it might lead him to a modification of the roughness penalty. The ball travels backwards and forwards between the Bayesian and non-Bayesian courts, the ball-game as a whole forming a plank of the Doogian platform.

It is easy to explain why the method of ML breaks down here. It was not designed for cases where there are very many parameters, and in this problem there is an infinite number of them, since the problem is nonparametric. (A nonparametric problem is one where the class of distribution functions cannot be specified in terms of a finite number of parameters, but of course any distribution can be specified in terms of an infinite number of parameters. My method of doing so is to regard the square root of the density function as a point in Hilbert space.)

To select a roughness penalty for multidimensional density functions, I find consistency appealing, in the sense that the estimate of densities that are known to factorize, such as $f(x)g(y)$ in two dimensions, should be the same whether f and g are estimated together or separately. This idea enabled me to propose a multidimensional roughness penalty but numerical examples of it have not yet been tried. [See also #1341.]

An interesting feature of the *subtractive roughness-penalty method* of density estimation, just described, is that it can be made invariant with respect to transformations of the $\mathbf{x}$ axes, even though such transformations could make the true density function arbitrarily rough. The method proposed for achieving invariance was to make use of the tensor calculus, by noticing that the elements of the matrix $\Delta(\theta)$ form a covariant tensor, which could be taken as the "fundamental tensor" g_{ij} analogous to that occurring in General Relativity. For "quadratic loss" this tensor becomes a constant, and, as in Special Relativity, it is then not necessary to use tensors. The same thing happens more generally if $u(\theta,\phi)$ is any function (with continuous second derivatives) of a quadratic.

(e) Type II Maximum Likelihood and the Type II Likelihood Ratio

The notion of a hierarchy of probabilities, mentioned earlier, can be used to produce a compromise between Bayesian and non-Bayesian methods, by treating hyperparameters in some respects as if they were ordinary parameters. In particular, a Bayes factor can be maximized with respect to the hyperparameters, and the hyperparameters so chosen (their "Type II ML" values) thereby fix the ordinary prior, and therefore the posterior distribution of the ordinary parameters. This *Type II ML method* could also be called the *Max Factor method*. This technique was well illustrated in #547. It ignores only judgments you might have about the Type III distributions, but I have complete confidence that this will do far less damage than ignoring all your judgments about Type II distributions as in the ordinary method of ML. Certainly in the reference just mentioned the Type II ML estimates of the physical probabilities were far better than the Type I ML estimates.

The same reference exemplified the *Type II likelihood Ratio*. The ordinary (Neyman-Pearson-Wilks) Likelihood Ratio (LR) is defined as the ratio of two maximum likelihoods, where the maxima are taken within two spaces corresponding to two hypotheses (one space embedded in the other). The ratio is then used as a test statistic, its logarithm to base $1/\sqrt{e}$ having asymptotically (for large samples) a chi-squared distribution with a number of degrees of freedom equal to the difference of the dimensionalities of the two spaces. The Type II Likelihood Ratio is defined analogously as

$$\max_{\theta\epsilon\omega} P\{E|H(\theta)\}/\max_{\theta\epsilon\Omega} P\{E|H(\theta)\}$$

where θ is now a hyperparameter in a prior $H(\theta)$, Ω is the set of all values of θ and ω is a subset of Ω. In the application to multinomial distributions this led to a new statistic called G having asymptotically a chi-squared distribution with one degree of freedom (corresponding to a single hyperparameter, namely the parameter of a symmetric Dirichlet distribution). Later calculations showed that this asymptotic distribution was accurate down to fantastically small tail-area probabilities such as 10^{-16}, see #862. In this work it was found that if the Bayes factor F, based on the prior selected in #547 [see also #1199] were used as a non-Bayesian statistic, in accordance with the Bayes/non-Bayes compromise, it was almost equivalent to the use of G in the sense of giving nearly the same significance levels (tail-area probabilities) to samples. It was also found that the Bayes factor based on the (less reasonable) Bayes postulate was roughly equivalent in the same sense, thus supporting my claims for the Bayes/non-Bayes compromise.

(f) The Non-Uniqueness of Utilities

For some decision problems the utility function can be readily measured in monetary terms; for example, in a gamble. In a moderate gamble the utility can reasonably be taken as proportional to the relevant quantities of money. Large insurance companies often take such "linear" gambles. But in many other decision problems the utility is not readily expressible in monetary terms, and can also vary greatly from one person to another. In such cases the Doogian, and many a statistician who is not Doogian or does not know that he is, will often wish to keep the utilities separate from the rest of the statistical analysis if he can. There are exceptions because, for example, many people might assume a squared loss function, but with different matrices, yet they will all find expected values to be the optimal estimates of the parameters.

One implication of the recognition that utilities vary from one person to another is that the expected benefit of a client is not necessarily the same, *nor even of the same sign*, as that of the statistical consultant. This can produce ethical problems for the statistician, although it may be possible to reward him in a manner that alleviates the problems. (See, for example, ##**26**, 690a.)

One example of this conflict of interests relates to the use of confidence-interval estimation. This technique enables the statistician to ensure that his interval estimates (*asserted* without reference to probability) will be correct say 95% of the time in the long run. If he is not careful he might measure his utility gain by this fact alone (especially if he learns his statistics from cookbooks) and it can easily happen that it won't bear much relation to his client's utility on a specific occasion. The client is apt to be more concerned with the final probability that the interval will contain the true value of the parameter.

Neyman has warned against dogmatism but his followers do not often give nor heed the warning. Notice further that there are *degrees* of dogmatism and that greater degrees can be justified when the principles involved are the more certain. For example, it seems more reasonable to be dogmatic that 7 times 9 is 63 than that witches exist and should be caused not to exist. Similarly it is more justifiable to be dogmatic about the axioms of subjective probability than to insist that the probabilities can be sharply judged or that confidence intervals should be used in preference to Bayesian posterior intervals. (Please don't call them "Bayesian confidence intervals," which is a contradiction in terms.)

Utilities are implicit in some circumstances even when many statisticians are unaware of it. Interval estimation provides an example of this; for it is often taken as a criterion of choice between two confidence intervals, both having the same confidence coefficient, that the shorter interval is better. Presumably this is because the shorter interval is regarded as leading to a more economical search or as being in general more informative. In either case this is equivalent to the use of an informal utility or quasiutility criterion. It will often be possible to improve the interval estimate by taking into account the customer's utility function more explicitly.

An example of this is when a confidence interval is stated for the position of a ship, in the light of direction finding. If an admiral is presented with say an elliptical confidence region, I suspect he would reinterpret it as a posterior probability density surface, with its mode in the center. (#618; Good, 1951.) The admiral would rationally give up the search when the expense per hour sank below the expected utility of locating the ship. In other words, the client would sensibly ignore the official meaning of the statistician's assertion. If the statistician knows this, it might be better, at least for his client, if he went Bayesian (in some sense) and gave the client what he wanted.

(g) Tail-Area Probabilities

Null hypotheses are usually known in advance to be false, and the point of significance tests is usually to find out whether they are nevertheless approximately true (p. 90 of #13). In other words *a null hypothesis is usually composite even if only just*. But for the sake of simplicity I shall here regard the null hypothesis as a simple statistical hypothesis, as an approximation to the usual real-life situation.

I have heard it said that the notion of tail-area probabilities, for the significance test of a null hypothesis H_0 (assumed to be a simple statistical hypothesis), can be treated as a primitive notion, not requiring further analysis. But this cannot be true irrespective of the test criterion and of the plausible alternatives to the null hypothesis, as was perhaps originally pointed out by Neyman and E. S. Pearson. A value X_1 of the test criterion X should be regarded as "more extreme" than another one X_2 only if the observation of X_1 gives "more evidence" against the null hypothesis. To give an interpretation of "more evidence" it is necessary to give up the idea that tail-areas are primitive notions, as will soon be clear. One good interpretation of "more evidence" is that the weight of evidence against H_0 provided by X_1 is greater than that provided by X_2, that is

$$\log \frac{\text{P.D.}\ (X_1 | H_1)}{\text{P.D.}\ (X_1 | H_0)} > \log \frac{\text{P.D.}\ (X_2 | H_1)}{\text{P.D.}\ (X_2 | H_0)},$$

where H_1 is the negation of H_0 and is a composite statistical hypothesis, and P.D. stands for "probability density." (When H_0 and H_1 are both simple statistical hypotheses there is little reason to use "tail-area" significance tests.) This interpretation of "more extreme" in particular provides a solution to the following logical difficulty, as also does the Neyman-Pearson technique if all the simple statistical hypotheses belonging to H_1 make the simple likelihood ratio monotonic increasing as x increases.

Suppose that the probability density of a test statistic X, given H_0, has a known shape, such as that in Figure 1a. We can transform the x axis so that the density function becomes any density function we like, such as that illustrated in Figure 1b. We then might not know whether the x's "more extreme" than the observed one should be interpreted as all the shaded part of 1(b), where the ordinates are smaller than the one observed. Just as the tail-area probability wallah points out that the Bayes postulate is not invariant with respect to transformations of the x axis, the Bayesian can say *tu quoque*. (Compare, for example, p. 53, of #750; Kalbfleisch, 1971, §7, 1-8.) Of course Doogians and many other modern Bayesians are not at all committed to the Bayes postulate, though they often use it as an approximation to their honest judgment, or marginalistically.

When tail-areas are used for significance testing, we need to specify what is meant by a "more extreme" value of the criterion. A smaller ordinate might

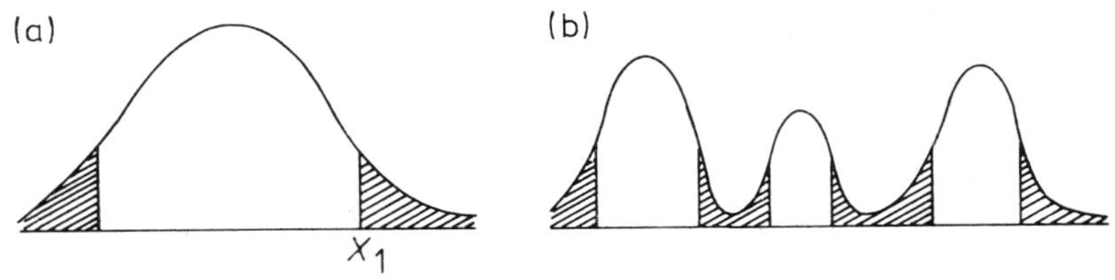

Figure 1.

not be appropriate, as we have just seen. I believe it is a question of ordering the values of the ordinate according to the weight of evidence against the null hypothesis, as just suggested. (Sometimes this ordering is mathematically independent of the relative initial probabilities of the simple statistical hypotheses that make up the composite non-null hypothesis H_1. In this case the interpretation of "more extreme" is maximally robust modulo the Bayesian assumptions.) This or similar fact is often swept UTC, although a special case of it is often implicit when it is pointed out that sometimes a single tail should be used and sometimes a double tail, depending on the nature of the non-null hypotheses.

For some problems it would not be appropriate to interpret "more extreme" to mean "further to the right" nor "either further to the right of one point or further to the left of another" (i.e. for "double tails"). For example, the null hypothesis might be a bimodal distribution with mean zero, the rivals being unmodal also with mean zero. Then we might need to regard values of the random variable close to the origin as significant, in addition to large positive and negative values. We'd be using a "triple tail" so to speak. All this comes out in the wash when "more extreme" is interpreted in terms of weight of evidence.

It is stimulating to consider what is "more extreme" in multivariate problems. It will be adequate to think of bivariate problems which are enough to bring out all the philosophical [or logical] aspects, which are more important than the mathematical ones. We might first ask what is the analogue of being "further to the right." One analogue is being "further to the north and east." This analogue is often dubious (unless the two independent varaiables are like chalk and cheese, or like oil and water) even without reference to any Bayesian or Neymanian-Pearsonian ideas. For under a linear transformation of the independent variables, such as an orthogonal transformation, there are a continuous infinity of different regions that are further to the north and east. The corresponding ambiguity in one dimension refers merely to the question of whether a single tail is more or less appropriate than a double tail.

The previously mentioned elucidation of "more extreme" in terms of weight of evidence applies just as much to multivariate problems as to univariate ones, and provides and answer to this "north-east" difficulty.

Even when a sensible meaning is ascribed to the expression "more extreme," my impression is that small tail-areas, such as 1/10000, are by no means as strong evidence against the null hypothesis as is often supposed, and this is one reason why I believe that Bayesian methods are important in applications where small tail areas occur, such as medical trials, and even more in ESP, radar, cryptanalysis, and ordinary life. It would be unfortunate if a radar signal were misinterpreted through overlooking this point, thus leading to the end of life on earth! *The more important a decision the more "Bayesian" it is apt to be.*

The question has frequently been raised of how the use of tail-area significance tests can be made comformable with a Bayesian philosophy. (See, for example, Anscombe [1968/69].) An answer had already appeared on p. 94 of #13, and I say something more about it here. (See also p. 61 of **#603B**.)

A reasonable informal Bayesian interpretation of tail-area probabilities can be given in some circumstances by treating the criterion X as if it were the whole of the evidence (even if it is not a sufficient statistic). Suppose that the probability density f_0 of X given H_0 is known, and that you can make a rough subjective estimate of the density f_1 given $\bar{H}_0$. (If you cannot do this at all then the tail area method is I think counterintuitive.) Then we can calculate the Bayes factor against H_0 as a ratio of ordinates $f_1(X)/f_0(X)$. It turns out that this is often the order of magnitude of $(1/\sqrt{N})\int_X^\infty f_1(x)dx/\int_X^\infty f_0(x)dx$, where N is the sample size, and this in its turn will be somewhat less than $1/(P\sqrt{N})$ where P is the right-hand tail-area probability on the null hypothesis. (See p. 863 of #127; improved on p. 416 of #547; and still further in #862.) Moreover, this argument suggests that, for a fixed sample size, there should be a roughly monotonic relationship and a *very* rough proportionality between the Bayes factor F against the null hypothesis and the reciprocal of the tail-area probability, P, provided of course that the non-null hypothesis is not at all specific. (See also p. 94 of #13; #547.)

Many elementary textbooks recommend that test criteria should be chosen before observations are made. Unfortunately this could lead to a data analyst's missing some unexpected and therefore probably important feature of the data. There is no existing substitute for examining the original observations with care. This is often more valuable than the application of formal significance tests. If it is easy and inexpensive to obtain new data then there is little objection to the usual advice, since the original data can be used to formulate hypotheses to be tested on later sample. But often a further sample is expensive or virtually impossible to obtain.

The point of the usual advice is to protect the statistician against his own poor judgment.

A person with bad judgment might produce many far-fetched hypotheses on the basis of the first sample. Thinking that they were worth testing, if he were non-Bayesian he would decide to apply standard significance tests to these hypotheses on the basis of a second sample. Sometimes these would pass the test, but some one with good judgment might be able to see that they were still improbable. It seems to me that the ordinary method of significance tests makes some sense because experimenters often have reasonable judgment in the formulation of hypotheses, so that the initial probabilities of these hypotheses are not usually entirely negligible. A statistician who believes his client is sensible might assume that the hypotheses formulated in advance by the client are plausible, without trying to produce an independent judgment of their initial probabilities.

Let us suppose that data are expensive and that a variety of different non-null hypotheses have been formulated on the basis of a sample. Then the Bayesian analyst would try, in conjunction with his client, to judge the initial probabilities $q_1, q_2, \ldots$ of these hypotheses. Each separate non-null hypothesis might be associated with a significance test if the Bayesian is Doogian. These tests might give rise to tail-area probabilities $P_1, P_1, P_3, \ldots$. How can these be combined into a single tail-area probability? (#174.)

Let us suppose that the previous informal argument is applicable and that we can interpret these tail-area probabilities as approximate Bayes factors C/P_1, C/P_2, $C/P_3, \ldots$ against the null hypothesis, these being in turn based on the assumption of the various rival non-null hypotheses. ("Significance tests in parallel.") By a theorem of weighted averages of Bayes factors, it follows that the resulting factor is a weighted average of these, so that the equivalent tail-area probability is about equal to a weighted harmonic mean of $P_1, P_2, P_3, \ldots$, with weights $q_1, q_2, q_3, \ldots$. This result is not much affected if C is a slowly decreasing function of P instead of being constant, which I believe is often the case. Nevertheless the harmonic-mean rule is only a rule of thumb.

But we could now apply the Bayes/non-Bayes compromise for the invention of test criteria, and use this weighted harmonic mean as a non-Bayes test criterion (p. 863 of #127; ##547, 862).

The basic idea of the Bayes/non-Bayes compromise for the invention of test criteria is that you can take a Bayesian model, which need not be an especially good one, come up with a Bayes factor on the basis of this model, but then use it as if it were a non-Bayesian test criterion. That is, try to work out or "Monte Carlo" its distribution based on the null hypothesis, and also its power relative to various non-null hypotheses.

An example of the Bayes/non-Bayes compromise arises in connection with discrimination between two approximately multinomial distributions. A crude Bayesian model would assume that the two distributions were precisely multinomial and this would lead to a linear discriminant function. This could then be used in a non-Bayesian manner or it might lead to the suggestion of using a linear discriminant function optimized by some other, possibly non-Bayesian, method. Similarly an approximate assumption of multinormality for two hypotheses leads to a quadratic discriminant function with a Bayesian interpretation but which can then be interpreted non-Bayesianwise. (See pp. 49-50 of #397 where there are further references.)

Let us now consider an example of an experimental design. I take this example from Finney (1953, p. 90) who adopts an orthodox (non-Bayesian) line. Finney emphasizes that, in his opinion, you should decide in advance how you are going to analyze the experimental results of a designed experiment. He considered an experimental design laid out as shown in Figure 2. The design consists of ten plots, consisting of five blocks each divided into two plots. We decide to apply treatment A and treatment B in a random order within each block, and we happen to get the design shown. Now this design could have arisen by another process: namely by selecting equiprobably the five plots for the application of treatment A from the $10!/(5!)^2 = 252$ possibilities. Finney then says, "The form of analysis depends not on the particular arrangement of plots and varieties in

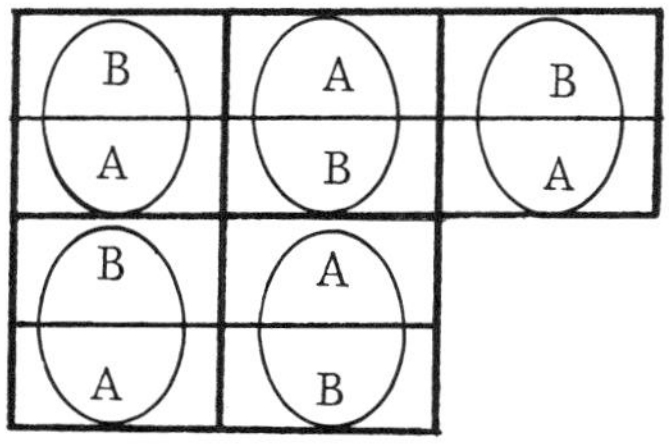

Figure 2. An agricultural experiment.

the field [I have been talking about *treatments* instead here but it does not affect the argument] but on the process of randomization from which the particular one was selected." (Perhaps one should talk of a stochastic or random design *procedure* and a *realization of the procedure.*) For one design procedure we would perhaps use the comparison within the five pairs, and for the other procedure we would compare the set of five yields from treatment A with the set of five yields from treatment B. Leaving aside the analysis of variance, we might find that every plot A did better than every plot B, thus bringing off a distribution-free chance of 1/252; but we are "permitted" to say merely that the chance is 1/32 if the design procedure was based on the five blocks. Suppose the statistician hadn't said which was his design and then he'd dropped dead after the experiment and suppose this is an important experiment organized by the government to decide whether a certain big expensive and urgent food production method was to be put into effect. Would it be reasonable to search the statistician's papers carefully to find out what his intentions had been? Or would it on the other hand be reasonable to call in agriculturalists to look at the plots in the field in order to try to decide which design would have been more reasonable? There are of course reasons for choosing one design rather than another one. So, if you entirely accept the Fisherian logic (as exemplified by Finney) you are whole-heartedly trusting the original judgment of choice of design: this is what the mystique recommends. My own feeling is that you would do better to judge the prior probabilities that each of the two designs is to be preferred, and then use these probabilities as weights in a procedure for combining significance tests (#174 and p. 83 of #750).

A living agriculturalist might examine the field and say that the design corresponding to the tail-area probability of 1/32 deserved twice as much weight as the other design. Then the harmonic-mean rule of thumb would suggest that the equivalent tail-area probability from the observations is

$$\frac{1}{\frac{2}{3} \times 32 + \frac{1}{3} \times 252} \approx \frac{1}{105}.$$

Of course we might do better by using the analysis of variance in a similar manner. I have used a distribution-free approach for the sake of simplicity. This imprecise result is better than either of the precise ones, 1/32 and 1/252. I predict that lovers of the "precision fallacy" will ignore all this.

It is often said that non-Bayesian methods have the advantage of conveying the evidence in an experiment in a self-contained manner. But we see from the example just discussed that they depend on a previous judgment; which in the special case of the dead-dropping of the statistician, has to be a posterior judgment. So it's misleading to tell the student he must decide on his significance test in advance, although it is correct according to the Fisherian technique.

(h) Randomness, and Subjectivism in the Philosophy of Physics

I would have included a detailed discussion on the use of random sampling and random numbers, but have decided not to do so because my views on the subject are explained, for example, on p. 255 of **#85A** and on pp. 83-90. The use of random sampling is a device for obtaining apparently precise objectivity but this precise objectivity is attainable, *as always*, only at the price of throwing away some information (by using a *Statistician's Stooge* who knows the random numbers but does not disclose them). But the use of sampling without randomization involves the pure Bayesian in such difficult judgments that, at least if he is at all Doogian, he might decide, by Type II rationality, to use random sampling to save time. As Cornfield (1968/70, p. 108) points out, this can be covered within the Bayesian framework.

Since this conference is concerned with physics as well as with statistics I should like to mention a connection between something I have been saying and a point that is of interest in the philosophy of physics. (This point is also discussed in #815.)

When discussing the probability that the millionth digit of π is a 7, I could have pointed out that similar statements can be made about pseudorandom numbers. These are deterministic sequences that are complicated enough so that they appear random at least superficially. It would be easy to make them so complicated that it would be practically impossible to find the law of generation when you do know it. Pseudorandom numbers are of value in computer applications of the so-called Monte Carlo method. They are better than random numbers for this purpose because it is possible to check a program exactly and conveniently when pseudorandom numbers are used. One could well say that the probabilities of the various sequences are the same as for a random sequence until the law of generation is known. This in an example of the shifting or evolving [or dynamic] probabilities that I mentioned before. The probabilities would all become 1 or 0 if the law of generation were discovered.

Now it seems to me that this is precisely the situation in classical statistical mechanics. Since the exact positions and velocities of the molecules are not known there is no practical difference between the assumptions of determinism and indeterminism in classical statistical mechanics.

Let us imagine that the world really is physically deterministic. The subjective probability (of the physicist) is almost equal to 1 that the entropy of a closed system will not decrease. He tends to believe that this subjective probability is also a physical probability but it is difficult to see how it can be on the assumption of determinism, any more than it can be a physical probability that the millionth digit of π is a 7. The world is so complicated that pseudo-indeterminism is indistinguishable in practice from strict physical indeterminism. Thus the physicist is "safe" when he says that certain of his subjective probabilities are physical, because he cannot be refuted. Yet he is really sweeping his subjective probabilities UTC. I claim that von Mises (1957, pp. 184-186) made a logical error when he argued that the effectiveness of classical statistical mechanics proves that the world is really indeterministic. (Compare Good, 1958; and p. 72.) Of course quantum mechanics is something else.

V. DISCUSSION

I have allowed for some of the discussion already in this write-up. Here I mention a point raised in the discussion by Peter Finch. He asked in what way initial probabilities come into statistical practice. Apart from the fact that Bayesian methods are gradually coming to the fore, I mentioned the following example.

Fisher (1959, p. 47) argued that, at the time he wrote, there was little evidence that smoking was a cause of lung cancer. The research by Doll *et al.* had shown a tail-area probability of about 1/80 suggesting such a relationship, but had also shown a similar level of significance suggesting that inhalers less often get lung cancer than non-inhalers. Fisher's ironical punch line then was that therefore the investigators should have recommended smokers to inhale! I believe that the investigators (unconsciously?) assumed that the initial probability here was too low to justify giving this advice. (#228; but see also pp. 365-367 of #339.)

Peter Finch also said that the purpose of avoiding far-fetched hypotheses was to get one's work published. My reply was that I had frequently used initial probabilities in my own classified statistical work to help decide what hypotheses I should follow up, with no question of publication.

Part II. Probability

CHAPTER 5

Which Comes First, Probability or Statistics (#85A)

The title of this note was selected so as to provide an excuse for discussing some rather general matters. Let us first consider the question of which of probability and statistics came first historically. This question is like the one about eggs and chickens. The question whether eggs or chickens came first could in principle be given a meaning by using arbitrarily precise definitions of eggs and chickens, and even then probably nobody would be able to answer the question. . . .

Let us consider some examples of statistical principles. For each of them we shall run into trouble by regarding them as golden rules leading to precise probability statements or decisions.

(i) *Maximum likelihood.* I recently won a one-cent bet by guessing the name of the last entry in a dictionary of 50,000 American scientists. (It was Zygmund.) The maximum-likelihood estimate of the number of names of American scientists known to me, on this evidence, is 50,000—clearly an unreasonable estimate. Fisher would recommend that the principle of maximum likelihood should be used with common sense. Another way of saying the same thing would be that initial probabilities and utilities should be taken into account. (For an example where maximum likelihood gets into trouble even for large samples see Lindley, 1947.)

(ii) *Tail-area probabilities.* One of the earliest attempts to avoid the use of more than the minimum of judgment was the use of tail-area probabilities (the so-much-or-more method). A typical example is the use of χ^2 by Karl Pearson in 1900. An earlier use was by Laplace in 1773 in a memoir on the inclination of the orbits of comets. There was an earlier, but rather trivial, example by Arbuthnot in 1712. But presumably gamblers must have used the method in a rough and ready way, even before 1654, for deciding whether to draw swords on their opponents for cheating.

To prove that the use of tail-area probabilities as the final summary of statistical evidence is controversial, it is sufficient to refer to a 1928 paper by Neyman & E. S. Pearson, in which it was emphasized that likelihoods on non-null hypotheses are relevant as well as those on the null hypothesis. It is possible to regard this emphasis as constituting a slight swing back to the Bayes-Laplace philosophy. (An example to show that the probability distribution of a statistic on the null hypothesis is not enough for determining the choice of which statistic to use is that the reciprocal of Student's t has the same distribution as t itself when the sample is of size 2.)

(iii) *Large-sample theory, or asymptotic properties of statistics.* A good deal of modern statistical theory is concerned with the asymptotic properties of statistics. One controversial question is how large samples have to be in order to make these asymptotic properties relevant.

(iv) *The likelihood-ratio method.* In this method a statistic is chosen that is equal to the ratio of maximum likelihoods among the class of simple statistical hypotheses being tested and among the class of all simple statistical hypotheses entertained. Though intuitively appealing and having desirable large-sample properties, a small-sample example was produced by Stein in which the method leads to absurd conclusions (see, for example, Neyman, 1952).

(v) *Unbiased statistics.* Unbiased statistics can take values outside the range of what is possible. For example, if a multinomial distribution has category chances $p_1, p_2, \ldots, p_m$, and if in a sample of size N the frequencies of the m classes are $n_1, n_2, \ldots, n_m$, then an unbiased estimate of Σp_i^2 is

$$\Sigma n_i(n-1)/N(N-1).$$

This estimate would vanish if each n_i were either 0 or 1, but the minimum possible value of the population parameter is $1/m$. It is tempting to replace by $1/m$ those values of the statistic that turn out to be less than $1/m$. Some statisticians would do this without noticing that they were now using a biased statistic.

It is sometimes argued that unbiased statistics have an advantage if it is intended to average over a number of experiments. Two questions then arise: (*a*) How many experiments? (*b*) Would a modified Bayes-Laplace philosophy do just as well if not better? (By the "modified Bayes-Laplace philosophy" we mean the philosophy described in [omitted parts of] the present note. It differs from the ordinary Bayes-Laplace philosophy in that it leaves room for individual judgment instead of assuming uniform initial distributions, i.e. Bayes postulates.) No modified Bayes-Laplace estimate can lie outside the possible range of values of the population parameter. Applications of the modified Bayes-Laplace philosophy do not yet belong to orthodox statistics. They are not intended to lead to precise results.

(vi) *Deciding on significance tests before taking a sample.* In elementary textbooks the advice is often given to decide on one's tests of significance before taking a sample. This may be good advice to those whose judgment you do not trust. Or a statistician may use the principle for himself as a precaution against wishful thinking, or as a guarantee against accusations of prejudice rather than judgment. But consider the following example. A sample of 100 readings is taken from some distribution for which the null hypothesis is that the readings are independently distributed with a normal distribution of zero mean and unit variance. It is decided in advance of sampling to divide this normal distribution up into ten equal areas, and to apply the χ^2 test to the ten-category equiprobable multinomial distribution of frequencies with which the readings fall into the ten areas. This would appear to be a very reasonable statistic. But what if it leads to a non-significant result even though one of the 100 readings was 20 standard deviations above the mean?

(vii) *Confidence intervals.* (Developed mainly by Neyman & Pearson [1930-33].) Suggested by E. B. Wilson (1927) [and Laplace (1878/1912), 286-287]. . . . One of the intentions of using confidence intervals and regions is to protect the reputation of the statistician by being right in a certain proportion of cases in the long run. Unfortunately, it sometimes leads to such absurd statements, that if one of them were made there would not be a long run. One objection, similar to the one above concerning unbiased estimates, was given by M. G. Kendall (1949). For others see the discussion on Daniels, "The theory of position-finding" (1951). A further objection is admitted in Neyman (1952), namely, that it can lead to absurdly long confidence intervals. Stein introduced a sequential sampling procedure to overcome this last objection, but it can lead to absurdly large samples.

A statistician can arrange to make confidence pronouncements that are correct in at least 95% of cases in the long run (if there is a long run). But if his customer decides to separate off the pronouncements that relate to a subclass of the possible experimental results (such as those in which a random variable is large), then it is no longer true that 95% of the subclass will be correct in general. In fact the judgment that the random variable is large is an indirect statement about the initial probability distribution, and it will imply that for this subclass the proportion of correct confidence interval statements will probably fall below 95%. This argument shows what is perhaps the main reason why the confidence method is a confidence trick, at least if used too dogmatically.

(viii) *Fiducial distributions.* The use of fiducial distributions in statistical inference is controversial if only because these distributions need not be unique. (See Mauldon, 1955. There is similar unpublished work by J. W. Tukey.)

(ix) *Errors of the first and second kinds.* The notion of the minimization of sampling costs for a given consumer's risk was used by Dodge & Romig (1929), and the subject was expanded by Neyman & Pearson in 1933. . . . As pointed out by Barnard at a recent British Mathematical Colloquium, the notion of errors of the first and second kinds ignores questions of "robustness" of a significance test. One wants a test to be sensitive in detecting certain types of departure from the null hypothesis but insensitive to other types of departure, a compromise between robustness and sensitivity.

(x) *The point of a significance test.* What is the point of a significance test anyway? A large enough sample will usually lead to the rejection of almost any null hypothesis (cf. p. 90 of #13). Why bother to carry out a statistical experiment to test a null hypothesis if it is known in advance that the hypothesis cannot be exactly true? The only answer is that we wish to test whether the hypothesis is in some sense approximately true, or whether it is rejectable on the sort of size of sample that we intend to take. These points are not usually made clear in text-books on statistics, and in any event they have never been formulated precisely.

(xi) *Is every significance test also an estimation problem?* This is another question on which there is controversy, but for our present purposes it is rather a side issue.

(xii) *On the use of random sampling numbers.* In avoiding the use of the Bayes-Laplace philosophy or of the modified Bayes-Laplace philosophy, orthodox statisticians attempt to make use of only two types of probability. These are (*a*) the tautological ones that occur in the definition of simple statistical hypotheses, and (*b*) probabilities obtained from random sampling numbers or roulette wheels or in some other way that does not lead in practice to much dispute concerning the numerical values of the probabilities. (In effect, nearly everybody accepts the null hypothesis that the random sampling numbers are at least approximately equiprobably random. This is an example where the distinction, made, for example, by Fisher, between non-rejection and acceptance seems to disappear.) The usefulness of the method of randomization in the design of an experiment is indisputable; nevertheless, it becomes controversial if it is put forward as absolutely precise.

We consider the famous tea-testing experiment (see Fisher [1949], p. 11). . . . [See #**815**.]

Thus the precision obtained by the method of randomization can be obtained only by ignoring information; namely, the information of what particular random numbers (or Latin square, etc.) occurred.

(xiii) *Does decision theory cover ordinary inference?* Just as there is fairly general agreement about the direct probabilities arising from random sampling numbers, there is also fairly general agreement within firms concerning certain utilities that occur in industrial processes. This is so when the utilities can be expressed in monetary terms and when the amounts of money are not large compared with the total capital of the firm. But in purely scientific matters there is much less agreement; in fact, the utilities as judged by a single individual will probably be bounded by upper and lower bounds that are very unequal. In other words the utilities are vague. For this reason the application of decision theory to scientific research is controversial (cf. p. 40 of #13). . . .

This note is based on a lecture given to the American Statistical Association and to the Society of Industrial Applied Mathematics, New York. The copyright is held by General Electric Company, who have kindly granted permission to publish. The present version gives effect to improvements suggested by Mr. Wilfred Perks.

CHAPTER 6

Kinds of Probability (#182)

The mathematician, the statistician, and the philosopher do different things with a theory of probability. The mathematician develops its formal consequences, the statistician applies the work of the mathematician, and the philosopher describes in general terms what this application consists in. The mathematician develops symbolic tools without worrying overmuch what the tools are for; the statistician uses them; the philosopher talks about them. Each does his job better if he knows something about the work of the other two.

What is it about probability that has interested philosophers? Principally, it is the question whether probability can be defined in terms of something other than itself, and, if not, how the idea of probability is used, what is its meaning, what are the shades of meaning. Can we verify that probability exists, or must we be satisfied to say how it is used? Is the "use" theory of meaning more appropriate than the "verification" theory? It seems to me that the philosopher's job is mainly to describe what a man does or thinks at the precise moment that he uses the idea of probability.

Our main question is this: are there different kinds of probability? The question is analogous to the one "Are there different kinds of life?" In a sense there are two kinds of life: animal and vegetable [this ignores some small organisms]; in another sense there are as many as there are genera or species; in yet another sense there is only one kind of life, since life is indivisible, and even the distinction between animals and vegetables is misleading in some contexts. . . . Much of the controversy about the theory of probability is like this. From some points of view there are at least five kinds of probability; from another point of view they can all be defined in terms of a single kind. I shall elaborate this remark and begin by describing some different kinds of probability. Classification of different kinds of probability is half the problem of the philosophy of probability.

THE CLASSICAL DEFINITION

Some billion years ago, an anonymous speck of protoplasm protruded the first primitive pseudopodium into the primeval slime, and perhaps the first state of uncertainty occurred. Thousands of years ago words such as *maybe, chance, luck,* and *fate* were introduced into languages. If a theory is a method of using language, we could say that theories of probability are thousands of years old. But often a usage of language is not dignified by the name *theory* unless a real effort has been made to describe this usage accurately: a theory, then, is not just talk, but is also talk about talk. (Philosophers of science talk about talk about talk.) So when Aristotle (about 300 B.C.) said "the probable is what usually happens," and when Cicero (about 60 B.C.) described probability as the "guide of life," they had formulated primitive theories of probability and of rational behavior. We can hardly tell whether these theories had any practical results; at any rate, the ancient Romans later practised insurance, and Domitius Ulpianus drew up a table of life expectancies (about A.D. 200.)

Mathematical ideas, however, date back only a few hundred years. A commentary in 1477 on Dante's *Purgatorio* gives the probabilities of various totals when three dice are thrown. Perhaps the application was to cleromancy (divination by dice). In the 16th century, Cardan, an inveterate gambler, made several simple probability calculations of use to gamblers. He defined probability as a "proportion of equally probable cases"; for example, of the 36 possible results of throwing two dice, three give a total of 11 or more points, so the probability of this event is defined as 1/12 if the 36 possible results are equally probable. The definition by equally probable cases is usually called the "classical definition."

The origin of the mathematical theory of probability is not usually ascribed to Cardan, but rather to Pascal in 1654, who, in correspondence with Fermat, solved the first mathematically nontrivial problems. The first book on the subject, of any depth, was published soon afterwards by Huygens.

All of these authors were concerned with games of chance, and although they defined probability as a proportion of equally probable cases, their purpose must have been to explain why certain long-run proportional frequencies of success occurred. Without being explicit about it they were trying to explain one kind of probability in terms of another kind. James Bernoulli was much more explicit about it, in his famous work *Ars Conjectandi*, published in 1713, eight years after his death. His "law of large numbers" states that in n "trials," each with probability p of success, the number of successes will very probably be close to pn if n is large. For example, if a coin has probability exactly 1/2 of coming down heads, and if it is tossed a thousand times, then the number of heads . . . will very probably lie between 470 and 530. . . . These results are based on the assumption that the probability of heads is 1/2 at each throw, no matter what the results of previous throws may have been. In other words, the trials must be "causally independent." Bernoulli did not make it clear that the trials must be causally independent and that pn must be large. If your probability, p, of winning a sweepstake is 1/1,000,000, then Bernoulli's theorem would not be applicable until you had entered several million sweepstakes by which time you would be too old to care.

Bernoulli proved his theorem on the assumption that the probability, p, was defined as a proportion of equally probable cases. But he tried to apply the theorem to social affairs in which this definition is hardly appropriate. Worse yet: the probability is likely to be variable.

SUBJECTIVE PROBABILITY

Even in games of chance the classical definition is not entirely satisfactory, for the games may not be "fair." A fair game of chance is one in which the apparently equal probabilities "really are" equal. In order to give this definition of a fair game any substance we must again distinguish between two kinds of probability. Consider, for example, the probability that cutting an ordinary pack of playing cards will put a red card at the bottom of the pack, an event that I shall call a "success." Since half the cards are red and half are black, the probability would seem to be 1/2 if the pack of cards, and its shuffling, are fair. But if all the red cards have dirty, sticky faces, then a black card is more likely to be brought to the bottom. If we knew the red cards had sticky faces we would prefer to bet on a black card, in a "level bet." But if we did not know it, then the probability would still be 1/2 *for us.* Even if we allowed for the possibility of stickiness, the black cards are as likely to be more sticky as to be less so, unless we have some further information. For us the *first* cut has probability 1/2 of being successful.

We may have an opponent who knows that the red cards are stickier. For him the probability is not the same as it is for us. This example shows that personal, or subjective, or logical probability depends on the given information as well as on the event whose probability is to be estimated. This is the reason for notation of the form

$$P(\mathrm{E}|\mathrm{F})$$

read from left to right (like all good notations) "the probability of E given F." For the sake of generality, E and F may be interpreted as propositions. This notation (or equivalent ones) has become standard during the present century. In this notation the probabilities we have just been discussing are

$$P(\text{bottom card is red}|\text{the cards have been well shuffled})$$

and

$$P(\text{bottom card is red}|\text{the cards have been well shuffled by normal standards, but the red ones have sticky faces}).$$

The use of the vertical stroke, or equivalent notation, is likely to save us from the errors that may arise through talking simply about the "probability that the bottom card will be red," without reference to the "given" (= assumed) information.

PHYSICAL PROBABILITY

Suppose that our opponent has carried out a very extensive experiment and has decided that the long-run proportion of successes is 0.47 (instead of ½). We may be tempted to call this the "true probability," or "physical probability," or "material probability," or "chance," or "propensity," and to regard it as having an impersonal, public, or objective significance. Whether or not physical probability is regarded as distinct from personal, private, intuitive, subjective, or logical probability, it is often convenient to talk as if it were distinct. I shall, however, argue later on that its numerical value can be defined in terms of subjective probability.

A physical probability is the probability of a "success" given the "experimental setup." So for physical probabilities, too, it is convenient to have a notation of the form $P(E|F)$. We can distinguish between true and hypothetical probabilities, depending on whether the experimental setup is true or hypothetical. For example, we can take an actual pack of cards and we can discuss the probability that the bottom card will be red "given" (= on the assumption that) all the clubs have been omitted. This probability makes sense even if the clubs have not in fact been omitted, and the probability will then be "hypothetical" and not "true." It so happens that it is decidedly useful to talk about hypothetical probabilities as well as true ones.

We could *imagine* a physical chemist who could analyze the chemicals on the faces of the cards and then compute the probability of success by quantum theory. But this would be a far cry from the simple physical symmetry that led Cardan and Pascal to judgments of equal probability, or from the logical symmetry that caused us to consider black and red to be equally likely to be the stickier. It is perhaps clear by now that the classical definition, however suggestive, is by no means general enough to cover all the uses of the word *probability*.

INVERSE PROBABILITY

Most applications of the theory of probability to the social sciences are more like unfair games of chance than fair ones. If n smokers are sent questionnaires and r of them refuse to fill them out, what is the probability, p, that the next smoker selected will refuse to fill out his questionnaire? And what is the proportion of all smokers who will refuse? Whereas Bernoulli's theorem works from a knowledge of p to information about the number of "successes" in the sample, the answer here seems to require the inverse process. A simple estimate of p is r/n, but if r is small this may be a poor estimate, especially if $r = 0$. (To say that the probability of an event is 0 is to say that the event is infinitely unlikely. Such an assertion is not justified merely by 100 percent failures in the past.) Sometimes r/n is taken as a definition of probability; it may be called the "naive" definition.

A better attempt at "inverting Bernoulli's theorem" was made by Thomas Bayes in a paper published posthumously in 1763. The method is known as "inverse probability," and was given a prominent place in Laplace's *Théorie analytique des probabilités* (1812). It may also be described as the Bayes-Laplace method of statistical inference. In modern terminology the principle of inverse probability can be expressed in terms of "initial" probabilities, "final" probabilities, and "likelihoods." The initial probability (also called the "prior" probability) of a hypothesis is its probability before some experiment is performed. (There may or may not have been previous experiments or evidence, so the description "a priori" is inappropriate.) The final probability is the probability after the experiment is performed. These probabilities are different, in general, because the given information is different. The *likelihood* of a hypothesis is the probability, given that hypothesis, of the actual result of the experiment.

For example, suppose we have two hypotheses about a coin, either that the coin is fair or that it is double-headed, and suppose that the initial probabilities of these two hypotheses are equal, that is, each is ½. Suppose now that the coin is tossed ten times and comes down heads every time. The likelihoods of the two hypotheses are then $2^{-10} = 1/1024$ and 1.

Bayes's theorem is, in effect, that the final probability of a hypothesis is proportional to its initial probability times its likelihood. In our example, the final probabilities are therefore proportional to 1/1024 and 1. Therefore, the final probability that the coin is double-headed is 1024/1025, or nearly certain.

Although Laplace's exposition was clearer than Bayes's, he blatantly assumed that initial probabilities were always equal, whereas Bayes was more modest. Laplace assumed, for example, that an unknown physical probability, p, was initially (that is, before any observations were taken) equally likely to "take any value" between 0 and 1; he assumed, for example, that each of the intervals (0, 0.01), (0.01, 0.02), . . . , (0.99, 1.00) initially had probability 0.01. In the applications, p is what we are calling a physical probability existing "out there," whereas the probability 0.01 is a more subjective kind of probability. By making this assumption of a "uniform distribution" of probability between 0 and 1, Laplace proved his so-called "law of succession." This states that after r "successes" in n "trials," p can be estimated as $(r + 1)/(n + 2)$. For example, after one success in two trials, p is estimated as ½; after one success in one trial, p is estimated as 2/3; after no success in one trial, p is estimated as 1/3; after no success in no trials, p is estimated as 1/2. The formula is open to dispute and has often been disputed. It leads, for example, to the conclusion that anything that has been going on for a given length of time has probability [close to] ½ of going on for the same length of time again. This does not seem to me to be too bad a rule of thumb if it is applied with common sense.

Inverse probability is by no means the only method of statistical inference. There is, for example, an important method known as "maximum likelihood," used at times by Daniel Bernoulli in 1777, Gauss in 1823, and especially by Fisher in 1912. In this method, that hypothesis is selected whose likelihood is a maximum, where "likelihood" is defined as it is above. For the simple sampling experiment mentioned above, the method of maximum likelihood leads to the naive estimate r/n, which in my opinion is not as good as the result given by Laplace's law of succession.

A familiar objection to the use of inverse probability is that the initial probabilities cannot usually be determined by clear-cut rules. The method of maximum likelihood is clear-cut, and does not lend itself so easily to conscious or unconscious cheating. But for small samples it can lead to absurd conclusions. The method of inverse probability, although more arbitrary, need never lead to absurdity unless it is dogmatically combined with an assumption that the initial probabilities of alternative hypotheses are invariably equal.

DEFINITION BY LONG-RUN FREQUENCY

One of Laplace's tricks was to use the expression "equally possible cases" instead of "equally probable cases," and thereby to pretend that he had defined probability completely. Not many people today are taken in by this verbal trick.

Leslie Ellis in 1843, A. Cournot in 1843, G. Boole in 1854, and J. Venn (in a full-length treatise, 1866), were not taken in. They asked, for example, how you could prove that a die was unloaded except by throwing it a great number of times. They proposed to solve the problem of inverting Bernoulli's theorem by simply defining physical probability in terms of long-run frequency ("frequentism").

If a roulette wheel is spun 300 times and there is no occurrence of a 7 should we regard the probability of a 7 on the next spin as 1/37 (its "official" value), or as 0, or as some intermediate value? This simple question exposes the weakness both of Laplace's position and of pure frequentism. The frequentist would perhaps refuse to make any estimate and would say "spin the wheel another few hundred times." Owing to lack of space I shall leave this question and consider an even simpler one.

Suppose that a coin-spinning machine is set to work and produces the sequence

HTHTHTHTHTHTHTHTHTHT

The proportion of heads is precisely 1/2 and it seems reasonable to predict that the "Venn limit," that is, the limiting proportion of heads if the sequence is indefinitely continued, will also be 1/2. Yet no one would say that the spinning was fair. This type of difficulty was recognized by Venn but was not adequately met. R. von Mises in 1919 proposed a new frequentist theory of probability based on the notion of infinitely long random sequences—what he called "irregular collectives." The main property of an irregular collective is that the proportion of "successes" (say heads) is the same for every sub-sequence selected in advance. This property is closely related to the impossibility of a successful gambling system. An irregular collective is an abstraction like a point in Euclidean geometry. Von Mises drew a clear distinction between the mathematical or abstract theory and the problem of application of that theory. He was perhaps the first person to make this distinction explicit for the theory of probability, in other words, to advocate Euclid's method, the "axiomatic method." But having made the distinction, he virtually ignored the philosophical problem of application. He stated, like the 19th-century frequentists, that in the applications the sequences must be long, but he did not say how long; just as the geometer might say that dots must be small before they are called points, without saying how small. But the modern statistician often uses small samples; he is like a draftsman with a blunt pencil. He would like to know how long is a long run. As J. M. Keynes said, "In the long run we shall all be dead."

If a frequentist is cross-examined about how long is a long run, it is possible to deduce something about the implicit initial probabilities that he is using. This can be done algebraically, by assuming that the initial probabilities exist as "unknowns," applying the theory of probability, including Bayes's theorem, making use of the frequentist's judgments, and finally solving for the unknown initial probabilities (or getting upper and lower bounds for them). In this way the frequentist may be seen, in spite of hot denials, to be behaving *as if* he had judgments concerning initial probabilities of hypotheses. Or he may be caught in a contradiction.

Like Venn, von Mises deliberately restricted the generality of the theory to situations where the long-run frequency definition seemed to be reasonable. He was entitled to do this but he was not justified in being intolerant of theories that try to achieve more, and especially those that concern themselves more with the philosophical problem of applicability.

Among other brilliant mathematicians since von Mises who have developed the mathematical theory, perhaps Kolmogorov deserves special mention. Most of these mathematicians have been concerned both with the mathematical theory and with its applications, but much less with the philosophical problem of applicability. Among those who have been so concerned were the philosopher W. E. Johnson, his pupil Keynes (1921), Jeffreys (1939/61), Ramsey (1926/64), B. de Finetti (1937/64), B. O. Koopman (1940a, b), R. Carnap (1950), B. Russell (1948), ##13, **26**, **43**, **85A**, 174, and L. J. Savage (1954).

NEOCLASSICAL DEFINITION

Some of these writers are dualists and hold that one should talk about two kinds of probability. Others put most emphasis on the subjectivistic or logical interpretation. Here I shall merely summarize some of my own views, which in one respect or another are closely related to those of the other authors just mentioned. The theory may reasonably be called "neoclassical" or "neo-Bayesian," since its opponents are primarily frequentists, and since Bayes's theorem is restored to a primary position from which it had been deposed by the orthodox statisticians of the second quarter of the 20th century, especially by R. A. Fisher.

1) The function of the theory of subjective probability is to introduce as much objectivity (impersonality) as possible into "your" subjective body of beliefs, not to make it completely impersonal, which may be impossible. With the help of a mathematical theory, based on a few axioms, a body of beliefs can be enlarged and inconsistencies in it can be detected. A subjective probability is a degree of belief that belongs to a body of beliefs from which the worst inconsistencies have been removed by means of detached judgments.

2) Subjective probabilities are not usually precise but are circumscribed by inequalities ("taking inequalities seriously" or "living with vagueness").

3) Probability judgments are plugged into a sort of black box (the abstract or mathematical theory) and discernments are fed out; the judgments can be of very varied type, so that nothing of value in frequentism, classicism, or any other theory is lost.

4) Many orthodox statistical techniques achieve objectivity only by throwing away information, sometimes too much. One way this can happen is if the observations supplied by a very expensive experiment support a hypothesis not thought of in advance of the experiment. In such circumstances, it will often happen that the experimenter will be thrown back on his personal judgment.

5) The theory can be extended to become a theory of rational behavior, by introducing "utilities" (value judgments).

6) All this is important for statistical practice and for the making of decisions.

7) A theory of subjective probability is general enough to cover physical probabilities, but not conversely. Although a physical probability can be regarded as something that is not subjective, its numerical value can be equated to the limiting value of a subjective probability when an experiment is repeated indefinitely under essentially constant circumstances.

KINDS OF PROBABILITY

Since this article is concerned mainly with subjective and physical probability, it would be inappropriate to discuss other kinds in great detail. Perhaps a mere list of various kinds will be of interest:

1) Degree of belief (intensity of conviction), belonging to a highly self-contradictory body of beliefs. (This hardly deserves to be called a probability.)

2) Subjective probability (personal probability, intuitive probability, credence). Here some degree of consistency is required in the body of beliefs.

3) Multisubjective probability (multicredence). The name here is self-explanatory.

4) Credibility (logical probability; impersonal, objective, or legitimate intensity of conviction).

5) Physical probability (material probability, chance, propensity; this last name was suggested by K. R. Popper).

6) Tautological probability. In modern statistics it is customary to talk about ideal propositions known as "simple statistical hypotheses." If, for each possible result, E, of an experiment, $P(E|H)$ is equal to a number that is specified as part of the *definition* of H, then the probability $P(E|H)$ may be called a "tautological probability," and H is a "simple statistical hypothesis."

Much of statistics is concerned with testing whether a simple statistical hypothesis is "true" (or approximately true) by means of sampling experiments. If we regard this as more than a manner of speaking, then, for consistency, we must believe in the existence of physical probabilities. For example, the proposition that a coin in unbiased is a simple statistical hypothesis, H, part of whose *definition* is that $P(\text{heads}|H) = 1/2$, a tautological probability. But if we say or believe that this proposition is *true*, then we are committed to saying or believing also that his tautological probability is a physical probability. It is at least a matter of linguistic convenience or consistency, and it may be more.

A full discussion of the relationships between the various kinds of probability would take us too far afield. I shall merely repeat dogmatically my opinion that although there are at least five different kinds of probability we can get along with just one kind, namely, subjective probability. This opinion is analogous to the one that we can know the world only through our own sensations, an opinion that does not necessarily make us solipsists, nor does it prevent us from talking about the outside world. Likewise, the subjectivist can be quite happy talking about physical probability, although he can measure it only with the help of subjective probability.

BEARING ON INDETERMINISM

On the face of it, the assumption that physical probabilities exist seems to imply the metaphysical theory of indeterminism. I shall conclude by trying to analyze this opinion.

When I say that a theory is "metaphysical," I mean that there is no conceivable experiment that can greatly change the logarithm of its odds. (The odds corresponding to probability p are defined as $p/(1 - p)$. It lies between 0 and plus infinity, and its logarithm lies between $-\infty$ and $+\infty$.) No theory is metaphysical if it can be virtually either proved or falsified, because its log-odds would then become very large, positive or negative. According to this definition, it is a question of degree whether a theory is metaphysical.

For example, the theory of determinism is less credible than it was a hundred years ago, but is by no means disproved and never will be. A statistician can never prove that "random numbers" are not "pseudo-random," and likewise "pseudo-indeterminism" cannot be disproved (#153).

We can consistently talk about physical probability without committing ourselves to the metaphysical theory that the universe is indeterministic, but only if we accept the existence of subjective probability or credibility. For if we assume determinism we can get physical probabilities only by having an incompletely specified physical setup. In this incomplete specification there must be probabilities. If we are determinists we must attribute these latter probabilities to our own ignorance and not merely to something basic in nature "out there." Whether or not we assume determinism, every physical probability *can* be interpreted as a subjective probability or as a credibility. If we do assume determinism, then such an interpretation is forced upon us.

Those philosophers who believe that the only kind of probability is physical must be indeterminists. It was for this reason that von Mises asserted indeterminism before it became fashionable. He was lucky.

CHAPTER 7

Subjective Probability as the Measure of a Non-measurable Set (#230)

1. INTRODUCTION

I should like to discuss some aspects of axiom systems for subjective and other kinds of probability. Before doing so, I shall summarize some verbal philosophy and terminology. Although the history of the subject is interesting and illuminating, I shall not have time to say much about it.

2. DEFINITION

In order to define the sense in which I am using the expression "subjective probability" it will help to say what it is not, and this can be done by means of a brief classification of kinds of probability (Poisson, 1837; Kemble, 1941; #**182**).

Each application of a theory of probability is made by a communication system that has apparently purposive behavior. I designate it as "you." It could also be called an "org," a name recently used to mean an organism or organization. "You" may be one person, or an android, or a group of people, machines, neural circuits, telepathic fields, spirits, Martians and other beings. One point of the reference to machines is to emphasize that subjective probability need not be associated with metaphysical problems concerning mind (compare #183).

We may distinguish between various kinds of probability in the following manner.

(i) Physical (material) probability, which most of us regard as existing irrespective of the existence of orgs. For example, the "unknown probability" that a loaded, but symmetrical-looking, die will come up 6.

(ii) Psychological probability, which is the kind of probability that can be inferred to some extent from your behavior, including your verbal communications.

(iii) Subjective probability, which is psychological probability modifed by the attempt to achieve consistency, when a theory of probability is used combined with mature judgment.

(iv) Logical probability (called "credibility" by Russell, 1948, for example), which is hypothetical subjective probability when you are perfectly rational, and therefore presumably infinitely large. Credibilities are usually assumed to have unique numerical values, when both the proposition whose credibility is under consideration and the "given" proposition are well defined. I must interrupt myself in order to defend the description "infinitely large."

You might be asked to calculate the logical probabilities of the Riemann, Fermat, and Goldbach conjectures. Each of these probabilities is either 0 or 1. It would be cheating to wait for someone else to produce the answers. Similarly, as pointed out by Popper (1957), you cannot predict the future state of society without first working out the whole of science. The same applies even if you are satisfied with the logical probabilities of future states of society. Therefore a rational being must have an infinite capacity for handling information. It must therefore be infinitely large, or at any rate much larger than is practicable for any known physical org. In other words, logical probabilities are liable to be unknown in practice. This difficulty occurs in a less acute form for subjective probability than for logical probability.

Attempts have been made (Carnap, 1950; Jeffreys, 1939) to define logical probability numerically, in terms of a language or otherwise. Although such a program is stimulating and useful, the previous remarks seem to show that it can never be completed and that there will always remain domains where subjective probability will have to be used instead.

(In Carnap's contribution to this Congress he has shifted his position, and now defines logical probability to mean what I call numerically completely consistent subjective probability. He permits more than one consistent system of probabilities. Thus his present interpretation of logical probability is a consistent system within a "black box" in the sense of Section 3 below.)

Physical probability automatically *obeys* axioms, subjective probability *depends* on axioms, psychological probability neither obeys axioms nor depends very much on them. There is a continuous gradation, depending on the "degree of consistency" of the probability judgments with a system of axioms, from psychological probability to subjective probability, and beyond, to logical probability, if it exists. Although I cannot define "degree of consistency," it seems to me to have very important intuitive significance. The notion is indispensable.

In my opinion, every *measure* of a probability can be interpreted as a subjective probability. For example, the physical probability of a 6 with a loaded die can be estimated as equal to the subjective probability of a 6 on the next throw, after several throws. Further, if you can become aware of the value of a logical probability, you would adopt it as your subjective probability. Therefore a single set of axioms should be applicable to all kinds of probability (except psychological probability), namely the axioms of subjective probability.

Superficially, at least, there seems to be a distinction between the axiom systems that are appropriate for physical probability and those appropriate for subjective probability, in that the latter are more often expressed in terms of inequalities, i.e., comparisons between probabilities. Theories in which inequalities are taken seriously are more general than those in which each probability is assumed to be a precise number. I do not know whether physical probabilities are absolutely precise, but they are usually assumed to be, with a resulting simplification in the axioms.

3. A BLACK-BOX DESCRIPTION OF THE APPLICATION OF FORMALIZED THEORIES

I refer here to a "description," and not to a "theory," because I wish to avoid a discussion of the theory of the application of the black-box theory of the application of theories (##13, 26, 43). The description is in terms of the block diagram of Figure 1 in which observations and experiments have been omitted. It consists of a closed loop in which you feed judgments into a black box and feed "discernments" out of it. These discernments are made in the black box as deductions from the judgments and axioms, and also, as a matter of expediency, from theorems deduced from the axioms alone. If no judgments are fed in, no discernments emerge. The totality of judgments at any time is called a "body of beliefs." You examine each discernment, and if it seems reasonable, you transfer it to the body of beliefs. The purpose of the deductions, in each application of the theory, is to enlarge the body of beliefs, and to detect inconsistencies in it. When these are found, you attempt to remove them by means of more mature judgment.

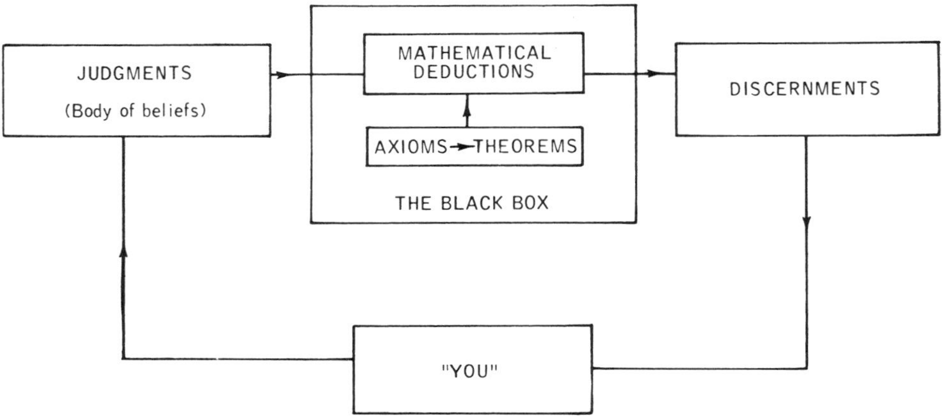

Figure 1. The black-box flow diagram for the application of formalized scientific theories.

The particular scientific theory is determined by the axioms and the rules of application.

The rules of application refer to the method of formalizing the judgments, and of "deformalizing" the mathematical deductions. For example, in a theory of subjective probability the standard type of judgment might be a comparison of the form

$$P'(\mathrm{E}|\mathrm{F}) \geqslant P'(\mathrm{G}|\mathrm{H}),$$

where $P'(\mathrm{E}|\mathrm{F})$ is the intensity of conviction or degree of belief that you would have in E if you regarded F as certain. The P''s are not necessarily numerical, and what is meaningful is not a P' by itself, but a comparison of intensities of conviction of the above type. These judgments are plugged into the black box by simply erasing the two dashes. Likewise, discernments can be obtained by taking an output inequality, $P(\mathrm{E}|\mathrm{F}) \geqslant P(\mathrm{G}|\mathrm{H})$, and putting dashes on it. The P's are assumed to be numbers, even if you can never discover their values at all precisely. This is the reason for the expression "black box." The black box may be entirely outside you, and used like a tame mathematician, or it may be partially or entirely inside you, but in any case you do not know the P's precisely.

Following Keynes and Koopman, I assume that the P''s are only partially ordered.

Apart from the axioms and rules, there are in practice many informal suggestions that you make use of, such as the need to throw away evidence judged to be unimportant, in order to simplify the analysis, and yet to avoid special selection of the evidence. (In spite of all the dangers, objectivistic methods in statistics invariably ignore evidence in order to achieve objectivity. In each application a subjective judgment is required in order to justify this device. Compare the idea of a "Statistician's Stooge" in ##199, 245.) But in this paper I am more concerned with axioms and rules of application than with "suggestions."

De luxe black boxes are available, with extra peripheral equipment, so that additional types of judgment and discernment can be used, such as direct judgments of odds, log-odds, "weights of evidence," numerical probabilities, judgments of approximate normality, and (for a theory of rational behavior) comparisons of utilities and of expected utilities (#26 or 43). (There are numerous aids to such judgments, even including black-box theorems, such as the central limit theorem, and a knowledge of the judgments of other orgs. All such aids come in the category of "suggestions.") But, as we shall see shortly, for the simplest kind of black box, a certain kind of output must not be available.

4. AXIOM SYSTEMS FOR SUBJECTIVE PROBABILITY

See, for example, Ramsey (1931), de Finetti (1937/64), Koopman (1940a, b), #13, Savage (1954), and, for similar systems for logical probability, Keynes (1921), Jeffreys (1939). The axioms of subjective probability can be expressed in terms of either

(i) comparisons between probabilities, or preferences between acts, or
(ii) numerical probabilities.

Koopman's system was concerned with comparisons between probabilities, without reference to utilities or acts. Although it is complicated it is convincing when you think hard. From his axioms he deduced numerical ones for what he called *upper* and *lower* probabilities, P^* and P_*. We may define $P_*(E|F)$ and $P^*(E|F)$ as the least upper bound and greatest lower bound of numbers, x, for which you can judge or discern that $P'(E|F) > x$ or $< x$. Here $P'(E|F)$ is not a number, although x is. The interpretation of the inequality $P'(E|F) > x$ is as follows. For each integer n, perfect packs of n cards, perfectly well shuffled, are imagined, so that for each rational number, $x = m/n (m < n)$, there exist propositions, G and H, for which $P(G|H)$ would usually be said to be equal to x. The inequality is then interpreted as $P'(E|F) > P'(G|H)$.

Note that $P_*(E|F)$ and $P^*(E|F)$ depend on the whole body of beliefs. Also note that $P_*(E|F)$ is not the least upper bound of all numbers, x, for which you can *consistently* state that $P'(E|F) > x$: to assume this interpretation for more than one probability would be liable to lead to an *in*consistency.

If $P_* = P^*$, then each is called P, and the usual axioms of probability are included in those for upper and lower probability. The analogy with inner and outer measure is obvious. But the axioms for upper and lower probability do not follow from the theory of outer and inner measure. It is a little misleading to say that probability theory is a branch of measure theory.

In order to avoid the complications of Koopman's approach, I have in the past adopted another one, less rigorous, but simpler. I was concerned with describing how subjective probability should be used in as simple terms as possible more than with exact formal justification. (I gave an informal justification which I found convincing myself.) This approach consisted in assuming that a probability *inside the black box* was numerical and precise. This assumption enables one to use a simple set of axioms such as the following set (axioms C).

C1. $P(E|F)$ *is a real number.* (Here and later, the "given" proposition is assumed not to be self-contradictory.)

C2. $0 \leqslant P(E|F) \leqslant 1$.

DEFINITION. *If $P(E/F) = 0$ (or 1), then we say that E is "almost impossible" (or "almost certain") given F.*

C3. *If E.F is almost impossible given H, then*

$$P(E \vee F|H) = P(E|H) + P(F|H) \text{ (addition axiom).}$$

C4. *If H logically implies E, then E is almost certain given H (but not conversely).*

C5. *If $H \cdot E$ and $H \cdot F$ are not self-contradictory and $H \cdot E$ implies F and $H \cdot F$ implies E, then*

$$P(E|H) = P(F|H) \text{ (axiom of equivalence).}$$

C6. $P(E \cdot F|H) = P(E|H) \cdot P(F|E \cdot H)$ (product axiom).

C7. (Optional.) *If* $E_i \cdot E_j$ *is almost impossible given* $H(i < j; i, j = 1, 2, 3, \ldots$ *ad inf.), then*

$$P(E_1 \vee E_2 \vee \ldots |H) = \sum_i P(E_i|H) \text{ (complete additivity).}$$

(The above axioms are not quite the same as axiom sets A and B of #13.)

C8. (The Keynes-Russell form of the principle of cogent reason. Optional.)

See p. 19 of Russell (1948), p. 4 of #13. *Let* ϕ *and* ψ *be propositional functions. Then*

$$P(\phi(a)|\psi(a)) = P(\phi(b)|\psi(b)).$$

I describe this axiom as "optional" because I think that in all those circumstances in which it is judged to be (approximately) applicable, the judgment will come to the same thing as that of the equation itself, with dashes on.

It follows from axiom C1, that $P(E|F) <, >$, or $= P(G|H)$, but we do not want to deduce that $P'(E|F)$ and $P'(G|H)$ are comparable. There is therefore an artificial restriction on what peripheral equipment is available with *de luxe* black boxes. This artificiality is the price to be paid for making the axioms as simple as possible. It can be removed by formulating the axioms in terms of upper and lower probabilities. To use axioms C is like saying of a non-measurable set that it really has an unknowable ("metamathematical") measure lying somewhere between its inner and outer measures. And as a matter of fact there is something to be said for this paradoxical-sounding idea. If you will bear with me for a moment I shall illustrate this in a nonrigorous manner.

Suppose A and B are two non-intersecting and non-measurable sets. Write m for the unknowable measure of a non-measurable set, and assume that

$$m(A + B) = m(A) + m(B).$$

Then

$$m(A + B) \leqslant m^*(A) + m^*(B), \; m(A + B) \geqslant m_*(A) + m_*(B).$$

Therefore (for elucidation, compare the following probability argument)

$$m^*(A + B) \leqslant m^*(A) + m^*(B), \; m_*(A + B) \geqslant m_*(A) + m_*(B),$$

and these inequalities are true. Similarly,

$$m(A) = m(A + B) - m(B) \leqslant m^*(A + B) - m_*(B).$$

Therefore

$$m^*(A) \leqslant m^*(A + B) - m_*(B),$$

which is also true.

The same metamathematical procedure can be used, more rigorously, in order to derive without difficulty, from axioms C together with the rules of application, a system of axioms for upper and lower probability. These are the axioms D listed below. As an example, I shall prove axiom D6(iii). We have

$$P(\mathrm{E}\cdot\mathrm{F}|\mathrm{H}) = P(\mathrm{E}|\mathrm{H})\cdot P(\mathrm{F}|\mathrm{E}\cdot\mathrm{H}).$$

Therefore, if $P(\mathrm{F}|\mathrm{E}\cdot\mathrm{H}) \neq 0$,

$$P(\mathrm{E}\cdot\mathrm{F}|\mathrm{H})/P(\mathrm{F}|\mathrm{E}\cdot\mathrm{H}) = P(\mathrm{E}|\mathrm{H}).$$

But $P^*(\mathrm{E}\cdot\mathrm{F}|\mathrm{H}) \geqslant P(\mathrm{E}\cdot\mathrm{F}|\mathrm{H})$, since, in this system, $P^*(\mathrm{E}\cdot\mathrm{F}|\mathrm{H})$ is defined as the greatest lower bound of numbers, x, for which it can be discerned that $x > P(\mathrm{E}\cdot\mathrm{F}|\mathrm{H})$. Similarly,

$$P_*(\mathrm{F}|\mathrm{E}\cdot\mathrm{H}) \leqslant P(\mathrm{F}|\mathrm{E}\cdot\mathrm{H}).$$

Therefore

$$P^*(\mathrm{E}\cdot\mathrm{F}|\mathrm{H})/P_*(\mathrm{F}|\mathrm{E}\cdot\mathrm{H}) \geqslant P(\mathrm{E}|\mathrm{H}).$$

Therefore

$$P^*(\mathrm{E}\cdot\mathrm{F}|\mathrm{H})/P_*(\mathrm{F}|\mathrm{E}\cdot\mathrm{H}) \geqslant P^*(\mathrm{E}|\mathrm{H}).$$

Q.E.D.

The main rule of application is now that the judgment or discernment $P'(\mathrm{E}|\mathrm{F}) > P'(\mathrm{G}|\mathrm{H})$ corresponds to the black-box inequality $P_*(\mathrm{E}|\mathrm{F}) > P^*(\mathrm{G}|\mathrm{H})$. Koopman derived most, but not all, of axioms D1-D6 from his non-numerical ones, together with an assumption that can be informally described as saying that perfect packs of cards can be imagined to exist. His derived axioms for upper and lower probability do not include axiom D6(iii) and (iv). (D7 and D9 were excluded since he explicitly avoided complete additivity.) I have not yet been able to decide whether it is necessary to add something to his non-numerical axioms in order to be able to derive D6(iii) and (iv). Whether or not it turns out to be necessary, we may say that the present metamathematical approach has justified itself, since it leads very easily to a more complete set of axioms for upper and lower probability than were reached by Koopman with some difficulty.

The axioms D will now be listed. I have not proved that the list is complete, i.e., that further independent deductions cannot be made from axioms C.

D1. P_**(E|F) and* P^**(E|F) are real numbers.* (Here and later the given proposition is assumed not to be self-contradictory.)

D2. $0 \leqslant P_*(\mathrm{E}|\mathrm{F}) \leqslant P^*(\mathrm{E}|\mathrm{F}) \leqslant 1$.

DEFINITION. *If* $\mathrm{P}^* = \mathrm{P}_*$, *each is called* P. *The previous definitions of "almost certain" and "almost impossible" can then be expressed as* $\mathrm{P}_* = 1$ *and* $\mathrm{P}^* = 0$.

D3. *If E · F is almost impossible given H, then* (addition axiom)

$$P_*(E|H) + P_*(F|H) \leqslant P_*(E \vee F|H) \leqslant P_*(E|H) + P^*(F|H) \\ \leqslant P^*(E \vee F|H) \leqslant P^*(E|H) + P^*(F|H).$$

D4. *If H logically implies E, then E is almost certain given H but not conversely.*

D5. *If H · E implies F, and H · F implies E, then* (axiom of equivalence)

$$P_*(E|H) = P_*(F|H), P^*(E|H) = P^*(F|H).$$

D6. (Product axiom.)

(i) $P_*(E \cdot F|H) \geqslant P_*(E|H) \cdot P_*(F|E \cdot H)$;
(ii) $P^*(E \cdot F|H) \geqslant P_*(E|H) \cdot P^*(F|E \cdot H)$;
(iii) $P^*(E \cdot F|H) \geqslant P^*(E|H) \cdot P_*(F|E \cdot H)$;
(iv) $P_*(E \cdot F|H) \leqslant P_*(E|H) \cdot P^*(F|E \cdot H)$;
(v) $P_*(E \cdot F|H) \leqslant P^*(E|H) \cdot P_*(F|E \cdot H)$;
(vi) $P^*(E \cdot F|H) \leqslant P^*(E|H) \cdot P^*(F|E \cdot H)$.

D7. (Complete super- and sub-additivity. Optional.)

If $E_i \cdot E_j$ *is almost impossible given* H ($i < j$; $i, j = 1, 2, 3, \ldots$ *ad inf.*), *then*

(i) $P_*(E_1|H) + P_*(E_2|H) + \cdots \leqslant P_*(E_1 \vee E_2 \vee \cdots |H)$;
(ii) $P^*(E_1|H) + P^*(E_2|H) + \cdots \geqslant P^*(E_1 \vee E_2 \vee \cdots |H)$.

D8. (Cogent reason. Optional.) *Let ϕ and ψ be propositional functions. Then*

$$P_*(\phi(a)|\psi(a)) = P_*(\phi(b)|\psi(b)), P^*(\phi(a)|\psi(a)) = P^*(\phi(b)|\psi(b)).$$

D9. (Complete super- and sub-multiplicativity. Optional.) *For any (enumerable) sequence of propositions, $E_1, E_2, \cdots$.*

(i) $P_*(E_1|H)P_*(E_2|E_1 \cdot H)P_*(E_3|E_1 \cdot E_2 \cdot H) \cdots \leqslant P_*(E_1 \cdot E_2 \cdots |H)$;
(ii) $P^*(E_1|H)P^*(E_2|E_1 \cdot H)P^*(E_3|E_1 \cdot E_2 \cdot H) \cdots \geqslant P^*(E_1 \cdot E_2 \cdots |H)$.

The corresponding result in the C system is a *theorem.* (See Appendix.)

I have not been able to prove that

$$P^*(E \vee F) + P^*(E \cdot F) \leqslant P^*(E) + P^*(F),$$

even though the corresponding property is true of Lebesgue outer measure, and I suspect that it does not follow by the above methods. It would be possible to prove it (compare p. 14 of Burkill, 1951) provided that we assumed the axiom:

D10. (Optional.) *Given any proposition, E, and a positive number, ϵ, there exists a proposition, G, which is implied by E, and has a precise probability* $P(G) < P^*(E) + \epsilon$. (This axiom may be made conditional on another proposition, H, in an obvious manner.)

I cannot say that D10 has much intuitive appeal, although the corresponding assertion is true in the theory of measure.

It seems that the theory of probability is not quite a branch of the theory of measure, but each can learn something from the other.

Incidentally, random variables can be regarded as *isomorphic* with arbitrary functions, not necessarily measurable. I understand that this thesis is supported by de Finetti. Also upper and lower expectations can be defined by means of upper and lower integrals in the sense of Stone (1948).

5. HIGHER TYPES OF PROBABILITY

A familiar objection to precise numerical subjective probability is the sarcastic request for an estimate correct to twenty decimal places, say for the probability that the Republicans will win the election. One reason for using upper and lower probabilities is to meet this objection. The objection is however raised, more harmlessly, against the precision of the upper and lower probabilities. In **#26** and in lectures at Princeton and Chicago in 1955, I attempted to cope with this difficulty by reference to probabilities of "higher type." When we estimate that a probability $P'(E|H)$ lies between 0.2 and 0.8, we may feel that 0.5 is *more likely to be rational* than 0.2 or 0.8. The probability involved in this expression "more likely" is of "type II." I maintain that we can have a subjective probability distribution concerning the estimate that a perfectly rational org would make for $P'(E|H)$ a subjective probability distribution for a credibility. *If* this probability distribution were sharp, then it could be used in order to calculate the expected credibility precisely, and this expectation should then be taken as our subjective probability of E given H. But the type II distribution is not sharp; it is expressible only in terms of inequalities. These inequalities themselves have fuzziness, in fact the fuzziness obviously increases as we proceed to higher types of probability, but it becomes of less practical importance.

It seems to me that type II probability is decidedly useful as an unofficial aid to the formulation of judgments of upper and lower probabilities of type I. I would not myself advocate even the unofficial use of type III probability for most practical purposes [but see ##398, 547, 862, 929, 1199], but the notion of an infinite sequence of types of probability does have the philosophical use of providing a rationale for the lack of precision of upper and lower probabilities.

APPENDIX. CONTINUITY

(See the remark following D9.) There is a well-known strong analogy between the calculus of sets of points and the calculus of propositions. In this analogy "E is contained in F" becomes "E implies F"; $E + F$ becomes $E \vee F$; $E - F$ becomes $E \cdot \bar{F}$; $E \cap F$ becomes $E \cdot F$; "E is empty" becomes "E is impossible"; "all sets are contained in E" becomes "E is certain"; $E_n \nearrow$ becomes $E_1 \supset E_2 \supset E_3 \supset \ldots$; $E_n \searrow$ becomes $\ldots E_3 \supset E_2 \supset E_1$.

Accordingly we can define, for an infinite sequence of propositions, $\{E_n\}$,

$$\limsup E_n = (E_1 \vee E_2 \vee \ldots) \cdot (E_2 \vee E_3 \vee \ldots) \cdot (E_3 \vee E_4 \vee \ldots) \ldots,$$
$$\liminf E_n = (E_1 \cdot E_2 \ldots) \vee (E_2 \cdot E_3 \ldots) \vee (E_3 \cdot E_4 \ldots) \vee \ldots.$$

If these are equal, each is called $\lim E_n$. The limit of a monotonic increasing (or decreasing) sequence of propositions is

$$E_1 \vee E_2 \vee \ldots \text{ (or } E_1 \cdot E_2 \ldots\text{).}$$

The other definitions and arguments given, for example, in pp. 84-85 of Loève, 1955, can be at once adapted to propositions, and we see that complete additivity is equivalent to continuity, i.e., $\lim P(E_n) = P(\lim E_n)$ if $\{E_n\}$ is a monotonic sequence of propositions. It can then be proved that, for example,

$$P(E_1 \cdot E_2 \ldots) = P(E_1) \cdot P(E_2 | E_1) \cdot P(E_3 | E_1 \cdot E_2) \ldots,$$

i.e., we have "complete multiplicativity." The axiom D9 is derived from this theorem by means of the metamathematical argument of Section 4.

The analogy between propositions and sets of points is imperfect. For the analogy of a point itself should be a logically possible proposition, E, that is not implied by any other distinct proposition that is logically possible. It is not easy to think of any such proposition E, unless the class of propositions has been suitably restricted. Fortunately the notion of a point is inessential for the above analysis: the algebra of sets of points makes little use of the notion of a point.

CHAPTER 8

Random Thoughts about Randomness (#815)

In this paper I shall bring together some philosophical and logical ideas about randomness many of which have been said before in scattered places. For less philosophical aspects see, for example #643.

When philosophers define terms, they try to go beyond the dictionary, but the dictionary is a good place to start and one dictionary definition of "random" is "having no pattern or regularity." This definition could be analyzed in various contexts but, at least for the time being, I shall restrict my attention to sequences of letters or digits, generically called "digits." These digits are supposed to belong to a generalized alphabet; perhaps a new word should be used, such as "alphagam," but I shall use the word "alphabet" in the generalized sense so as to avoid a neologism. For simplicity I shall assume that the alphabet is finite and consists of the t digits 0, 1, 2, . . ., $t-1$. From a philosophical point of view it would make little difference if the number of digits in the alphabet were countably infinite. It *is* important whether the sequence itself is finite or infinite in length. Finite sequences are more practical, but I shall briefly discuss infinite sequences first. Infinite random sequences belong more to mathematics than to the external world but, apart from their purely mathematical interest, they might stimulate practical procedures, that is, procedures applicable to finite sequences. Fortunately, the mathematical history of infinite random sequences, which turn out to be equivalent to irregular collectives, has been extremely well covered by Dr. Coffa in this conference so I shall take most of it for granted. I should just like to add what I think is the simplest convincing *intuitive* argument to show the consistency of the idea of an irregular collective. The argument uses the idea of *generalized decimals*. . . . Let me remind you that an irregular collective, spelt more impressively as *Irregular Kollektiv*, is an infinite sequence satisfying some postulates of von Mises, Wald, etc., relating to long-run frequencies. (For precise definitions and detailed references see Martin-Löf [1969] and

Coffa's paper.) The long-run frequencies of digits, digit-pairs, and "polynomes" (*n*-plets), are to apply to a certain countable number of subsequences of digits, these subsequences being determined by the so-called "place-selection" rules. (There can be only a countable number of such rules expressible in an unambiguous language.) . . .

[The omitted page gives the argument concerning generalized decimals. Anyone referring to the original printing should note that the diagram was somewhat misprinted.] Naturally an irregular collective cannot be constructed, that is, it cannot correspond to a computable number in the sense of Turing (1937). Existence proofs in mathematics cannot necessarily be made constructive. For irregular collectives they necessarily *cannot*. (For the "mathematical intuitionist" *in Brouwer's sense*, irregular collectives do *not* exist!)

An infinite random sequence can be directly defined in terms of probability: the probability is p_r that a digit in each specific place is r, conditional on *any* information about the other digits. It turns out that infinite random sequences are the same mathematical entities as irregular collectives. Once this is accepted there is no need to continue to use the expression "irregular collective."

Sometimes when we say that a *finite* sequence is random we mean that it has been obtained from a machine that is supposed to produce an infinite sequence of digits, this infinite sequence being random. Put otherwise, we sometimes mean that the finite sequence is a segment of an infinite random sequence, and moreover the position of this segment is itself in some sense chosen "at random" from the infinite random sequence. So we should consider also what is meant by selecting a segment at random. There is no difficulty here if the sequence is produced by a machine. For we can then produce a random segment of length N by merely switching the machine on and letting it produce N digits. If we select a finite segment from a *book* of random digits we have to select a random starting point, and this can be done "equiprobably" by means of devices such as roulette wheels and icosahedral dice. But it is philosophically unsatisfactory to use a printed book of random digits because of the possibility that the digits we select have been used by another experimenter. In a perfect but indeterministic world, when any one used a piece of his copy of the book the corresponding digits would vanish from all other copies of the book, and eventually all copies would become blank and could be used as notebooks. (In terms of magnetic tapes and computer networks, this point is not quite as ludicrous as it seems.) I shall return later to the subject of finite random sequences.

Apart from the distinction between finite and infinite sequences, we now have three interconnected concepts, that of randomness, regularity, and probability. Philosophers might wish to define or to describe any one of these by itself, or in terms of one of the other concepts or in terms of the other two. This gives rise to twelve distinct possible discussions, but do not worry, for I shall not try to tackle them all.

There is one minor paradox worth cleaning up concerning regularity. The laws of large numbers provide a kind of regularity which is not supposed to run counter to the randomness of a sequence. This does not give rise to a paradox within the rigorous interpretation of irregular collectives: it merely shows that the definition of randomness as "having no regularity" is not rigorous enough. There is even an idea that has been floating around for some time that all laws of nature are statistical and analogous to the laws of large numbers: "order out of chaos."

Limiting frequencies can be used, in the style of von Mises and his elaborators, as a basis for the definition of an infinite random sequence, but not of course directly for finite sequences. On the other hand both finite and infinite random sequences can be directly and easily defined in terms of probability: the definition given before for infinite sequences applies verbatim for finite sequences. If in this definition the probabilities are supposed to be physical then the definition is that of a physically random sequence; if the probabilities are mathematical or logical or subjective, then the sequence is mathematically or logically or subjectively a random sequence respectively. (For a discussion of kinds of probability see, for example, #182.) Since most of us believe that the ordinary calculus of probability is self-consistent, it follows that we should also believe that the notion of a finite or infinite random sequence is self-consistent. (We only need to assume that the Peano axioms of the integers are self-consistent.)

So why the emphasis on von Mises collectives? Presumably collectives are still of interest outside mathematics to those who regard randomness or limiting frequencies as primary and probability as secondary.

What can be meant by asking whether randomness or probability is primary? It could be a matter of which is the logically simpler concept, or the psychological question of which concept is formed first by a baby. Let's try some armchair psychology in which we guess how a baby learns these concepts.

Obviously a baby must have a latent ability to do some categorization of its sense impressions so as to form concepts of different kinds of objects. It must look for or notice regularities, otherwise it could not categorize. Presumably the baby enjoys looking for, and finding, *new* regularities, when they are not *too* new, just as an adult does. This is part of an explanation for the enjoyment of music. When the baby tries to find a regularity and fails to do so it would begin to arrive at the concept of randomness. But it could not find regularities without at the same time having expectations which involve some notion of probability. When one of its expectations is verified it begins to think it has found a regularity. Although I have not developed this line of thought in detail, it seems fairly obvious that the baby forms the concepts of regularity, randomness, and probability in an intertwined manner. The building of these concepts in the baby seems to be analogous to that of the building of concepts by scientists. As Popper once said, "Science is based on a swamp," which we can now parody by saying that scientists and babies are both based on swamps. To ask which of the concepts of regularity, randomness, and probability is psychologically prior might be like asking whether "the chicken is the egg's way of making another egg" [Samuel Butler]. It might be a fruitful field for psychological research. Such research would not be easy, as it is with older children, since the baby's lack of language is not fully compensated by its lungpower [or swamp-power].

I do not believe that the concept of randomness is logically either prior or secondary to that of probability, but randomness can be easily defined in terms of probability. It is also possible, along the lines of von Mises and Co., to define probability in terms of a suitable interpretation of randomness, although this is not so easy. It is less easy one way round than the other because the ordinary axioms of probability are understood and accepted as sensible by nearly all mathematicians since they can be interpreted in terms of lengths on a line, whereas the concept of a random sequence or collective (developed in the von Mises manner) ties up more with formal logic. For historical reasons, and perhaps also for economic reasons, measurement of a line has been a more traditional part of elementary mathematical education than has formal logic. Therefore, for most of us, the usual axioms of probability seem simpler than an adequate set of axioms for irregular collectives. But measurement of a line involves the concept of a real number which only seems simple because of its familiarity. So, as far as I can see, it is only a matter of taste which of probability and randomness is regarded as primary.

So far, I have been mixing up the real world with mathematics. I have suggested that the real world is what provokes the baby to form some more or less mathematical concepts, such as randomness and probability, before he has learnt the appropriate terminology. The implication is that there must be some connection between physical randomness in the world and the mathematical notion of randomness. But the existence of physical randomness is not strictly necessary as part of my argument. All I need is that the baby behaves *as if* he believed in physical randomness. His belief in physical randomness, if he has it, may or may not be correct, and it might be merely a projection of his own ignorance of fully deterministic laws of nature. It is only later, when he is no longer a baby, that he might argue whether the external world is deterministic, a question with a high degree of metaphysicality in my opinion, but which is stimulating to discuss. (For "degrees of metaphysicality" see ##**182**, 243.) If there is such a thing as physical randomness then there is also such a thing as physical probability and conversely. The question is one that can never be definitely settled, but physics might one day stabilize to a point where one of the two views, determinism or indeterminism, becomes the more convenient assumption, and the result might be to determine the predominant philosophical viewpoint at that time.

Let's now consider the place of randomness in statistics. Random sampling and random designs of experiments were introduced into statistics to achieve apparent precision and objectivity. I believe that the precision attained by objectivistic methods in statistics invariably involves the throwing away of information (p. 102 of #13), although the clients are seldom told this. In particular, the use of random sampling leads to precise results about the hypotheses of interest to the client only if the experimenter is unaware of his choice of random numbers. The comment dates back at least to 1953 (pp. 264-265 of Good, 1953a), and to some years earlier in conversations. This difficulty can be clearly seen in terms of the famous experiment for testing whether a certain lady had some ability to tell by the taste whether the milk was put into her tea first. R. A. Fisher begins Chapter 2 of his book on the design of experiments with a discussion of this experiment, and presents the official objectivistic line. Let us suppose that we decide to test the lady's ability by giving her twenty cups of tea to taste, in ten of which the milk was put in first. We choose one of the $20!/(10!)^2$ sequences equiprobably. Suppose that she gets all twenty judgments right. Then, on the assumption that she really has no ability at milk-in-first detection, a coincidence of 1/184756 has occurred and this is an objective (or at least a highly multisubjective) probability. But suppose that, on looking at the random selection of sequences, we find that the milk happened to be put in first in accordance with a suspiciously regular-looking pattern. Then we have difficulty in deciding the probability that the lady might have thought of this pattern independently, and the objective precision of the probability vanishes. Matters are really worse, for even if the sequence of milk first and last is not mathematically simple we still have to judge the probability that the lady might have thought of this sequence. (It might turn out to be the 5-unit teleprinter encoding of the word "milk.") We then have to make a judgment, which in the present state of knowledge would always be subjective (personal), of the probability that the lady would have thought of this sequence for reasons unrelated to the tastes of the various cups of tea. *This* for us is the *relevant* probability, not 1/184756 (p. 62). It must be for such a reason that Lindley suggested that, rather than generate random numbers mechanically, it is better to generate them by personal judgment. I think it is safer to use a compromise, that is, to generate them mechanically and then to use both statistical tests and personal pattern recognition to reject suspicious-looking sequences. But what is suspicious for one purpose might not be for another. When a statistical experiment involves a human, then any sequence that is a close approximation to one with a simple definition is suspicious, where the simplicity might be *relative* to existing human knowledge, but if the subject of an experiment is not a person this level of caution is less required. For example, in an agricultural experiment involving Latin squares, I would be suspicious of one having marked diagonal properties because the field might have once been ploughed in that direction, but I would not necessarily be worried if the Latin square somehow reminded me of a chess endgame composition by Troitsky.

A statistician who uses subjective probabilities is called a "Bayesian." As Herman Rubin said at the Waterloo conference on the foundations of statistical inference in 1970 (Godambe & Sprott, 1971), "A good Bayesian does better than a non-Bayesian, but a bad Bayesian gets clobbered." This is clear in the present context. Another name for a non-Bayesian is an objectivist. Fisher once mentioned (private communication) that many of his clients were not exceptionally bright so perhaps his emphasis on objectivism was provoked by a thought like the second half of Rubin's remark.

The problem of which finite random sequences are satisfactory occurs in a dramatic form in the science of cryptology. It might be supposed that there is no better method of enciphering a secret message than a random method. Yet suppose that, by extraordinarily bad luck, the enciphered message came out exactly the same as the original one: would you use it? Would you use it if you knew that your opponent knew you were using a random method?

The objectivist can try to avoid the difficulty that every finite random sequence can be found to have some apparently "non-random features" by engaging a *Statistician's Stooge* who selects the random numbers, carries out the experiment, and reports the number of successes out of twenty achieved by the Lady. If he reveals the sequence of random numbers to the experimenter he will be shot and knows it. The experimenter achieves his objectivity by saying "I don't wish to know that fact." Thus the experimenter who wishes to appear objective and precise can take his choice between two crimes: threatening to murder the Stooge or suppressing some of the evidence. He can strengthen his position further by crucifying the Bayesians who had not already been clobbered. I believe that, at least since 1956, many objectivists have been suppressing the fact that they suppress evidence.

M. G. Kendall (1941), after considering matters of "place selection" for infinite random sequences, and perhaps prompted by the infinite case, turns his attention to the more practical case of finite sequences and says that they can only be *random with respect to the tests of randomness used.* These tests are to some extent analogous to the place selection rules, but do not refer to limiting frequencies; rather they are ordinary statistical tests of significance used for rejecting "unsuitable" sequences. He points out that this cuts down on the number of usable sequences of length N. Since a doctored sequence is no longer strictly random, this procedure affects the interpretation of the experiment for which the sequence is used. The first paper known to me that tries to cope with the practical problem that this creates is a scathing book review by Christopher S. O'D Scott (1958). Since this was published in a journal not readily available, I shall mention Scott's result here. (See also p. 252 of #162.)

Suppose that we apply some statistical test for randomness to a finite sequence and we reject it with probability P_1. If it gets rejected, we generate another sequence and test it also. In this way we eventually find a sequence that passes the test and we use it in some experimental design. [Then the sequence that is used is not strictly random. Nevertheless statisticians are wont to regard such doctored sequences as strictly random, and, when used for some statistical experiment, the sequence might give rise on this assumption to a tail-area probability P_2. Scott points out that the correct tail area is P_2', where

$$\frac{P_2 - P_1}{1 - P_1} \leqslant P_2' \leqslant \frac{P_2}{1 - P_1}.$$

If P_1 is small, then the doctoring causes little harm since P'_2/P_2 is then known to be not much greater than 1. It might be much smaller than 1 unless P_1/P_2 is also small.

Note that if we were to apply a collection of tests, perhaps a countable infinity of them, with rejection probabilities Q_1, Q_2, Q_3, . . . , then the entire collection can be regarded as a single test with a rejection probability P_1 where

$$P_1 < 1 - (1 - Q_1)(1 - Q_2)(1 - Q_3) \ldots .$$

If we are sensible we would make the Q's small, for complicated tests, in such a manner that the product would have a value close to 1 (compare #191).

Kolmogorov (1963) sets himself the problem of defining what might be meant by a finite random sequence, without reference to its being a segment of an infinite random sequence. (See also Martin-Löf, 1969.) Basically Kolmogorov considers a finite sequence to be random if there is no appreciably shorter sequence that describes it fully, in some unambigous mathematical notation that uses the same alphabet as the sequence itself. (Simplicity was defined in terms of brevity by Valéry, 1921. See also ##599, 876.) Although this involves some philosophical difficulties, and is of mathematical interest, it does not yet solve the practical problems. Scott's book review comes closer to this goal in terms of objectivistic statistics.

In terms of strictly subjectivistic or Bayesian statistics it is necessary to make more explicit use of the degree of simplicity of the rule that could have generated the observed finite sequence or of approximately generating it (#162) and of the degree of approximation. Somehow we have to judge the probability that the Lady, or the agricultural plot, might have attained the observed level of apparent significance when the design of the experiment is exactly the one used, including the specific random numbers or Latin square etc., if we are not to suppress known evidence. Clearly this is apt to be a very difficult task, even when only interval judgments of probability are to be made (as I have long advocated). But in strict logic this difficulty would have to be faced if all the information were to be taken into account. I think that, from a practical point of view, a compromise between the subjectivistic and the objectivistic analysis should often be adopted. This still fits into a Bayesian framework in one of its interpretations, namely when the costs of thinking are taken into account. (For varieties of Bayesians, see **#765.**) The aim of a statistical experiment, in the Bayesian framework, is the maximization of the mathematical expectation of the utility; but, when allowance for the costs of thinking and calculation are taken into account, one reaches a very practical compromise which I call *rationality of the second type* (for example, **#659**). It is impossible to reconcile this fully with ordinary logic. It is, however, in my opinion, the only possible practical decision rule in applied statistics although the statistics textbooks do not yet mention it as far as I know, perhaps partly for "political" reasons. It is tied up with the notion of "evolving" or "shifting" [or "dynamic"] probabilities, which are subjective probabilities that change in virtue of thought alone, without obtaining new empirical information (#938). An example is when you estimate the probability of a mathematical theorem before trying to prove it, as every good mathematician must do at least informally, or you might for betting purposes estimate as 0.1 the probability that the millionth digit of π is a 7 (p. 49 of #13). In strict logic the probability must be 0 or 1, whereas, in the π example, the evolving subjective probability is 0.1 that the credibility is 1. I doubt if practical subjective probabilities can ever be other than evolving. Evolving probabilities are used to resolve an important paradox about simplicity in #599.

The need to throw away information to attain apparent objectivity occurs even in the familiar technique of random sampling. Suppose that we wish to find out the proportion of people who would support McGovernment. We take a random sample, we do the interviewing at the doors of the electorate, and we can record or notice many facts about the interviewees. Weather forecasts are also relevant since the weather can affect the result of an election. Only by throwing away most of this information can we compute an "objective" statement of apparent probability.

To return, for definiteness, to the lady tea-taster, suppose that we were to use pseudorandom numbers for the selection of the sequence. Recall that pseudorandom numbers appear random in accordance with a variety of tests for randomness, but in fact are generated according to a deterministic rule, preferably one that cannot be readily computed mentally. For example, we might use the binary expansion of $\sqrt{2}$ starting at the 777th digit. It is extremely unlikely that the lady will know these digits, although they have been published (#526), and we are subjectively sure that she cannot compute them mentally. The chance that she will have even partial knowledge of these digits would presumably be judged by the statistician as much less that 1/184756. Hence pseudorandom numbers can be just as useful as strictly random ones, in fact they have some important practical advantages in some contexts.

I think this point about pseudorandomness has repercussions for the philosophy of physics and for the problem of determinism in general. In classical (non-quantum) statistical mechanics (Bohm, 1952), it might be possible to assume consistently that the physical processes are strictly deterministic. But they are so complicated that, as far as we shall ever know, they might appear to be random. Thus they might be pseudorandom but indistinguishable from being strictly random by any method now known and possibly by any method that will ever be discovered (p. 15 of #13; #153). From an operational point of view then, a deterministic interpretation might be indistinguishable from an indeterministic one. It cannot make an operational difference whether we consider that the probabilities involved in our predictions are physical probabilities inherent in the external world, or are probabilities related to our permanent ignorance of some underlying determinism. It is in this sense that I disagree with an argument of Mach and of von Mises that statistical physical laws cannot be deduced from a deterministic physical theory (pp. 184-186 of von Mises, 1957; Good, 1958). If the reasoning of Mach and von Mises were correct on this point it would be even more unjustified to use pseudorandom numbers in simulation studies. The universe might be a pseudorandom study in preparation for an improved design.

Dr. Coffa has pointed out that the mathematical concept of randomness, due to von Mises and his elaborators, is not identical with that of physical randomness. I agree with this statement but we must not be misled by it. It is conformable with the fact that physical randomness is based on physical probability, whereas mathematical randomness is based on mathematical probability as mentioned before. A mathematically random sequence is one that cannot be effectively constructed by mathematical rules. A physically random sequence arising in some physical system is one that cannot be predicted with the help of any separate physical apparatus. Just as one mathematically random sequence can be independent of another, a physically random process can be physically independent of another. Although a mathematically random sequence is not the same as a physically random one, there is nevertheless a strong analogy between them. A probabilistic mathematical model might have good predictive value for a physical situation. It seems to me that this is what statistical mechanics is about. We can never prove that a physical system is deterministic so we can lose nothing practical by displacing the indeterminism from our shoulders to those of the physical system and by saying we are describing some properties of the system that might be true independently of our own existence. On the other hand, the Copenhagen interpretation of quantum mechanics has solipsistic tendencies (Bohm, 1952; Bunge, 1955; Wigner, 1962; p. 42 of #13), but I agree with Bohm and Bunge and I think with Coffa and Kyburg (in this symposium) that this tendency might be removed. Whether it is possible to remove the solipsistic element is a part of the main problem in the philosophy of science, the mind-body problem, even more important than the problem of determinism versus indeterminism, although the two problems happen to be closely associated especially in the philosophy of quantum mechanics. Bunge (1955), who argues powerfully for the existence of the external world, shows convincingly that the solipsism of "complementarity" is based primarily on faith, but physicists who accept it have been reluctant to take the next step of regarding the wave function of the whole universe as that of an observer outside the physical universe. This point has been made with the heading "And Good saw that it was God(d)" (Good, 1971; ##882, 1322A). Physicists, following Laplace, may "have no need of this hypothesis" but it is at any rate more economical than the branching universe theory of Everett (1957). (This branching theory had occurred in a primitive form in science fiction, and at a philosophical conference [Good, 1962b].) The only form of solipsism that appeals to common sense would be solipsism by God (if he [she or it] exists) and this would logically imply that the mind of any conscious entity would have to be a part of God. I shall return to the question of solipsism in connection with de Finetti's theorem.

In the discussion some one raised the question of whether the physical universe might be "infinitely random" in the sense that an infinite amount of information would be required to make an accurate prediction of a closed system. The following observation is relevant. Consider a model of classical statistical mechanics in which the various particles are assumed to be perfectly elastic spheres. Then, to predict a time T ahead, to a given tolerance, I think we would need to know the positions and velocities of the particles to a number of decimal places proportional to T. There must be some degree of accuracy that is physically impossible to measure, and in this sense classical statistical mechanics provides indeterminism arising out of determinism. On the other hand, quantum mechanics suggests the possibility that the universe is a finite-state machine. From this point of view quantum mechanics might provide determinism arising out of indeterminism!

The question of the existence of physical probabilities, and therefore of physical randomness, is analogous to that of the existence of the "outside world" (outside your mind) though somewhat less convincing. I find it much easier to make decisions by assuming that the world exists, and, to a much lesser extent, by assuming that physical probabilities exist. Regarding *logical* probabilities, that is, *credibilities,* I think it is mentally healthy to believe some of the time that they "exist" since this belief gives you some psychological support when estimating your own subjective probabilities. (Your judgment is helped by your imagination just as in the Device of Imaginary Results, pp. 35, 70, 81 of #13; pp. 19, 20, 45 of #398; and especially #547.) It leads naturally to the notion of a hierarchy of types of probability, which is useful in Bayesian statistics (##398, **1230**).

Let me turn now to one of the commonest problems both in life and in statistics, since it has been touched on in this symposium, namely that of estimating the probability that some event will occur on some specific occasion. This is always a single-case probability; it is always a matter of estimating the probability of an event that has never occurred before if enough background is taken into account (such as the time on the clock). The problem subsumes *all* other problems as special cases, so that a completely general solution is out of the question. In a paper (#83) rejected in 1953 (in its original form), I was concerned with the estimation of the probability, physical or subjective, corresponding to a cell in a contingency table, where the cell might be empty. I described this as the probability of an event that had never occurred, a description whose deletion was recommended by the referee on the grounds that it was "provocative" (which perhaps means that is was philosophical), thus suggesting that the concept was somewhat pioneering at that time. Jimmie Savage once wittily said, at the expense of the philistines, "philosophy" is a dirty ten-letter word.

An example of a contingency table is one where people are classified by their occupations and causes of death, a two-way classification. Some of the cells in the table might be empty, such as the cell corresponding to the number of professional chess-players kicked to death by a horse [the Bortkiewicz effect]. To make an estimate we have to look for some source of information that is not immediately obvious. One way is to try to lump together some of the rows or some of the columns. The method that gave an apparently sensible answer for the two contingency tables examined was based on the observation that the "amount of mutual information" between a row and a column, $\log(p_{ij}/(p_{i.}p_{.j}))$, seemed to have a normal distribution over the table as a whole. This was used as a kind of semi-initial distribution in a Bayesian argument. (Here p_{ij} denotes the physical probability corresponding to cell (i, j) and $p_{i.}$ and $p_{.j}$ denote row and column total probabilities.) Another example of an event that has never occurred before crops up in the sampling of species or vocabulary (##38, 86). Suppose that we sample N words of text at random and that the number of words (species) each of which is represented r times in the sample is denoted by n_r (the "frequency of the frequency" r). Then we ask what is the probability that the next word or animal sampled will belong to a species that has never occurred before. The answer is approximately n_1/N if n_1 is not too small, a result that is not obvious. The basic idea behind this, of using the frequencies of the frequencies as a source of information, is due to Turing (1940) and it anticipates the logic of the [sophisticated form of the] empirical Bayes method to a great extent.

Actuaries have been aware of the problem of estimating probabilities of unique events for more than a hundred years and they usually select some reference class that favors their insurance company at the expense of the customer (Vajda, 1959). But this method would not be used if the competition between insurance companies were perfect, and then actuaries would have to face the problem in greater generality, which is that of the estimation of a multidimensional mixed probability density and mass function. A solution is also needed for probabilistic medical diagnosis. Proposals for an approximate solution to the problem have been made (Dickey, 1968; p. 267 of #701).

Some of you might have expected me, as a confirmed Bayesian, to restrict the meaning of the word "probability" to subjective (personal) probability. That I have not done so is because I tend to believe that physical probability exists and is in any case a useful concept. I think that physical probability can be *measured* only with the help of subjective probability (##**182**, 617) whereas de Finetti (1937/64) believes that it can be *defined* in terms of subjective probability. (See also the index of #398.) De Finetti showed that if a person has a consistent set of subjective or logical probabilities, then he will behave *as if* there were physical probabilities, where the physical probability has an initial subjective probability distribution. It seems to me that, if we are going to act as if the physical probability exists, then we don't lose anything practical if we assume it really *does* exist. In fact I am not sure that existence means more than that there are no conceivable circumstances in which the assumption of existence would be misleading. But this is perhaps too glib a definition. The philosophical impact of de Finetti's theorem is that it supports the view that solipsism cannot be logically disproved. Perhaps it is the mathematical theorem with most potential philosophical impact.

In discussions of probabilistic causality the notion of physical probability is useful (#**223B**), but subjective probabilities are more relevant for explanation (##599, **1000**). (My work on explanation was quantitative, but is I think compatible with the qualitative idea of Hempel's that the explanation should be of a more general nature than what is to be explained.)

I should like to conclude my discussion with a reference to a book by McShane, and more especially to the anonymous review of this book in the *Times Literary Supplement*, which review I wrote (#697). The main idea of the book is that emergence of new things in the world and also the emergence of new sciences and new ideas and hypotheses about the world are all possible only in virtue of randomness. As an example, consider whether life can in principle be explained in terms of physics. The argument in favor of this reductionist hypothesis is that quantum mechanics seems to be sufficient to explain the whole of chemistry "in principle"; and most quantum physicists believe that physical randomness is responsible for mutations [and other chromosomal changes] and hence, when combined with natural selection, for the emergence of new species of life. McShane denies the possibility of reductionism because he says that randomness precludes prediction with anything approaching certainty. The study of biology is the study of life as it happens to have emerged in this world. This seems to me to be equivalent to saying that reductionism is false because biology is to some extent a study of a historical process and especially of the present results of that process: the millions of species of living organisms as they exist today. Let me continue with an exact quotation from the book review.

"But suppose we imagine a different kind of biology, a study of the class of all possible life that could exist: in fact this could be regarded as the ultimate aim of the emerging field of theoretical biology. Its completion would be a task of greater magnitude than the analysis of all possible chess games: utterly impossible in practice, but 'logically' possible, possible 'in principle.' For that matter, even the study of all possible machines that could be designed is a study that could never be completed. Machines, too, are emergent. Why pick on life? It is [perhaps] only because of the limitations of time, and because of the convention of studying only life forms that we meet, that reductionism appears to be false." In other words, the trouble with reductionism might be merely lack of time.

CHAPTER 9

Some History of the Hierarchical Bayesian Methodology (#1230)

SUMMARY

A standard technique in subjective "Bayesian" methodology is for a subject ("you") to make judgments of the probabilities that a physical probability lies in various intervals. In the hierarchical Bayesian technique you make probability judgments (of a higher type, order, level, or stage) concerning the judgments of lower type. The paper will outline *some* of the history of this hierarchical technique with emphasis on the contributions by I. J. Good because I have read every word written by him.

1. PHILOSOPHY

In 1947, when few statisticians supported a Bayesian position, I had a non-monetary bet with M. S. Bartlett that the predominant philosophy of statistics a century ahead would be Bayesian. A third of a century has now elapsed and the trend supports me, but I would now modify my forecast. I think the predominant philosophy wil be a Bayes/non-Bayes synthesis or compromise, and that the Bayesian part will be mostly hierarchical. But before discussing hierarchical methods, let me "prove" that my philosophy of a Bayes/non-Bayes compromise or synthesis is necessary for human reasoning, leaving aside the arguments for the specific axioms.

Proof. Aristotelean logic is insufficient for reasoning in most circumstances, and probabilities must be incorporated. You are therefore forced to make probability judgments. These subjective probabilities are more directly involved in your thinking than are physical probabilities. This would be even more obvious if you were an android (and you cannot prove you are not). Thus subjective probabilities are required for reasoning. The probabilities cannot be sharp, in general. For it would be only a joke if you were to say that the probability of rain tomorrow (however sharply defined) is 0.3057876289. Therefore a theory of partially ordered subjective probabilities is a necessary ingredient of rationality.

Such a theory is "a compromise between Bayesian and non-Bayesian ideas. For if a probability is judged merely to lie between 0 and 1, this is equivalent to making no judgment about it at all" (p. 137 of **#838**). Therefore a Bayes/non-Bayes compromise or synthesis is an essential ingredient of a theory of rationality. Q.E.D.

The notion of a hierarchy of different types, orders, levels, or stages of probability is natural (i) in a theory of physical (material) probabilities, (ii) in a theory of subjective (personal) probabilities, and (iii) in a theory in which physical and subjective probabilities are mixed together. I shall not digress to discuss the philosophy of kinds of probability. (See, for example, Kemble, 1941; **#182**; Chapter 2 of #398.) It won't affect what I say whether you believe in the real existence of physical (material) probability or whether you regard it as defined in terms of de Finetti's theorem concerning permutable (exchangeable) events.

I shall first explain the three headings, leaving most of the elaborations and historical comments until later.

(i) *Hierarchies of physical probabilities.* The meaning of the first heading is made clear merely by mentioning populations, superpopulations, and super-duper-populations, etc. Reichenbach (1949, Chapter 8) introduced hierarchies of physical probabilities in terms of random sequences, random sequences of random sequences, etc.

(ii) *Hierarchies arising in a subjective theory.* Most of the justifications of the axioms of subjective probability assume sharp probabilities or clear-cut decisions, but there is always some vagueness and one way of trying to cope with it is to allow for the confidence that you feel in your judgments and to represent this confidence by probabilities of a higher type.

(iii) *Mixed hierarchies.* The simplest example of a mixed hierarchy is one of two levels wherein a subjective or perhaps logical distribution is assumed for a physical probability. But when there are only two levels it is somewhat misleading to refer to a "hierarchy."

In Case (i), Bayes's theorem is acceptable even to most frequentists; see, for example, von Mises (1942). He made the point, which now seems obvious, that if, in virtue of previous experience, something is "known" about the distribution of a parameter θ, then Bayes's theorem gives information about the final probability of a random variable x whose distribution depends on θ. Presumably by "known" he meant "judged uncontroversially." In short he emphasized that a "non-Bayesian" can use Bayes's theorem in some circumstances, a point that was also implicit in Reichenbach's Chapter 8. The point was worth making in 1942 because statisticians had mostly acquired the habit of using Fisherian techniques which nearly always ignore the possibility that there might sometimes be uncontroversial approximate prior distributions for parameters. F. N. David (1949, pp. 71 & 72) even said that Bayes's *theorem* "is wholly fallacious except under very restrictive conditions" and ". . . at the present time there are few adherents of Bayes's theorem." von Mises (1942, p. 157) blew it by saying that the notion that prior probabilities are non-empirical "cannot be strongly enough refuted." He certainly failed to refute them strongly enough to stem the expansion of modern forms of subjectivistic Bayesianism.

Some people regard the *uncontroversial* uses of Bayes's theorem, that is, those uses acceptable to von Mises, as a case of the empirical Bayes method. Others, such as R. G. Krutchkoff, use the expression "empirical Bayes" only for the more subtle cases where the prior is assumed to exist but drops out of the formula for the posterior expectation of θ. It was in this sense that A. M. Turing used the empirical Bayes method for a classified application in 1941. I applied his method with many elaborations in a paper published much later (#38) which dealt with the population frequencies of species of animals or plants or words. . . . [see p. 28.]

Perhaps a statistical argument is not fully Bayesian unless it is subjective enough to be controversial, even if the controversy is between Bayesians themselves. Any subjective idea is bound to be controversial in spite of the expression "de gustibus non disputandum est" (concerning taste there is no dispute). Perhaps most disputes *are* about taste. We can agree to differ about subjective probabilities but controversies arise when communal decisions have to be made. The controversy cannot be avoided, though it may be decreased, by using priors that are intended to represent ignorance, as in the theories of Jeffreys and of Carnap. (Of course "ignorance" does not here mean ignorance about the prior.) All statistical inference is controversial in any of its applications, though the controversy can be negligible when samples are large enough. Some anti-Bayesians often do not recognize this fact of life. The controversy causes difficulties when a statistician is used as a consultant in a legal battle, for few jurymen or magistrates understand the foundations of statistics, and perhaps only a small fraction even of statisticians do. I think the fraction will be large by A.D. 2047 [and the effect on legal procedures may well be substantial].

Now consider heading (ii), in which at least two of the levels are logical or subjective. This situation arises naturally out of a theory of partially ordered subjective probabilities. In such a theory it is not assumed, given two probabilities p_1 and p_2, that either $p_1 \geqslant p_2$ or $p_2 \geqslant p_1$. Of course partial ordering requires that probabilities are not necessarily numerical, but numerical probabilities can be introduced by means of random numbers, shuffled cards etc., and then the theory comes to the same thing as saying that there are upper and lower probabilities, that is, that a probability lies in some interval of values. Keynes (1921) emphasized such a theory except that he dealt with logical rather than subjective probabilities. Koopman (1940a, b) developed axioms for such a theory by making assumptions that seemed complex but become rather convincing when you think about them. I think the simplest possible acceptable theory along these lines was given in #13, and was pretty well justified by C. A. B. Smith (1961). (See also **#230**.) Recently the theory of partially-ordered probability has often been called the theory of qualitative probability, though I think the earlier name "partially ordered" is clearer. When we use sharp probabilities it is for the sake of simplicity and it provides an example of "rationality of type 2" (**#679**).

If you can say confidently that a logical probability lies in an interval (a, b) it is natural to think it is more likely to be near to the middle of this interval than to the end of it; or perhaps one should convert to log-odds to express a clear preference for the middle. (Taking the middle of the log-odds interval is an invariant rule under addition of weight of evidence.) At any rate this drives one to contemplate the notion of a higher type of probability for describing the first type, even though the first type is not necessarily physical. This is why I discuss hierarchies of probabilities in my paper on rational decisions, **#26**. Savage (1954, p. 58) briefly discusses the notion of hierarchies of subjective probabilities, but he denigrates and dismisses them. He raises two apparent objections. The first, which he heard from Max Woodbury, is that if a primary probability has a distribution expressed in terms of secondary probabilities, then one can perform an integration or summation so as to evaluate a composite primary probability. Thus you would finish up with a sharp value for the primary probability after all. (I don't regard this as an objection.) The second objection that he raises is that there is no reason to stop at secondary probabilities, and you could in principle be led to an infinite hierarchy that would do you no good.

On p. 41 of **#13** I had said that higher types of probability might lead to logical difficulties but in **#26** I took the point of view that it is mentally healthy to think of your subjective probabilities as estimates of credibilities, that is, of logical probabilities (just as it is healthy for some people to believe in the existence of God). Then the primary probabilities might be logical but the secondary ones might be subjective, and the composite probability obtained by summation would be subjective also. Or the secondary ones might also be logical but the tertiary ones would be subjective. This approach does not deny Max Woodbury's point; in fact it might anticipate it. I regard the use of hierarchical chains as a technique helping you to sharpen your subjective probabilities. Of course if the subjective probabilities at the top of the hierarchy are only partially ordered (as they normally would be if your judgments were made fully explicit), the same will be true of the composite primary or type I probabilities after the summations or integrations are performed. A further use of the hierarchical approach in **#26** is in relation to minimax decision functions. Just as these were introduced to try to meet the difficulty of using ordinary Bayesian decisions, one can define a minimax decision function of type II, to avoid using Bayesian decision functions of type II. (The proposal was slightly modified in #80.) Leonid Hurwicz (1951) made an identical proposal simultaneously and independently. I still stand by the following two comments in my paper: ". . . the higher the type the woollier the probabilities . . . the higher the type the less the woolliness matters provided [that] the calculations do not become too complicated." (The hierarchical method must often be robust, otherwise, owing to the woolliness of the higher levels, scientists would not agree with one another as often as they do. This is why I claimed that the higher woolliness does not matter much.) Isaac Levi (1973, p. 23) says, "Good is prepared to define second order probability distributions . . . and third order probability distributions over these, etc. until he gets tired." This was funny, but it would be more accurate to say that I stop when the guessed expected utility of going further becomes negative if the cost is taken into account.

Perhaps the commonest hierarchy that deserves the name comes under heading (iii). The primary probabilities, or probabilities of type I, are physical, the secondary ones are more or less logical, and the tertiary ones are subjective. Or the sequence might be: physical, logical, logical [again], subjective. In the remainder of my paper I shall discuss hierarchies of these kinds.

2. SMALL PROBABILITIES IN LARGE CONTINGENCY TABLES

I used a hierarchical Bayesian argument in #83 for the estimation of small frequencies in a large pure contingency table with entries (n_{ij}). By a "pure" table I mean one for which there is no clear natural ordering for the rows or for the columns. Let the physical probabilities corresponding to the cells of the table be denoted by p_{ij}, and the marginals by $p_{i.}$ and $p_{.j}$. Then the amount of information concerning row i provided by seeing a column of class j can be defined as $\log[p_{ij}/(p_{i.}p_{.j})]$ and it seemed worth trying the assumption that this has approximately a normal distribution over the table as a whole. This turned out to be a readily acceptable hypothesis for two numerical examples that were examined. In other words it turned out that one could accept the loglinear model

$$\log p_{ij} = \log p_{i.} + \log p_{.j} + \epsilon$$

where ϵ has a normal distribution whose parameters can be estimated from the data. (This was an early example of a loglinear model. Note that if ϵ is replaced by ϵ_{ij} and its distribution is not specified, then the equation does not define a model at all.) If then a frequency n_{ij} is observed it can be regarded as evidence concerning the value of p_{ij}, where p_{ij} has a lognormal distribution. Then an application of Bayes's theorem provides a posterior distribution for p_{ij}, even when $n_{ij} = 0$. . . . The lognormal distribution was used as a prior for the parameter p_{ij} and the parameters in this distribution would now often be called hyperparameters. Perhaps this whole technique could be regarded as a non-controversial use of Bayes's theorem. Incidentally, if it is assumed that $p_{ij}/(p_{i.}p_{.j})$ has a Pearson Type III distribution the estimates turn out to be not greatly affected, so the method appears to be robust. (The calculation had to be iterative and was an early example of the EM method as pointed out by Dempster *et al.*, 1977, p. 19.)

3. MAXIMUM LIKELIHOOD/ENTROPY FOR ESTIMATION IN CONTINGENCY TABLES

For ordinary and multidimensional *population* contingency tables, with some marginal probabilities known, the method of maximum entropy for estimating the probabilities in the individual cells leads to interesting results (#322). (The principle of maximum entropy was interpreted by Jaynes [1957] as a method for selecting prior distributions. #322 interprets it as a method for formulating hypotheses; in the application it led to hypotheses of vanishing interactions of various orders. Barnard mentions that an early proposer [but not developer] of a principle of maximum entropy was Jean Ville in the Paris conference on the history and philosophy of science in 1949 but I have not yet been able to obtain this reference.) When there is a sample it is suggested on p. 931 of #322 that one might find the estimates by maximizing a linear combination of the log-likelihood and the entropy, that is, in the two-dimensional case, by maximizing an expression of the form $\Sigma(n_{ij} - \lambda p_{ij})\log p_{ij}$, subject to constraints if the marginal probabilities are assumed. (Here (n_{ij}) is the sample and (p_{ij}) the population contingency table.) This technique could be adopted by a non-Bayesian who would think of λ as a "procedure parameter." A Bayesian might call it a hyperparameter because the ML/E method, as we may call it, is equivalent to the maximization of the posterior density when the prior density is proportional to $\Pi p_{ij}^{-\lambda p_{ij}}$. This method has been investigated by my ex-student Pelz (1977). I believe that the best way to estimate the hyperparameter λ is by means of the method of cross-validation or predictive sample reuse, a method that could also be used for comparing the ML/E method with other methods (#1245). We intend to try this approach.

4. MULTINOMIAL DISTRIBUTIONS

Some hierarchical models that have interested me over a long period are concerned with multinomials and contingency tables, and these models received a lot of attention in my monograph on the estimation of probabilities from a Bayesian point of view (#398). (See also Good, 1964.) To avoid controversy about purely mathematical methods I there used the terminology of distributions of types I, II, and III without committing myself about whether the probabilities were physical, logical, or subjective. But, in a Bayesian context, it might be easiest to think of these three kinds of probability as being respectively of the types I, II, and III. My next few hundred words are based on #398 where more details can be found although the present discussion also contains some new points.

The estimation of a binomial parameter dates back to Bayes and Laplace, Laplace's estimate being known as "Laplace's law of succession." This is the estimate $(r + 1)/(N + 2)$, where r is the number of successes and N the sample size. This was the first example of a shrinkage estimate. It was based on the uniform prior for the binomial parameter p. The more general conjugate prior of beta form was proposed by the actuary G. F. Hardy (1889). De Morgan (1837/1853) (cited by Lidstone, 1920) generalized Laplace's law of succession to the multinomial case where the frequencies are (n_i) $(i = 1, 2, \ldots, t)$. (I have previously attributed this to Lidstone.) De Morgan's estimate of the ith probability

p_i was $(n_i + 1)/(N + t)$ which he obtained from a uniform distribution of $(p_1, p_2, \ldots, p_t)$ in the simplex $\Sigma p_i = 1$ by using Dirichlet's multiple integral. The estimate is the logical or subjective expectation of p_i and is also the probability that the next object sampled will belong to the ith category. The general Dirichlet prior, proportional to $\Pi p_i^{k_i - 1}$, leads to the estimate $(n_i + k_i)/(N + \Sigma k_j)$ for p_i. But if the information concerning the t categories is symmetrical it is adequate, at the first Bayesian level, to use the prior proportional to Πp_i^{k-1} which leads to the estimates $(n_i + k)/(N + tk)$. In fact we can formulate the Duns-Ockham hyper-razor as "What can be done with fewer (hyper)parameters is done in vain with more." ("Ockham's razor" had been emphasized about twenty years before Ockham by the famous medieval philosopher John Duns Scotus.) We can regard k [either] as a flattening constant or as the hyperparameter in the symmetric Dirichlet prior. The proposal of using a continuous linear combination of Dirichlet priors, symmetric or otherwise, occurs on p. 25 of #398. Various authors had previously proposed explicitly or implicitly that a single value of k should be used but I am convinced that we need to go up one level. (Barnard tells me he used a combination of two beta priors in an unpublished paper presented at a conference in Bristol in about 1953 because he wanted a bimodal prior.)

The philosopher W. E. Johnson (1932) considered the problem of what he called "multiple sampling," that is, sampling from a t-letter alphabet. He assumed permutability of the N letters of the sample (later called "exchangeability" though "permutability" is a slightly better term). Thus he was really considering multinomial sampling. He further assumed what I call his "sufficientness postulate," namely that the credibility (logical probability) that the next letter sampled will be of category i depends only on n_i, t, and N, and does not depend on the ratios of the other $t - 1$ frequencies. Under these assumptions he proved that the probability that the next letter sampled will be of category i is $(n_i + k)/(N + tk)$, but he gave no rules for determining k. His proof was correct when $t \geqslant 3$. He was presumably unaware of the relationship of this estimate to the symmetric Dirichlet prior. The estimate does not merely follow from the symmetric Dirichlet prior; it also implies it, in virtue of a generalization of de Finetti's theorem. (This particular generalization follows neatly from a purely mathematical theorem due to Hildebrandt & Schoenberg; see p. 22 of #398.) De Morgan's estimate is the case $k = 1$. Maximum Likelihood estimation is equivalent to taking $k = 0$. The estimates arising out of the invariant priors of Jeffreys (1946) and Perks (1947) correspond to the flattening constants $k = 1/2$ and $k = 1/t$. [S. Zabell tells me that Laplace anticipated De Morgan.]

Johnson's sufficientness assumption is unconvincing because if the frequencies $n_2, n_3, \ldots, n_t$ are far from equal it would be natural to believe that p_1 is more likely to be far from $1/t$ than if $n_2, n_3, \ldots, n_t$ are nearly equal. [This is an application of the Device of Imaginary Results because you don't need an actual experience to make this judgment.] Hence it seemed to me that the "roughness" of the frequency count (n_i) should be taken into account. Since roughness can be measured by a scalar I felt that k could be estimated from the sample (and approximately from its roughness), or alternatively that a hyperprior could be assumed for k, say with a density function $\phi(k)$. This would be equivalent to assuming a prior for the p_i's, with density

$$\int_0^\infty \frac{\Gamma(tk)\Pi p_i^{k-1}\phi(k)dk}{[\Gamma(k)]^t}.$$

Those who do not want to assume a hyperprior could instead estimate k say by Type II Maximum Likelihood or by other methods in which the estimate of k is related to $X^2 = (t/N)\Sigma(n_i - N/t)^2$. These methods were developed in ##398, 522, 547. #547 was mainly concerned with the Bayes factor, provided by a sample (n_i), against the null hypothesis $p_i = 1/t(i = 1, 2, \ldots, t)$. The estimation of the cell probabilities p_i was also covered. (It seems to me to be usually wrong in principle to assume distinct priors, given the non-null hypothesis, according as you are doing estimation or significance testing, except that I believe that more accurate priors are required for the latter purpose.) The null hypothesis corresponds to the complete flattening $k = \infty$ and we may denote it by H_∞. Let H_k denote the non-null hypothesis that the prior is the symmetric Dirichlet with hyperparameter k. Let $F(k)$ denote the Bayes factor in favor of H_k as against H_∞, provided by a sample (n_i). (See p. 862 of #127; or p. 406 of #547.) If k has a hyperprior density $\phi(k)$, then the Bayes factor F against H_∞ is

$$F = \textstyle\int_0^\infty F(k)\phi(k)dk.$$

$\phi(k)$ must be a proper density, otherwise F would reduce to 1, in other words the evidence would be killed. This is an interesting example where impropriety is a felony. One might try to be noncommittal about the value of k and the usual way of being noncommittal about a positive parameter k is to use the Jeffreys-Haldane density $1/k$ which is improper. This can be approximated by the log-Cauchy density which has the further advantage that its quantiles are related in a simple manner to its hyperhyperparameters (pp. 45-46 of **#603B**). One can determine the hyperhyperparameters by guessing the upper and lower quartiles of the repeat rate Σp_i^2, given the non-null hypothesis, and thereby avoid even a misdemeanor. The Bayes factor F is insensitive to moderate changes in the quartiles of the log-Cauchy hyperprior, and the estimates of the p_i's are even more robust. If you prefer not to assume a hyperprior then a type II or second order or second level Maximum Likelihood method is available because $F(k)$ has a unique maximum F_{max} if $X^2 > t - 1$. This was conjectured on p. 37 of #398, largely proved in #860 and completely proved by Levin & Reeds (1977). Other methods of estimating k are proposed on pp. 27, 33, and 34 of #398 and by Bishop, Fienberg, and Holland (1975, Chapter 12). When a hyperparameter is estimated the latter authors call the method "pseudo-Bayesian." It is an example of a Bayes/non-Bayes compromise.

F_{max} is an example of a Type II (or second order or second level) Likelihood Ratio defined in terms of the hyperparametric space which is one-dimensional.

Hence the asymptotic distribution of F_{max} is proportional to a chi-squared with one degree of freedom. In 1967 the accuracy of this approximation was not known but it was found to be fairly accurate in numerous examples in #862, even down to tail-area probabilities as small as 10^{-16}. We do not know why it should be an adequate approximation in such extreme tails.

5. INDEPENDENCE IN CONTINGENCY TABLES

#398 began the extension of the multinomial methods to the problem of testing independence of the rows and columns of contingency tables, and this work was continued in #929 where extensions to three and more dimensions were also considered. But I shall here consider only ordinary (two-dimensional) tables with r rows and s columns. The case $r = s = 2$ is of especial interest because 2×2 tables occur so frequently in practice.

As is well known outside Bayesian statistics, there are three [main] ways of sampling a contingency table, known as sampling Models 1, 2, and 3. In Model 1, sampling is random from the whole population; in Model 2, the row totals (or the column totals) are fixed in advance by the statistician; and in Model 3 both the row and column totals are fixed. Model 3 might seem unreasonable at first but it can easily arise. Denote the corresponding Bayes factors against the null hypothesis H of independence by F_1, F_2, and F_3. But in our latest model it turns out that $F_1 = F_2$ because in this model the fixing of the row totals alone provides no evidence for or against H. The model also neglects any evidence that there might be in the order of rows or of columns; in other words we restrict our attention in effect to "pure" contingency tables. This is of course also done when X^2 or the likelihood-ratio statistic is used.

The basic assumption in the analysis is that, given the non-null hypothesis $\overline{\text{H}}$, the prior for the physical probabilities p_{ij} in the table is a mixture of symmetric Dirichlets. (Previous literature on contingency tables had discussed symmetric Dirichlet distributions but not mixtures.) From this assumption F_1 and F_3 can be calculated. We can deduce FRACT (the factor against H provided by the row and column totals alone, in Model 1) because FRACT = F_1/F_3. A large number of numerical calculations have been done and were reported in #1199. We found that FRACT usually lies between ½ and 2½ when neither of the two sets of marginal totals is very rough and the two sets are not both very flat, and we gave intuitive reasons for these exceptions. We did not report the results for 2×2 tables in that paper but we have done the calculations for this case with the sample size $N = 20$. We find, for example, with our assumptions, that FRACT = 1.48 for the table with margins [5,15;7,13]; FRACT = 2.5 for [10,10;10,10]; FRACT = 2.56 for [1,19;2,18]; and FRACT = 8.65 for the extreme case [1,19; 1,19].

If the mixture of Dirichlets is replaced by a single symmetrical Dirichlet with hyperparameter k, then F_3 is replaced by $F_3(k)$, and $\max_k F_3(k)$ is a Type II Likelihood Ratio. Its asymptotic distribution again turns out to be fairly accurate in the extreme tail of the distribution, even down to the tail-area probabilities such as 10^{-40}. The unimodality of $F_3(k)$ . . . has yet to be proved, but is well supported by our numerical results.

I noticed only as recently as May 1978 that the consideration of contingency tables sheds light on the hyperprior ϕ for multinomials. This was first reported in #1160. We write $\phi(t, k)$ instead of $\phi(k)$ to indicate that it might depend on t as well as k. The prior for a t-category multinomal is then $D^*(t)$ where

$$D^*(t) = \int_0^\infty D(t, k)\phi(t, k)dk$$

and where $D(t, k)$ denotes the symmetric Dirichlet density. Our assumption of $D^*(rs)$, given $\bar{\mathrm{H}}$ and Model 1, implies the prior $\int_0^\infty D(r, sk)\phi(rs, k)dk$ for the row totals. But, if the row totals alone contain no evidence concerning H, this must be mathematically independent of s and it can be deduced that $\phi(t, k)$ must be of the form $\psi(tk)/k$. Strictly therefore some of the calculations in #862 should be repeated, but of course the distribution of the Type II Likelihood Ratio is unaffected, and we have reason to believe the remaining results are robust. This example shows how logical arguments can help to make subjective probabilities more logical. Logical probabilities are an ideal towards which we strive but seldom attain.

A spin-off from the work on contingency tables has been the light it sheds on the classical purely combinatorial problem of the enumeration of rectangular arrays of integers (##974, 1173). This problem had not previously been treated by statistical methods as far as I know.

T. Leonard has used hierarchical Bayesian methods for analyzing contingency tables and multinomial distributions, but since he has attended this conference I shall leave it to him to reference his work in the discussion of the present paper.

6. PROBABILITY DENSITY ESTIMATION AND BUMP-HUNTING

Probability density estimation has been a popular activity since at least the nineteenth century, but bump-hunting, which is closely related to it, is I think comparatively recent. There is a short discussion of the matter on pp. 86-87 of #13 where the "bumpiness" of a curve is defined as the number of points of inflection, though half this number is a slightly better definition. The number of bumps was proposed as one measure of complexity, and the greater the number the smaller the initial probability of the density curve *ceteris paribus.* [See also p. 45.]

In the 1970 Waterloo conference, Orear and Cassel (1971) said that bump-hunting is "one of the major current activities of experimental physicists." In the discussion I suggested the idea of choosing a density function f by maximizing $\Sigma \log f(x_i) - \beta R$, that is, log-likelihood minus a roughness penalty proportional to a measure R of roughness of the density curve. (Without the penalty term one gets $1/N$ of a Dirac function at each observation.) It was pointed out that the problem combines density estimation with significance testing. In #699 the argument is taken further and it is mentioned that $\exp(-\beta R)$ can be regarded as the prior density of f in function space. In this Bayesian interpretation β is a hyperparameter. (There were 21 misprints in this short article, owing to a British dock strike.) The method was developed in considerable detail in ##701 and 810 and applied to two real examples, one relating to high-energy physics and the other to the analysis of chondrites (a common kind of meteorite containing round pellets) in #1200. In the latter work, the estimation of the hyperparameter was made by means of non-Bayesian tests of goodness of fit so as to avoid controversies arising out of the use of hyperpriors.

Leonard (1978, p. 129) mentions that he hopes to report a hierarchical form of his approach to density estimation. Also he applies his method to the chondrite data, and he brought this data to my attention so that our methods could be compared.

7. INFERENCE ABOUT NORMAL DISTRIBUTIONS AND LINEAR MODELS

In 1969 I suggested to my student John M. Rogers that he might consider analogies of the multinomial work for the estimation of the parameters of multivariate normal distributions. It turned out that even the univariate problems were complicated and he completed his thesis without considering the multivariate problems. He considered the estimation of a (univariate) normal mean when the prior contains hyperparameters. The priors were of both normal and Cauchy form (Rogers, 1974) and the hyperparameters were estimated by type II maximum likelihood.

Meanwhile hierarchical Bayesian models with three or four levels or stages had been introduced for inferences about normal distributions and linear models by Lindley (1971) and by Lindley and Smith (1972). A survey of these matters could be better prepared by Lindley so I shall say no more about them.

[This lecture was followed by discussion and a response, omitted here.]

CHAPTER 10

Dynamic Probability, Computer Chess, and the Measurement of Knowledge (#938)

Philosophers and "pseudognosticians" (the artificial intelligentsia[1]) are coming more and more to recognize that they share common ground and that each can learn from the other. This has been generally recognized for many years as far as symbolic logic is concerned, but less so in relation to the foundations of probability. In this essay I hope to convince the pseudognostician that the philosophy of probability is relevant to his work. One aspect that I could have discussed would have been probabilistic causality (Good, #**223B**) in view of Hans Berliner's forthcoming paper "Inferring causality in tactical analysis," but my topic here will be mainly dymanic probability.

The close relationship between philosophy and pseudognostics is easily understood, for philosophers often try to express as clearly as they can how people make judgments. To parody Wittgenstein, what can be said at all can be said clearly *and it can be programmed.*

A paradox might seem to arise. Formal systems, such as those used in mathematics, logic, and computer programming, can lead to deductions outside the system only when there is an input of assumptions. For example, no probability can be numerically inferred from the axioms of probability unless some probabilities are assumed without using the axioms: *ex nihilo nihil fit.*[2] This leads to the main controversies in the foundations of statistics: the controversies of whether intuitive probability[3] should be used in statistics and, if so, whether it should be logical probability (credibility) or subjective (personal). We who talk about probabilities of hypotheses, or at least the relative probabilities of pairs of hypotheses (##13, **846**), are obliged to use intuitive probabilities. It is difficult or impossible to lay down precise rules for specifying the numerical values of these probabilities, so some of us emphasize the need for subjectivity, bridled by axioms. At least one of us is convinced, and has repeatedly emphasized for the last thirty years, that a subjective probability can usually be judged only to lie in some interval of values, rather than having a sharp numerical value (#13). This approach arose as a combination of those of Keynes (Keynes, 1921) and of Ramsey (Ramsey, 1931); and Smith's (Smith, 1961) proof of its validity based on certain desiderata was analogous to the work of Savage (Savage, 1954) who used sharp probabilities. [See also **#230.**]

It is unfortunately necessary once again to express this theory of "comparative subjective probability" in a little more detail before describing the notion of dynamic probability. The theory can be described as a "black box" theory, and the person using the black box is called "you." The black box is a formal system that incorporates the axioms of the subject. Its input consists of your collection of judgments, many of which are of the form that one probability is not less than another one, and the output consists of similar inequalities better called "discernments." The collection of input judgments is your initial *body of beliefs, B*, but the output can be led back into the input, so that the body of beliefs [usually] grows larger as time elapses. The purpose of the theory is to enlarge the body of beliefs and to detect inconsistencies in it. It then becomes your responsibility to resolve the inconsistencies by means of more mature judgment. The same black box theory can be used when utilities are introduced and it is then a theory of rationality (##13, **26**).

This theory is not restricted to rationality but is put forward as a model of *all* completed scientific theories.

It will already be understood that the black box theory involves a time element; but, for the sake of simplicity in many applications, the fiction is adopted (implicitly or explicitly) that an entirely consistent body of beliefs has already been attained. In fact one of the most popular derivations of the axioms of probability is based on the assumption that the body of beliefs, including judgments of "utilities" as well as probabilities, is consistent.[4]

One advantage of assuming your body of beliefs to be consistent, in a static sense, is that it enables you to use conventional mathematical logic, but the assumption is not entirely realistic. This can be seen very clearly when the subject matter is mathematics itself. To take a trivial, but very clear example, it would make sense for betting purposes to regard the probability as 0.1 that the millionth digit of π is a 7, yet we know that the "true probability" is either 0 or 1. If the usual axioms of intuitive probability are assumed, together with conventional static logic, it is definitely inconsistent to call the probability 0.1. If we wish to avoid inconsistency we must change the axioms of probability or of logic. Instead of assuming the axiom that $P(\mathrm{E}|\mathrm{H}) = 1$ when H logically implies E, we must assume that $P(\mathrm{E}|\mathrm{H}) = 1$ when we have *seen* that H logically implies E. In other words probabilities can change in the light of calculations or of pure thought without any change in the *empirical* data (cf. p. 49 of #13, where the example of chess was briefly mentioned). In the past I have called such probabilities "sliding," or "evolving," but I now prefer the expression *dynamic probability.*[5] It is difficult to see how a subjective probability, whether of a man or of a machine, can be anything other than a dynamic one. We use dynamic probability whenever we make judgments about the truth or falsity of mathematical theorems, and competent mathematicians do this frequently, though usually only informally. There is a naive view that mathematics is concerned only with rigorous logic, a view that arises because finished mathematical proofs *are* more or less rigorous. But in the process of finding and conjecturing theorems every real mathematician is guided by his judgments of what is probably true.[6] This must have been known for centuries, and has been much emphasized and exemplified by Polya (Polya, 1941, 1954). A good "heuristic" in problem solving is one that has a reasonable chance of working.[7]

Once the axioms of probability are changed, there is no genuine inconsistency. We don't have to say that $P(E|H)$ has more than one value, for we can denote its value at time t by $P_t(E|H)$, or we can incorporate a notation for the body of beliefs B_t if preferred. There is an analogy with the FORTRAN notation, as in $x = x + 3$, where the symbol x changes its meaning during the course of the calculation without any real inconsistency.[8]

Believing, as I did (and still do), that a machine will ultimately be able to simulate all the intellectual activities of any man, if the machine is allowed to have the same mechanical help as the man,[9] it used to puzzle me how a machine could make probability judgments. I realized later that this is no more and no less puzzling than the same question posed for a man instead of a machine. We *ought* to be puzzled by how judgments are made, for when we know how they are made we don't call them judgments (#183).[10] If judgments ever cease then there will be nothing left for philosophers to do. For philosophical applications of dynamic probability see Appendix A.

Although dynamic probability is implicitly used in most mathematical research it is even more clearly required in the game of chess.[11] For in most chess positions we cannot come as close to certainty as in mathematics. It could even be reasonably argued that the sole purpose of analyzing a chess position, in a game, is for improving your estimate of the dynamic probabilities of winning, drawing, or losing. If analysis were free, it would pay you in expectation to go on analyzing until you were blue in the face, for it is known that free evidence is always of non-negative expected utility (for example, **#508**, but see also #855). But of course analysis is not free, for it costs effort, time on your chess clock, and possibly facial blueness. In deciding formally how much analysis to do, these costs will need to be quantified.

In the theory of games, as pioneered mainly by von Neumann (von Neumann, 1944/47), chess is described as a "game of perfect information," meaning that the rules involve no reference to dice and the like. But in practice most chess positions cannot be exhaustively analyzed by any human or any machine, *present or future.*[12] Therefore play must depend on probability even if the dependence is only implicit. Caissa is a cousin of the Moirai after all.

Against this it can be argued that the early proposals for helping humans and computers to play chess made use of evaluation functions (for quiescent positions) and did not rely on probability, dynamic or otherwise. For example, the beginner is told the value of the pieces, $P = 1$, $B = 3.25$, etc. and that central squares are usually more valuable than the others.[13] But an evaluation function can be fruitfully interpreted in probabilistic terms and we now recall a conjectured approximate relationship that has been proposed (##183, 521) by analogy with the technical definition of weight of evidence.

The weight of evidence, provided by observations E, in favor of one hypothesis H_1, as compared with another one H_2, is defined as

$$\log \frac{O(H_1/H_2|E)}{O(H_1/H_2)} = \log \frac{P(E|H_1)}{P(E|H_2)}$$

where P denotes probability and O denotes odds. In words, the weight of evidence, when added to the initial log-odds, gives the final log-odds. . . . The conjecture is that *ceteris paribus* the weight of evidence in favor of White's winning as compared with losing, in a given position, is roughly proportional to her advantage in material, or more generally to the value of her evaluation function, where the constant of proportionality will be larger for strong players than for weak ones. The initial log-odds should be defined in terms of the playing strengths of the antagonists, and on whether the position is far into the opening, middle-game, or end-game, etc. Of course this conjecture is susceptible to experimental verification or refutation or improvement by statistical means, though not easily; and at the same time the conjecture gives additional meaning to an evaluation function.[14] As an example, if an advantage of a pawn triples your odds of winning as compared with losing, then an advantage of a bishop should multiply your odds by about $3^{3.25} = 35.5$. This quantitative use of probability is not in the spirit of Polya's writings, even if interval estimates of the probabilities are used.

If dynamic probability is to be used with complete seriousness, then it must be combined with the principle of rationality (see Appendix A). First you should decide what your utilities are for winning, drawing, and losing, say u_W, u_D, and u_L. More precisely, you do not need all three parameters, but only the ratio $(u_W - u_D)/(u_D - u_L)$. Then you should aim to make the move, or one of the moves, that maximize the mathematical expectation of your utility, in other words you should aim to maximize

$$p_W u_W + p_D u_D + p_L u_L \tag{1}$$

where p_W, p_D, and p_L are your dynamic probabilities of winning, drawing, or losing. When estimating (1) you have to allow for the state of the chess clock so that the "costs of calculation," mentioned in Appendix A, are very much in the mind of the match chess player.[15] This is not quite the whole picture because you might wish to preserve your energy for another game: this accounts for many "grandmaster draws."

Current chess programs all depend on tree analysis, with backtracking, and the truncation of the tree at certain positions. As emphasized in #521 it will [perhaps] eventually be necessary for programs to handle descriptions of positions[16] if Grandmaster status is to be achieved, and the lessons derived from this work will of course change the world, but we do not treat this difficult matter in this paper. [But the latest indications are that computers may soon beat Grandmasters by tactical brute force.]

For the moment let us suppose that the problem has been solved of choosing the nodes where the tree is to be truncated. At each such node the probabilities p_W, p_D, and p_L are a special kind of dynamic probability, namely *superficial* or *surface* probabilities, in the sense that they do not depend on an analysis in depth. The evaluation function used at the end-nodes, which is used for computing these three probabilities, might depend on much deep cogitation and statistical analysis, but this is not what is meant here by an "analysis in depth." The the minimax backtracking procedure can be used; or expectimaxing if you wish to allow for the deficiencies of your opponent,[17] and for your own deficiencies. In this way you can arrive at values of the dynamic probabilities $p_W{}^0$, $p_D{}^0$, and $p_L{}^0$ corresponding to the positions that would arise after each of your plausible moves in the *current* position, π_0. Of course these probabilities depend on the truncation rules (pruning or pollarding).

Some programs truncate the analysis tree at a fixed depth but this is very unsatisfactory because such programs can never carry out a deep combination. Recognizing this, the earliest writers on chess programming, as well as those who discussed chess programming intelligently at least ten years earlier,[18] recognized that an important criterion for a chess position π to be regarded as an endpoint of an analysis tree was *quiescence.* A quiescent position can be defined as one where the player with the move is neither threatened with immediate loss, nor can threaten his opponent with immediate loss. The primary definition of "loss" here is in material terms, but other criteria should be introduced. For example, the advance of a passed pawn will often affect the evaluation non-negligibly. We can try to "materialize" this effect, for example, by regarding the value of a passed pawn, not easily stopped, as variable. My own proposals are 1¼ on the fourth rank, 1½ on the fifth rank, 3 on the sixth, 5 on the seventh, and 9 on the eighth!, but this is somewhat crude.[19]

An informal definition of a turbulent position is a combinative one. For example, the position: White K at e3, R at c8; Black K at a1, R at a4; is turbulent. But if Black also has a Q at f6, then White's game is hopeless, so the turbulence of the position does not then make it much worth analyzing.

Hence in #521 I introduced a term *agitation* to cover both turbulence and whether one of p_W, p_D, and p_L is close to 1. Apart from considering whether to threaten to take a piece, in some potential future position π, we should consider whether the win of this piece would matter much. Also, instead of considering one-move threats, it seems better to consider an analysis of unit cost, which might involve several moves, as, for example, when checking whether a pawn can be tackled before it touches down in an end-game. The definition of the *agitation* $A(\pi)$ was the expected value of $|U(\pi|\$) - U(\pi)|$ where $U(\pi)$ is the superficial utility of π and $U(\pi|\$)$ is the utility of π were a unit of amount of analysis to be done. ($U(\pi|\$)$ is a subjective random variable before the analysis is done.)

But the depth from the present position π_0 to π is also relevant in the decision whether to truncate at π. More exactly, the dynamic probability $P(\pi|\pi_0)$ that position π will be reached from π_0 is more relevant than the depth. We could even reasonably define the probabilistic depth as proportional to $-\log P(\pi|\pi_0)$ and the effective depth of the whole analysis as $-\Sigma P(\pi|\pi_0) \log P(\pi|\pi_0)$, summed over all endpoints π, as suggested in #521. But the most natural criterion for whether to treat π as an endpoint in the analysis of π_0 is obtained by setting a threshold on $P(\pi|\pi_0)A(\pi)$. The discussion of agitation and allied matters is taken somewhat further on pp. 114-115 of #521.

As a little exercise on dynamic probability let us consider the law of multiplication of advantage which states that "with best play on both sides we would expect the rate of increase of advantage to be some increasing function of the advantage." This might appear to contradict the conjecture that the values of the pieces are approximately proportional to weights of evidence in favor of winning rather than losing. For we must have the "Martingale property" $E(p_t|p_0) = p_0$, where p_0 and p_t are the probabilities of winning at times 0 and t. This only sounds paradoxical if we forget the elementary fact that the expectation of a function is not usually equal to that same function of the expectation. For example, we could have, for some $\epsilon > 0$,

$$E(\log \frac{p_t}{1-p_t}) \approx (1+\epsilon)^t \log \frac{p_0}{1-p_0} \tag{2}$$

without contradicting the Martingale property, and (2) expresses a possible form of the law of multiplication of advantage, though it cannot be very accurate.

An idea closely associated with the way that dynamic probabilities can vary is the following method for trying to improve any given chess program. Let the program starting in a position π_0 play against itself, say for the next n moves, and then to quiescence, at say π_1. Then the odds of winning from position π_1, or the expected utility, could be used for deciding whether the plan and the move adopted in position π_0 turned out well or badly. This information could be used sometimes to change the decision, for example, to eliminate the move chosen before revising the analysis of π_0. This is not the same as a tree analysis alone, starting from π_0, because the tree analysis will often not reach the position π_1. Rather, it is a kind of learning by experience. In this procedure n should not be at all large because non-optimal moves would introduce more noise the larger n was taken. The better the standard of play the larger n could be taken. If the program contained random choices, the decision at π_0 could be made to depend on a *collection* of sub-games instead of just one. This idea is essentially what humans use when they claim that some opening line "appears good in master practice."

To conclude this paper I should like to indicate the relevance of dynamic probability to the quantification of knowledge, for which Michie proposed a non-probabilistic measure.[20] As he points out, to know that $12^3 = 1728$ can be better than having to calculate it, better in the sense that it saves time. His discussion was non-probabilistic so it could be said to depend, at least implicitly, on dynamic *logic* rather than on dynamic *probability*. In terms of dynamic probability, we could describe the *knowledge* that $12^3 = 1728$ as the ascribing of dynamic probability $p = 1$ to this mundane fact. If instead p were less than 1, then the remaining dynamic information available by calculation would be $-\log p$ (p. 75 of #13; p. 126 of #599). This may be compared with Michie's definition of amount of knowledge, which is [or was] based on Hartley's non-probabilistic measure of information (Hartley, 1928).

Amount of knowledge can be regarded as another quasiutility of which weight of evidence and explicativity are examples. A measure of knowledge should be usable for putting programs in order of merit.

In a tree search, such as in chess, in theorem-proving, and in medical diagnosis, one can use entropy, or amount of information, as a quasi-utility, for cutting down on the search (##592, 755, 798) and the test for whether this quasi-utility is sensible is whether its use agrees reasonably well with that of the principle of rationality, the maximization of expected utility. Similarly, to judge whether a measure of knowledge is a useful quasi-utility it should ultimately be compared with the type 2 principle of rationality (Appendix A). So the question arises what form this principle would take when applied to computer programs.

Suppose we have a program for evaluating a function $f(x)$ and let's imagine for the moment that we are going to make one use of the program for calculating $f(x)$ for some unknown value of x. Suppose that the probability that x will be the value for which we wish to evaluate the function is $p(x)$ and let's suppose that when we wish to do this evaluation the utility of the calculation is $u(x,\lambda)$ where λ is the proportional accuracy of the result. Suppose further that the cost of obtaining this proportional accuracy for evaluating $f(x)$, given the program, is $c(x,\lambda)$. Then the total expected utility of the program, as far as its next use is concerned, is given by the expression

$$U = \int p(x) \max\{0, \max_{\lambda}[u(x,\lambda) - c(x,\lambda)]\} dx$$

or (3)

$$\sum_{x} p(x) \max\{0, \max_{\lambda}[u(x,\lambda) - c(x,\lambda)]\}.$$

The notion of dynamic probability (or of rationality of type 2) is implicit in the utilities mentioned here, because, if the usual axioms of probability are assumed, the utilities would be zero because the costs of calculation are ignored. Anything calculable is "certain" in ordinary logic and so conveys no logical information, only dynamic information.

If the program is to be applied more than once, then formula (3) will apply to each of its applications unless the program is an adaptive one. By an *adaptive program* we could mean simply that *the costs of calculation tend to decrease when the program is used repeatedly*. This will be true for example, in the adaptive rote learning programs that Donald Michie described in lectures in Blacksburg in 1974. To allow for adaptability would lead to severe complications and I suspect that similar complications would arise if Donald Michie's definition of amount of knowledge were to be applied to the same problem.

I expect his definition will usually be considerably easier to use than the expression (3), but I do not know which is the better definition on balance.

Example 1. Suppose that (i) all accurate answers have a constant utility a, and all others have zero utility. Then

$$u(x,\lambda) = \begin{cases} a \text{ if } \lambda = 0 \\ 0 \text{ otherwise;} \end{cases}$$

(ii) $c(x,0) = b$, a constant, when x belong to a set X, where $a > b > 0$, and that $c(x,0) > a$ if x does not belong to X; (iii) all values of x are equally likely a priori, that is, $p(x)$ is mathematically independent of x. Then (3) is proportional to the number of elements in X, that is, to the number of values of x that can be "profitably" computed.

Example 2.

$$u(x,\lambda) = \begin{cases} a \text{ if } \lambda < \lambda_0 \\ 0 \text{ otherwise.} \end{cases}$$

The analysis is much the same as for Example 1 and is left to the reader.

Example 3. $u(x,\lambda) = -\log\lambda(\lambda < 1)$; then the utility is approximately proportional to the number of correct significant figures.

Example 4. In the theory of numbers we would often need to modify the theory and perhaps use a utility $u(x,\mu)$, where μ is the number of decimal digits in the answer.

Example 5: knowledge measurement in chess. Let x now denote a chess position instead of a number. Let $u(x)$ denote the expected utility of the program when applied in position x, allowing this time for the costs. Then $\nu = \Sigma_x p(x)u(x)$ measures the expected utility of the program per move, where $p(x)$ is the probability of the occurrence of position x. The dependence between consecutive positions does not affect this formula because the expectation of a sum is always the sum of the expectations regardless of dependence. A measure of the knowledge added to the program by throwing the book on opening variations at it, can be obtained by simply subtracting the previous value of ν from the new value.

It should now be clear that dynamic probability is fundamental for a theory of practical chess, and has wider applicability. Any search procedure, such as is definitely required in non-routine mathematical research, whether by humans or by machines, *must* make use of subgoals to fight the combinatorial explosion. Dynamic utilities are required in such work because, *when you set up subgoals, you should estimate their expected utility as an aid to the main goal* before you bother your pretty head in trying to attain the subgoals.

The combinatorial explosion is often mentioned as a reason for believing in the impracticability of machine intelligence, but if this argument held water it would also show that human intelligence is impossible. Perhaps it is impossible for a human to be intelligent, but the real question is whether machines are necessarily equally unintelligent. Both human problem-solvers and pseudognostical machines must use dynamic probability.

APPENDIX A. PHILOSOPHICAL APPLICATIONS OF DYNAMIC PROBABILITY

An interesting application of dynamic probability is to a fundamental philosophical problem concerning simplicity. Many of us believe that of two scientific laws that explain the same facts, the simpler is usually the more probable. Agassi, in support of a thesis of Popper, challenged this belief by pointing out that, for example, Maxwell's equations imply Fresnel's optical laws and must therefore be not more probable, yet Maxwell's equations appear simpler. This difficulty can be succinctly resolved in terms of dynamic probability, and I believe this is the only possible way of resolving it. For the clarification of these cryptic remarks see ##599, **846**. . . .

A further philosophical application of dynamic probability arises in connection with the principle of rationality, the recommendation to maximize expected utility. It frequently happens that the amount of thinking or calculation required to obey this principle completely is very great or impracticably large. Whatever its size, it is rational to allow for the costs of this effort (for example, #**679**), whatever the difficulties of laying down rules for doing so. When such allowance is made we can still try to maximize expected utility, but the probabilities, and sometimes the utilities also, are then dynamic. When a conscious attempt is made to allow for the costs we may say we are obeying the principle of rationality of type 2. This modified principle can justify us in using the often convenient but apparently *ad hoc* and somewhat irrational methods of "non-Bayesian" statistics, that is, methods that officially disregard the use of subjective probability judgments. But such judgments are always at least implicit: all statisticians are implicit Bayesians whether they know it or not, except sometimes when they are making mistakes. (Of course Bayesians also sometimes make mistakes.)

Thus dynamic probability and dynamic utility help us to achieve a Bayes/non-Bayes synthesis. Inequality judgments rather than sharp probability judgments also contribute to this synthesis: a strict non-Bayesian should choose the interval (0,1) for all his subjective probabilities! For interesting examples of Bayes/non-Bayes syntheses see ##547, 862, 929, 1199.

NOTES

1. Lighthill's joke, cracked in a BBC TV debate. Jokes don't wear well for long, however risible they were originally, so I have invented a neologism that just might replace the clumsy and ambiguous "workers in A. I." The "g" of "pseudognostics" belongs to the

third syllable! Michie's expression "knowledge engineering" might be preferred in some contexts, but it will tend to prevent A. I. work in any university department outside engineering. Engineering departments already tend to take the universities over.

2. Each axiom merely *relates* probability values. *Suggestions,* such as the "principle of sufficient reason," are not axioms and they require judgments about the real world.

3. By "intuitive probability" I mean either logical or subjective probability (Koopman, 1940) as contrasted with the physical probabilities that arise, for example, in quantum mechanics, or the tautological probabilities of mathematical statistics (#**182**).

4. More precisely, it must be "coherent" in the sense that a "Dutch book" cannot be made against it in a gambling situation. A Dutch book is a proposed set of bets such that you will lose whatever happens (Savage, 1954).

5. Donald Michie expressed a preference for this term in conversation in 1974, since he thought that "evolving probability," which I have used in the past, was more likely to be misunderstood.

6. (i) A real mathematician, by definition, cannot do all his work by low-level routine methods; but one man's routine is another man's creativity. (ii) Two famous examples of the use of *scientific* induction in mathematics were Gauss's discoveries of the prime number theorem and of the law of quadratic reciprocity. He never succeeded in proving the first of these results.

7. Polya's writings demonstrate the truth of the aphorism in Note 6. Polya's use of probability in mathematical research is more qualitative than mine. A typical theorem in his writings is "The more confidence we placed in an incompatible rival of our conjecture, the greater will be the gain of faith in our conjecture when that rival is refuted" (Polya, 1954, vol 2, p. 124). His purely qualitative approach would prevent the application of the principle of rationality in many circumstances.

8. Presumably the ALGOL notation $x := x + 3$ was introduced to avoid the apparent inconsistency.

9. It is pointless to make such judgments without some attached dynamic probabilities, so I add that I think there is a probability exceeding 1/2 that the machine will come in the present century. But a probability of only 1/1000 would of course justify the present expenditures.

10.

Judgments are never formalized
You can sign that with your blood gents
For when they are formalized
No one dare call them judgments.

Drol Doog (With apologies to Sir John Harrington.)

11. In case this seems too obvious the reader is reminded that it was not explicit in the earlier papers on chess programming, and there is no heading "Probability" in (Sunnucks, 1970).

12. Even if every atom in the moon examined 10^{24} *games* per second (light takes about 10^{-24} sec. to traverse the diameter of an electron), it would take ten million times the age of the universe to examine 10^{100} games, which is a drop in the mare.

13. The values of the pieces also vary with the position, in anyone's book. There is much scope for conjectures and statistical work on evaluation functions. For example, it was suggested in #521 that the "advantage of two bishops" could be explained by assuming that it is "in general" better to control two different squares than to control one square twice, although "overprotection of the center" might be an exception. For example, the contribution to the total "score" from the control n times of one square might be roughly proportional to $(n + 1)^{\alpha} - (n + 1)^{-\beta}$ $(0 < \alpha < 1, \beta > 0)$.

14. Perhaps the odds of a draw are roughly the geometric mean of those of winning and losing.

15. The International Chessmaster and senior Civil Servant, Hugh Alexander, once

remarked that it is more important for a Civil Service administrator to make his mind up promptly than to reach the best decision. He might have had in mind that otherwise the administrator would "lose on the clock."

16. To be precise I said that natural language should be used, and John McCarthy said from the floor that descriptions in symbolic logic might be better.

17. This is known as psychological chess when Emanual Lasker does it, and trappy chess when I do it.

18. By definition of "intelligently."

19. The difficulty of evaluating unblocked passed pawns is one for the human as well as for the machine, because it is often in the balance whether such pawns *can* be blocked. This might be the main reason for the difficulty of formalizing end-game play. It is said that mathematicians have an advantage in the end-game but I do not know the evidence for this nor clearly why it should be true.

20. This part of the paper is based on my invited discussion of Michie's public lecture on the measurement of knowledge on October 30, 1974 in Blacksburg: see Michie (1977).

Part III.
Corroboration, Hypothesis Testing, Induction, and Simplicity

CHAPTER 11

The White Shoe Is a Red Herring (#518)

Hempel's paradox of confirmation can be worded thus, "A case of a hypothesis supports the hypothesis. Now the hypothesis that all crows are black is logically equivalent to the contrapositive that all non-black things are non-crows, and this is supported by the observation of a white shoe."

The literature of the paradox is large and I have myself contributed to it twice (##199, 245). The first contribution contained an error, but I think the second one gave a complete resolution. The main conclusion was that it is simply not true that a "case of a hypothesis" necessarily supports the hypothesis; and an explanation was also given for why it seems to be true.

In the present note we show in six sentences, and even without reference to the contrapositive, that a case of a hypothesis does not necessarily support it.

Suppose that we know we are in one or other of two worlds, and the hypothesis, H, under consideration is that all the crows in our world are black. We know in advance that in one world there are a hundred black crows, no crows that are not black, and a million other birds; and that in the other world there are a thousand black crows, one white one, and a million other birds. A bird is selected equiprobably at random from all the birds in our world. It turns out to be a black crow. This is strong evidence (a Bayes-Jeffreys-Turing factor of about 10) that we are in the second world, wherein not all crows are black. Thus the observation of a black crow, in the circumstances described, undermines the hypothesis that all the crows in our world are black. Thus the initial premise of the paradox of confirmation is false, and no reference to the contrapositive is required.

In order to understand why it is that a case of a hypothesis seems to support it, note that

$$W(\mathrm{H}:\mathrm{Black}|\mathrm{Crow}) > 0,$$

where $W(\text{H:E}|\text{G})$, the weight of evidence, support, or log-factor, for H provided by E given G, is the logarithm of the Bayes-Jeffreys-Turing factor, $P(\text{E}|\text{G and H})/P(\text{E}|\text{G and not H})$. The above inequality is clear from the fact that $P(\text{Black}|\text{H and Crow}) = 1$, and a similar inequality will follow for all other explicata of corroboration (#211; App. ix of Popper, 1959). On the other hand $W(\text{H:Crow}|\text{Black})$ can be negative.

It is formally interesting to recall that

$$W(\text{H:Black Crow}) = W(\text{H:Crow}) + W(\text{H:Black}|\text{Crow}),$$

and that only the last of these three terms needs to be positive. The first two terms can both be negative.

CHAPTER 12

The White Shoe qua Herring Is Pink (#600)

Hempel (1967) points out that I (#**518**) had misunderstood him and that in his context [of no background knowledge] a white shoe is not a red herring. But in my context (#245) it was a red herring; so in its capacity as a herring it seems to be pink. I shall now argue its redness even within Hempel's context.

Let H be a hypothesis of the form that class A is contained in class B, for example, "all crows are black." Let E be what I call a "case" of H, that is, a proposition of the form "this object is in both class A and B." I showed by means of a clear-cut succinct example that when there is background or given knowledge G it is possible for E to undermine H. In fact, in a technical sense, the weight of evidence concerning H provided by E given G, denoted by $W(H:E|G)$, can be numerically measurable and negative. Hempel's thesis however is unaffected unless we can show that $W(H:E)$ can be negative, where $W(H:E)$ is in no sense an abbreviation for $W(H:E|G)$ at any rate if G has any empirical content.

Since the propositions H and E would be meaningless in the absense of empirical knowledge, it is difficult to decide whether $W(H:E)$ is necessarily positive. The closest I can get to giving $W(H:E)$ a practical significance is to imagine an infinitely intelligent newborn baby having built-in neural circuits enabling him to deal with formal logic, English syntax, and subjective probability. He might now argue, after defining a crow in detail, that it is initially extremely likely that all crows are black, that is, that H is true. "On the other hand," he goes on to argue, "if there are crows, then there is a reasonable chance that they are of a variety of colors. Therefore, if I were to discover that even a black crow exists I would consider H to be less probable than it was initially."

I conclude from this that the herring is a fairly deep shade of pink.

CHAPTER 13

A Subjective Evaluation of Bode's Law and an "Objective" Test for Approximate Numerical Rationality (#603B)

[The original version of this publication was preceded by a summary.]

1. INTRODUCTION

This paper is intended in part to be a contribution to the Bayesian evaluation of physical theories, although the main law discussed, Bode's law or the Bode-Titius law, is not quite a theory in the usual sense of the term.

At the suggestion of the Editor [of *JASA*], a brief discussion of the foundations of probability, statistics, and induction, as understood by the author, is given in Part 2 [of the paper], in order to make the paper more self-contained. In Part 3 Bode's law is given a detailed subjective evaluation. At one point the argument makes use of a new philosophical analysis of "explanation." In the course of Part 3 the question whether some real numbers are "close to being rational " arises. A non-Bayesian test for this is given in Part 4, and it is applied to the periods of the solar bodies. Blagg's law, which is a complicated modification of Bode's law, is discussed in Part 5, and a law concerning the saturnine satellites, due to a "dabbler in astronomy," is shown in Part 6 to have had predictive value, a fact that seems to have been previously overlooked. This law resembles Bode's law in form and presumably strongly supports Bode's law indirectly although we do not try to evaluate the magnitude of this support in this paper. Properties of the log-Cauchy distribution, which is used in Part 3, are listed in the appendix. [In this volume, Parts 4, 5, and 6 of the paper have been omitted.]

2. PROBABILITY, STATISTICS, AND INDUCTION

2.1 *Seven kinds of probability.* It is important to recognize that the word "probability" can be used in an least seven senses. A dendroidal categorization was given in #522. The main kinds, all of interest even if they do not all exist, are *tautological* (mathematical), *physical*, and *intuitive*, and the latter splits up into *logical, subjective, multisubjective, evolving* [or *dynamic*], and *psychological.* Some further relevant references are Poisson (1837), Jeffreys (1939), Kemble (1942), Carnap (1950), ##13, **182**, 398, 617.

Subjective probability is psychological probability to which some canons of consistency have been applied. Logical probability is the subjective probability in the mind of a hypothetical perfectly rational man, and is often called *credibility* (Edgeworth, 1961; Russell, 1948; #398). "Your" (sing. or pl.) subjective probabilities can be regarded as estimates of credibilities if these exist, and they can be regarded as estimates of physical probabilities when these in their turn exist. An evolving [dynamic] probability is one that changes in the light of reasoning alone without the intervention of strictly new evidence, as in the game of chess or when estimating the probability of a mathematical theorem (p. 49 of #13, ##521, 592, 599, **938**).

Evolving probabilities are in my opinion the most important kind in practical affairs and in scientific induction even though they are not completely consistent. It is convenient to distinguish between two types of rationality: rationality of Type 1 (the classical kind) in which complete consistency is demanded with the usual antique axioms of utility and conditional probability (such as the axioms in ##13, **26**); and of Type 2, when no contradiction has been found *as far as the reasoning to date is concerned.* (See #290.) For both types of rationality, the "principle of rationality" is the recommendation to maximize expected utility. (We shall also discuss the rationality of real numbers in this paper!)

My view is that physical probabilities exist in much the same sense that the physical world exists (that is, it is not misleading to say they exist), but that they can be measured only with the help of a theory of subjective (personal) probability. (See, for example, #617.)

Further I believe that subjective probabilities are only partially ordered. In this respect I follow the views of Keynes (1921) and Koopman (1940a, b), except that Koopman was concerned with intuitive probability, and Keynes with credibility, although he admitted later (p. 243 of Keynes, 1921) that subjective probabilities were primary. The assumption of partial ordering leads in a natural way to the ideas of lower and upper probabilities, these being the ends of an interval in which any given probability is judged to lie. Nevertheless it is convenient to make use of the usual axioms for well-ordered probabilities as if the mathematical theory were a black box into which judgments of inequalities can be plugged and discernments fed out (##13, **26**, **230**). A justification of this extremely simple black-box approach, along the lines of a previous justification of the axioms of rationality when preferences are well ordered (Savage, 1954), has been given by Smith (1961, 1965).

2.2. *Terminology.* I have tried for decades to use terminology that is as close as possible to the ordinary English uses of the terms, as an aid to intuition, instruction and the unity of knowledge, and for aesthetic reasons. I prefer "initial" to "prior" and use it as an adjective only, "final" to "posterior," and "intermediate" to "preposterior." Improper priors are excluded from my temple but improper initial distributions are entertained. [More recently I have followed the masses in the use of "prior."] The odds corresponding to a probability p are defined as $p/(1-p)$, so that, for example, odds of 3 mean that betting odds of "3 to 1 on" could just be given. Let H denote a hypothesis and E an event, experimental result, or evidence. I adopt Turing's suggestion (1940) that $P(\mathrm{E}|\mathrm{H})/P(\mathrm{E}|\overline{\mathrm{H}})$, where $\overline{\mathrm{H}}$ is Hilbert's notation for the negation of H, should be called the "factor in favor of H provided by E" (or the "Bayes factor" or the "Bayes-Jeffreys-Turing factor"), since it is equal to $O(\mathrm{H}|\mathrm{E})/O(\mathrm{H})$, where O denotes odds. It is denoted by $F(\mathrm{H}:\mathrm{E})$ or by $F(\mathrm{H}:\mathrm{E}|\mathrm{G})$ if G is given throughout. Also $F(\mathrm{H}:\mathrm{E}|(\mathrm{H} \vee \mathrm{H}').\mathrm{G})$, where v denotes "or," is sometimes denoted by $F(\mathrm{H}/\mathrm{H}':\mathrm{E}|\mathrm{G})$, pronounced "the factor in favor of H as against H$'$ provided by E given G." A factor is equal to the "likelihood ratio" only when H and $\overline{\mathrm{H}}$ are both simple statistical hypotheses. The expression "likelihood ratio" should be reserved exclusively for the ratio of maximum likelihoods, since otherwise the expression "likelihood ratio test" is made obscure, a pity in view of its great importance in statistics. The term "confirmation" which is used by Carnap (1950) for logical probability should be abandoned [in this sense] even by philosophers: the older term "credibility" does the job perfectly.

The expression "weight of evidence," abbreviated to W, is used for the logarithm of a Bayes factor: the term is highly appropriate owing to the properties $\text{log-odds}(\mathrm{H}|\mathrm{E}) = \text{log-odds}(\mathrm{H}) + W(\mathrm{H}:\mathrm{E})$ and $W(\mathrm{H}:\mathrm{E}.\mathrm{F}|\mathrm{G}) = W(\mathrm{H}:\mathrm{E}|\mathrm{G}) + W(\mathrm{H}:\mathrm{F}|\mathrm{E}.\mathrm{G})$, where logical conjunction is denoted by a period. $W(\mathrm{H}:\mathrm{E}|\mathrm{G})$ is pronounced "the weight of evidence in favor of H provided by E given G." It is related to the historically later concept of the amount of information concerning H provided by E (given G), $I(\mathrm{H}:\mathrm{E}|\mathrm{G})$, which is defined as the logarithm of the "association factor" (Keynes's term) $P(\mathrm{E}.\mathrm{H}|\mathrm{G})/(P(\mathrm{E}|\mathrm{G})P(\mathrm{H}|\mathrm{G}))$. The relationship is $W(\mathrm{H}:\mathrm{E}|\mathrm{G}) = I(\mathrm{H}:\mathrm{E}|\mathrm{G}) - I(\overline{\mathrm{H}}:\mathrm{E}|\mathrm{G})$. The unit in terms of which weights of evidence and amounts of information are measured depends on the base of the logarithms. Turing (1940) suggested the term "ban" when the base is 10. This term was derived from "Banbury," a town where half a million square feet of forms were printed for the application of the concept. One tenth of a ban was called a "deciban" by analogy with the acoustic decibel. The term "centiban" is then self-explanatory. A deciban is about the smallest weight of evidence perceptible to the intuition, so the analogy holds up. When the base is e Turing suggested the expression "natural ban." When the base is 2 the unit is the "bit" (see Shannon and Weaver, 1949 [1948]; this term was suggested by J. W. Tukey). It is usually sufficient to estimate a weight of evidence to within 5 decibans, corresponding to a factor of about 3, but the smaller units are useful for cumulation.

Weight of evidence appears to be the best explication for "corroboration" (##211, 599). The expected weight of evidence in favor of a correct hypothesis is always non-negative. I have discussed the properties and applications of Bayes factors or weights of evidence on numerous occasions (for example, ##13, 174, 211, 221, **223B**, 245, 398, **508**, **518**, 521, 522, 547, 570, 574, 592, 599, 617, 618, 622). The expression "weight of evidence" was independently suggested by Minsky and Selfridge (1961) and by Peirce (1878) in a similar sense. (See #1382.) (Peirce also drew the analogy with the Weber-Fechner law.) I hope that statisticians and philosophers of science will try to decide for good *scientific and philosophical* reasons whether they should adopt this and other scientific and philosophical terminology. [Logically, terminology should not matter; but, in practice, it does.]

2.3. *Hypothesis testing.* A Bayesian test of a hypothesis or theory H requires that both H and its negation be formulated with some accuracy. Suppose, for example, that H is a simple statistical hypothesis. Then "in principle," its negation $\overline{H}$ should be expressed as a logical disjunction of all possible simple statistical hypotheses that contradict H and each of these must be attributed an initial probability. Since this is impracticable it is necessary to make use of judgment in deciding which of these simple statistical hypotheses should be entertained at all. When making this judgment it is essential to allow informally both for the complexities of the hypotheses and for whether they could be rejected out of hand by the evidence. For this reason it is often necessary for a Bayesian to look at the observations before deciding on the full specification of $\overline{H}$. This has its dangers if detached judgment is not used, but the non-Bayesian is in the same boat (p. 60). It is necessary also to make some further simplifying assumptions in order that the calculations should not get out of hand. But the Bayesian is in no worse a position than his opponents in this respect: in both a Bayesian and a non-Bayesian analysis we have to guess that some mathematical model of reality is adequate. The main difference is that in a non-Bayesian analysis more is swept under the carpet. This makes non-Bayesian methods politically expedient. The Bayesian is forced to put his cards on the table instead of up his sleeve. He thus helps others to improve his analysis, and this is always possible in a problem concerning the real world.

2.4. *Degrees of sensitivity of Bayesian methods to the assumed initial distributions.* It has often been remarked (see, for example, p. 146 of Jeffreys, 1961) that, when testing a hypothesis by Bayesian methods, the results are not sensitive to "ordinary" changes in the assumed initial distribution of the parameters. This is true enough when there are only a few parameters. It is by no means true when there are many parameters: see #547. In any given application it is always possible to test the effect of such changes, but sometimes we are too busy to do this, just as the non-Bayesian is often too busy to test his procedures for robustness (or to read the literature!) and to measure the power of his significance tests against many alternatives. Deep down inside we are all subjectivists and, when dealing with a complicated problem, we are especially prone to indulge in incomplete arguments. As S. S. Wilks (1963) said, "statistics is infinite." Judgment is necessary when deciding whether a set of judgments is sufficient.

We shall not consider the philosophy of Bayesian estimation procedures here. These procedures can be regarded as a special case of Bayesian hypothesis testing since every statement of the form that a vectorial parameter belongs to a region is itself a hypothesis [but estimates are less often formulated before making observations].

2.5. *Scientific induction.* By "scientific induction" I understand the activity of trying to judge the subjective probabilities of theories and future events. It has come under attack in recent years under the lead of Popper (1959) who believes that the logical probability of any general scientific theory is zero, a view that if adopted would kill induction stone dead and is intended to do so. I think that one reason he came to this view is that he overlooked that the sum of an infinite number of positive numbers can be finite (see #191). If his view is adopted, then $P(E) = P(E|H_0)P(H_0) + P(E|H_1)P(H_1) + P(E|H_2)P(H_2) + \ldots = P(E|H_0)$, where $H_1, H_2, \ldots$ is an enumerable sequence of possible probabilistically verifiable mutually exclusive and exhaustive theories, E is a future event whose probability is of interest, and H_0 is the disjunction of all unspecifiable simple statistical hypotheses (that is, ones that cannot be specified in a finite number of words). Since H_0 is unspecifiable, there is no basis for a theoretical calculation of the physical probability of E. Hence Popper's view kills science, not just scientific induction [but see #956]. Other arguments against Popper's pilosophy of induction are given in ##599, **838**.

My own view on induction is close to that of Jeffreys (1939) in that I think that the initial probability is positive for every self-consistent scientific theory with consequences verifiable in a probabilistic sense. No contradiction can be inferred from this assumption since the number of statable theories is at most countably infinite (enumerable). It is very difficult to decide on numerical values for the probabilities, but is not quite as difficult to judge the *ratio* of the subjective initial probabilities of two theories by comparing their complexities. This is one reason why the history of science is scientifically important. Little progress has been made towards objectivity in this matter, although in practice there is often a reasonable measure of agreement concerning which of two theories is preferable. This is often because there is a very large weight of evidence for one of the theories as against the other. In some cases it should be possible to make some progress in the estimation of relative initial probabilities by assuming that the current judgments are correct and by using Bayes's theorem *backwards*. This idea is closely related to the "device of imaginary results" in which judgments are made concerning the values that the final probabilities would have after some imaginary results are obtained; then Bayes's theorem is used backwards to estimate what the consistent initial probabilities must be assumed to be; and lastly Bayes's theorem is used (forwards) to obtain estimates of the final probabilities in the light of the actual observations. (See ##13, 398, and especially #547.) This device takes *seriously* the view that a theory of probability or of rationality is no more than a theory of consistency.

2.6. *Bayesian statistics.* It is ambiguous to describe oneself, or a statistical technique as "Bayesian" *tout court.* I am a Bayesian in the sense that I believe it is useful to talk about the probabilities of hypotheses and that statisticians should make a reasonable attempt to avoid holding self-contradictory judgments. I am an eclectic in the sense that I see little objection to using any of the tricks of the statistical trade as an aid to judgment, so long as these tricks do not violate the axioms of probability. Perhaps the most fundamental forms of judgment are of the form of inequalities between subjective probabilities or between expected utilities of two different actions, but there is no need to restrict judgments to these forms. In particular it is possible with practice to make direct inequality judgments concerning "weights of evidence." Magistrates and medical diagnosticians do it all the time without quite knowing it. Statisticians should be more like magistrates and magistrates more like Bayesian statisticians.

3. A SUBJECTIVE ANALYSIS OF BODE'S LAW

3.1. The evaluation of scientific theories is a thorny problem in the philosophy of science. For each theory or set of conflicting theories the evaluation is often made by each scientist as an overall judgment in which, partly for economic reasons and partly out of reasonable respect, the authority of other more powerful scientists is taken into account. This is the sociological element of truth in the contention (Popper, 1959) that induction is irrelevant to the progress of science. It is a challenge to find some way of breaking these overall judgments down in order to increase their objectivity. All judgments are subjective, but some are more subjective than others. To increase the objectivity of judgments is the only function of theories of subjective probability (p. 4 of #13). The Bayesian considers that one important breakdown is obtained by separating the initial probability from the Bayes factor, since the initial probability is often more subjective than the factor. In this paper this approach will be used for the evaluation of Bode's law concerning the "mean distances" of the planets from the sun. Of novel philosophical interest is the use made of the "sharpened razor" in Section 3.7, although this is not the main issue.

Bode's law is an example of physical "numerology." A physical theory is "numerological" if it provides a formula for the calculation of physical quantities with little or no attempt to explain the formula. The word "numerology" was perhaps first suggested in this sense in conversation in 1947 by Blackett in connection with his paper of that year in which he discussed the magnetic moment and the angular momentum of large bodies. For some further examples of physical numerology see Good (1962c). The evaluation of numerological theories is more clearly a problem for the statistician than is the evaluation of non-numerological ones. But there is no clear-cut division between the two kinds of theory; all science is numerological to some extent, and we shall see that the evaluation of Bode's law requires both statistical and ordinary scientific reasoning. Ordinary scientific reasoning is more subjective than is sometimes appreciated.

If we were convinced that Bode's law *could not be ascribed to chance,* its importance for theories of the origin of the solar system would be increased. The need for a statistical evaluation is clear because different astronomers hold opposing views about it. In fact the distinguished astronomer Cecelia Payne-Gaposchkin (1961, pp. 177 & 253) even apparently contradicts herself . . . when she says it is "probably an empirical accident" and that "the fact that the asteroids move in a zone near to which Bode's law predicts a planet suggests that they represent such a planet." Davidson (1943) says "it cannot be considered a mere coincidence." Fath (1955) says " . . . it seems likely that it is merely a chance arrangement of numbers which breaks down completely for Neptune and Pluto." Edgeworth (1961) says " . . . it is reasonable to say that there is a certain measure of regularity in the spacing of the planetary orbits, and this regularity cannot be entirely without significance." Young (1902) manages to contradict himself in a single sentence: "For the present, at least, it must therefore be regarded as a mere coincidence rather than a real 'law,' but it is not unlikely that its explanation may ultimately be found . . . ." He wins, or loses, both ways.

3.2. The "mean distance" of a planet from the sun is the conventional misnomer for the semimajor axis of its orbit. In 1766 Johann Daniel Titius of Wittenberg announced, and in 1772 Johann Elert Bode published an approximate empirical "law" for the relative mean distances from the sun of the planets then known. (See Newcomb, 1910, where earlier references are given.) The law was later found to fit the mean distance of Uranus, and the average mean distances of the asteroids, but it failed for Neptune and for Pluto. When expressed in its most natural manner it fails also for Mercury. The usual expression is that the distances are approximately proportional to $4 + 2^n 3$, where $n = -\infty, 0, 1, 2,$ etc., but the first of these values is artificial, and the fair value to associate with Mercury is $n = -1$, since this increases the simplicity of the law and makes it less *ad hoc,* although it then does not give good agreement with Mercury's actual mean distance from the sun. . . .

[Much of the article and all the controversial discussion and reply are omitted.]

CHAPTER 14

Some Logic and History of Hypothesis Testing (#1234)

ABSTRACT

It is familiar that the probability of the outcome of an experiment is usually exceedingly small, especially when described in detail, and cannot be used by itself for hypothesis testing because it has no intuitive appeal. Three methods for producing a criterion of more sensible size are the uses of (i) tail-area probabilities, (ii) Bayes factors, or their logarithms the weights of evidence, and (iii) surprise indexes. The Jeffreys-Good[-Robbins]-Lindley paradox constitutes a serious logical objection to the first method for large samples, if the Likelihood Principle is accepted, and there are some difficulties in applying the second and third methods. The author believes that it is best to use the "Doogian" philosophy of statistics, which is a Bayes/non-Bayes compromise or synthesis, and examples of this synthesis are described.

1. INTRODUCTION

The foundations of statistics are controversial, as foundations usually are. The main controversy is between so-called Bayesian methods, or rather neo-Bayesian, on the one hand and the non-Bayesian, or "orthodox," or sampling-theory methods on the other.[1] The most essential distinction between these two methods is that the use of Bayesian methods is based on the assumption that you should try to make your subjective or personal probabilities more objective, whereas anti-Bayesians act as if they wished to sweep their subjective probabilities under the carpet. (See, for example, #838.) Most anti-Bayesians will agree, if asked, that they use judgment when they apply statistical methods, and that these judgments must make use of intensities of conviction,[2] but that they would prefer not to introduce numerical intensities of conviction into their formal and documented reports. They regard it as politically desirable to give their reports an air of objectivity and they therefore usually suppress some of the background judgments in each of their applications of statistical methods, where these judgments would be regarded as of potential importance by the

Bayesian. Nevertheless, the anti-Bayesian will often be saved by his own common sense, if he has any. To clarify what I have just asserted, I shall give some examples in the present article.

My own philosophy of probability and statistics is a Bayes/non-Bayes compromise. I prefer to call it the Doogian philosophy rather than the Good philosophy because the latter expression might appear self-righteous. Although I have expounded this philosophy on several previous occasions (for example, pp. 31-32 of #13; pp. 6-11 of #398; ##**26**, **230**, **679**) it can be roughly expressed succinctly enough to justify describing it yet again. In fact, the theory can be expressed so succinctly that it is liable to be ignored. . . . [The details are omitted to avoid repetition.]

A similar theory incorporating partially ordered utilities can then be constructed in a natural, almost an obvious, way (#**26**). Yet the obvious is often overlooked.

A theory of partially ordered probabilities is in a sense a compromise between a "strict" or "sharp" Bayesian philosophy in which all probabilities are precise, and non-Bayesian philosophies in which they are assumed merely to lie between 0 and 1 (p. 30).

Any sufficiently complete theory can be expressed in terms of axioms, rules and suggestions. I stated a set of axioms carefully on pp. 19 and 49 of #13 and will not repeat them here except to mention the product axiom. This is

$$P(\mathrm{E}\ \&\ \mathrm{F}|\mathrm{H}) = P(\mathrm{E}|\mathrm{H}) \cdot P(\mathrm{F}|\mathrm{E}\ \&\ \mathrm{H})$$

and its meaning is that if it is assumed that any two of the probabilities belong to certain intervals, then the third probability can be inferred to belong to some interval by using the equation. In my opinion this single assumption discriminates between Bayesians and non-Bayesians as effectively as any other equally simple assumption.

For a codification of the rules and suggestions see #**679**. Unlike a system of axioms, such a codification cannot be complete, and it does not readily lead to mathematical theorems, but I believe it is very useful.

As a psychological aid to introspection, for eliciting your own probabilities, I advocate the use of probability distributions of probabilities. This gives rise to a hierarchical theory the history of which is reviewed in some detail in #**1230**. It shows how a good philosophical point of view leads to practical statistical procedures, a possibility that might surprise many philosophers and statisticians. Like mathematics, philosophy can be either pure, applied, or applicable.

This survey of "Doogianism" has of course been highly succinct and the reader with an open mind who would like to see more details, if such a reader exists, will presumably examine some of the references.

2. EVIDENCE

Does your degree of belief (= intensity of conviction = subjective probability = personal probability) in some hypothesis or theory or proposition H depend only on the evidence E? It depends what you mean by the evidence. Your belief that 43917 = 43917 depends on logic and conventions rather than on the result of an experiment. Similarly, your judgment of the probability of a hypothesis depends on its prior probability as well as on the results of some experimental trial. If someone guesses a single decimal digit correctly you wouldn't seriously believe that he could always do so, for you would judge the prior probability of this hypothesis as being too small.

The usual way to change initial or prior probabilities into final or posterior probabilities is by means of Bayes's theorem. For a convenient formulation of this theorem it is customary to introduce the term "likelihood" in a sense first used by R. A. Fisher in a special case. The probability of E given H, denoted by $P(\mathrm{E}|\mathrm{H})$, is also called the likelihood of H given E. The special case considered by Fisher is when the value of $P(\mathrm{E}|\mathrm{H})$ is uncontroversial, in fact "tautological"; that is, where the numerical value of $P(\mathrm{E}|\mathrm{H})$ is determined by the *definition* of H (in which case H is called a *simple statistical hypothesis*). For example, H might state that the probability of r successes in some experiment is $e^{-r}a^r/r!$; then of course the observation of r successes has this probability, given H, by definition. A Bayesian is prepared to use the notation $P(\mathrm{E}|\mathrm{H})$ for probabilities that are not tautological, even for subjective (personal) probabilities, so, for a Bayesian, the concept of likelihood is more widely applicable than for a non-Bayesian such as Fisher. (At least he *thought* he was a non-Bayesian.) It might be advisable to call $P(\mathrm{E}|\mathrm{H})$ the "Bayesian likelihood" of H given E when this probability is not tautological. (One reason for this caution is that the expression "likelihood ratio" is usually used by statisticians in a technical sense as a ratio of maximum likelihoods.)

The concept of likelihood or Bayesian likelihood is used only when there are at least two hypotheses, so let's suppose we have n mutually exclusive hypotheses $\mathrm{H}_1, \mathrm{H}_2, \ldots, \mathrm{H}_n$. Then we have n likelihoods $P(\mathrm{E}|\mathrm{H}_1), \ldots, P(\mathrm{E}|\mathrm{H}_n)$. In some usages any set of probabilities $kP(\mathrm{E}|\mathrm{H}_1), \ldots, kP(\mathrm{E}|\mathrm{H}_n)$ is called a set of likelihoods, where k is any positive constant. With either of these definitions we can express Bayes's theorem, in the form suggested by Jeffreys as "final probability is proportional to initial probability $\times$ (Bayesian) likelihood" or, in symbols

$$P(\mathrm{H}_i|\mathrm{E}) \propto P(\mathrm{H}_i) \cdot P(\mathrm{E}|\mathrm{H}_i).$$

Note the immediate deduction

$$\frac{P(\mathrm{H}_i|\mathrm{E})}{P(\mathrm{H}_j|\mathrm{E})} = \frac{P(\mathrm{H}_i)}{P(\mathrm{H}_j)} \cdot \frac{P(\mathrm{E}|\mathrm{H}_i)}{P(\mathrm{E}|\mathrm{H}_j)}.$$

In particular, if $\overline{\mathrm{H}}$ denotes the negation of H, we have

$$O(\mathrm{H}|\mathrm{E}) = O(\mathrm{H}) \cdot P(\mathrm{E}|\mathrm{H})/P(\mathrm{E}|\overline{\mathrm{H}}),$$

where $O(\mathrm{H}|\mathrm{E})$ means $P(\mathrm{H}|\mathrm{E})/P(\overline{\mathrm{H}}|\mathrm{E})$ and is called the *odds* of H given E. The ratio $P(\mathrm{E}|\mathrm{H})/P(\mathrm{E}|\overline{\mathrm{H}})$ is called the *(Bayes) factor in favor of* H *provided by* E and may be denoted by $F(\mathrm{H}:\mathrm{E})$. Jeffreys, 1939, denoted it by K but did not give it a name. The logarithm of the Bayes factor was independently called the *weight of evidence* in favor of H provided by E by C. S. Peirce (1878), by Good in #13 (and in about forty other publications by him) and by Minsky and Selfridge (1961). [For a correction, see #1382.] The weight of evidence can be added to the initial log-odds of the hypothesis to obtain the final log-odds. If the base of the logarithms is 10, the unit of weight of evidence was called a *ban* by Turing (1941) who also called one tenth of a ban a *deciban* (abbreviated to db). . . . I hope that one day judges, detectives, doctors and other earth-ones will routinely weigh evidence in terms of decibans because I believe the deciban is an intelligence-amplifier.

If someone guesses one digit correctly the hypothesis that he can always do so gains a Bayes factor of ten and, as I said before, you still don't believe the hypothesis. But if a new typewriter prints the digit corresponding to a key you press, you would tend to believe it will always do so until it breaks down. Therefore there is simply no question of the likelihoods alone determining your states of belief in the various possible hypotheses. They can only modify the beliefs you had before you made the relevant observation.

Nevertheless, there is something called the "likelihood principle," and, if it is correctly stated, it is a trivial consequence of Bayes's theorem. It can be stated in the following manner: *Let E and E′ be two distinct experimental results or observations. Suppose that they do not affect the utilities (if true) of hypotheses* $H_1, H_2, \ldots, H_n$ *under consideration. Suppose further that E and E′ provide the same likelihoods to all the hypotheses, that is, that* $P(E|H_i) = P(E'|H_i)$ $(i = 1, 2, \ldots, n)$. *Then E and E′ should affect your beliefs, recommendations, and actions concerning* $H_1, H_2, \ldots, H_n$ *in the same way.* Clearly this principle is built into Bayes's theorem. When $n = 2$ the principle is built into the terminology "weight of evidence."

The likelihood principle has been supported by some non-Bayesians who want to avoid the use of probabilities that are neither apparently physical (material) nor tautological (mathematical) but are subjective (personal) or are credibilities (logical probabilities). Unfortunately members of the likelihood brotherhood have sometimes given the impression that the likelihoods by themselves always enable you to choose between hypotheses.[3] We have already seen, by the example of the guessed digit, that the likelihoods by themselves are clearly not enough for this purpose (although they are "sufficient statistics" in Fisher's technical sense). They tell you all you need to know about the experimental results, if your model of the situation is not to be questioned, but the result of an experiment is not by itself enough for choosing between hypotheses. If it were, there would be no need for the Duns-Ockham razor. As de Finetti (1975, p. 248) said,

> they ignore one of the factors (the prior probability) altogether, and treat the other (the likelihood) as though it . . . meant something other than it actually does. This is the same mistake as is made by someone who has scruples about measuring the arms of a balance

> (having only a tape-measure at his disposal . . .), but is willing to assert that the heavier load will always tilt the balance (thereby implicitly assuming, although without admitting it, that the arms are of equal length!).

We never reject a hypothesis H merely because an event E of very small probability (given H) has occurred although we often carelessly talk as if that were our reason for rejection. If the result E of an experiment or observation is described in sufficient detail its probability given H is nearly always less than say one in a million. As Jeffreys (1939, p. 315) said,

> If mere probability of the observations, given the hypothesis, was the criterion, any hypothesis whatever would be rejected. Everybody rejects the conclusion [of rejecting hypotheses because of improbability of the evidence] but this can only mean that the improbability of the observations, given the hypothesis, is not the criterion, and some other must be provided.

If we want to be able to say that H should be rejected "because the observation is too improbable" given H we have to do more than compute $P(\mathrm{E}|\mathrm{H})$ even when this probability can be computed. Let us consider various approaches to this problem.

3. THE TESTING OF HYPOTHESES

The Bayesian Approach

A Bayesian believes it is meaningful to talk about the (Bayes) factor against (or in favor of) H, or about its logarithm the weight of evidence. In statistical applications H is often (at least as an approximation) a simple statistical hypothesis in the sense that $P(\mathrm{E}|\mathrm{H})$ is a tautological probability while $\overline{\mathrm{H}}$ is often composite, that is, it is a logical disjunction of a set of simple statistical hypotheses. For the sake of simplicity of exposition I shall suppose that $\overline{\mathrm{H}}$ is the logical disjunction "H_1 or H_2 or H_3 or . . . " (although the number of such hypotheses is often non-countable in statistical models), where H_1, H_2, H_3, . . . are mutually exclusive. (In other words $\overline{\mathrm{H}}$ is true if and only if one of H_1, H_2, H_3, . . . is true.) The probabilities $P(\mathrm{H}_i|\overline{\mathrm{H}})$ and $P(\mathrm{E}|\mathrm{H}_i)$ are assumed to exist and $P(\mathrm{E}|\overline{\mathrm{H}})$ can then "in principle" be calculated by the formula

$$P(\mathrm{E}|\overline{\mathrm{H}}) = P(\mathrm{E}|\mathrm{H}_1)P(\mathrm{H}_1|\overline{\mathrm{H}}) + P(\mathrm{E}|\mathrm{H}_2)P(\mathrm{H}_2|\overline{\mathrm{H}}) + \dots .$$

Then the Bayes factor $P(\mathrm{E}|\overline{\mathrm{H}})/P(\mathrm{E}|\mathrm{H})$, against the null hypothesis, can in principle be calculated, and its logarithm is the weight of evidence against the "null hypothesis" H provided by E, say $W(\overline{\mathrm{H}}:\mathrm{E})$. Note that this is mathematically independent of the initial or prior probability $P(\mathrm{H})$. The main objection to this Bayesian approach is of course that it is usually difficult to specify the probabilities $P(\mathrm{H}_i|\overline{\mathrm{H}})$ with much precision. The Doogian reply is that we cannot dispense with judgments of probabilities and of probability inequalities; also that non-Bayesian methods also need Bayesian judgments and merely sweep them under the carpet.

The Tail-Area-Probability Approach

In order to judge H in terms of a probability that is not necessarily microscopic, "tail-area probabilities" (sometimes called *P*-values) are often used. A tail-area probability is, by definition, the probability that the outcome of an experiment or observation would have been "at least as extreme" as the actual outcome E.

Different statisticians have interpreted a tail-area probability P in different ways. Some statisticians say that if P is small enough, then the null hypothesis H should be rejected. Braithwaite (1951) argued that rejection of a hypothesis is always provisional and that there are "degrees of provisionality" of rejection. Many statisticians decide in advance of an experiment what threshold P_0 should be used such that if $P < P_0$ then H should be "rejected," and they don't mention whether the rejection is only provisional. Typical values of P_0 are 0.05, 0.02, 0.01, 0.005, 0.002, and 0.001 because Fisher had issued tables (for some random variables) corresponding to these thresholds. I understand that he issued the tables in this form partly to avoid having to use Karl Pearson's tables. The threshold would be chosen depending partly on the initial probability that H is true, or is an adequate approximation, and partly on the seriousness of rejecting H when it is true or accepting it when false. Some statisticians choose a threshold without being fully conscious of why they chose it. Others will put one asterisk against outcomes that have $0.05 > P > 0.01$, two asterisks when $0.01 > P > 0.001$, and three asterisks when $P < 0.001$. There is no special justification for doing this except that it is conventional and saves work if tables like Fisher's are being used. If P can be calculated at little cost, which is increasingly possible, it seems better to state its actual value.

Let us consider a null hypothesis that is a (sharp) simple statistical hypothesis H. This is often done in statistical practice although it would usually be more realistic to lump in with H a small "neighborhood" of close hypotheses as on p. 97 of #13. The non-null hypothesis $\overline{H}$ is usually a logical disjunction of a continuous infinity of simple statistical hypotheses which I shall call the *components* of $\overline{H}$. These components will be at various "distances" from H and some of them will be very close to it. Hence it is often sensible to talk of testing H *within* $\overline{H}$ although statisticians often use the word "against." An experiment or observation usually consists of a sample and when we refer to the size of an experiment we mean the size of the sample. (Neyman calls the α-level of a test the size of the test but I prefer to call it the α-level. We define α-levels later.) To distinguish H from the close components of $\overline{H}$ requires a large experiment. If no prior distribution is assigned to the components of $\overline{H}$ (given $\overline{H}$), then one could never obtain evidence in favor of H even if it is true, because there will be components of $\overline{H}$ close enough to H to be indistinguishable from it given any sample of a specified size. If, however, H is false, then a large enough sample should demonstrate its falsity to a Fisherian. On the other hand if a prior over the components of $\bar{H}$ is assumed, then a Bayesian or Doogian can obtain evidence in favor of H, though typically this evidence is not very great. Of course if H is redefined to be a composite hypothesis by including within H a small "neighborhood" of the "sharp" null hypothesis, then it becomes possible to obtain much more evidence in favor of H even without assuming a prior over the components of $\bar{H}$. Similarly if by the truth of Newtonian mechanics we mean that it is approximately true in some appropriate well defined sense we could obtain strong evidence that it is true; but if we mean by its truth that it is exactly true then it has already been refuted.

Very often a statistician doesn't bother to make it quite clear whether his null hypothesis is intended to be sharp or only approximately sharp. He also often has in mind an experiment of moderate size but does not state this explicitly. It is hardly surprising then that many Fisherians (and Popperians) say that "you can't get [much] evidence in favor of a null hypothesis but can only refute it." Regarding this statement itself as a kind of null hypothesis (!) the Fisherian's experience tends to support it, as an approximation, Doogianwise, so the Fisherian (and Popperian) comes to believe it, because he is to some extent a Doogian without knowing it.

Perhaps the simplest example of a significance test is a test for whether a binomial physical probability p is equal to 1/2. I am assuming the sample to be definitely binomial and that only the value of p is uncertain. Suppose there are r "successes" in n "trials" so that the "sigmage" is $x = (r - \frac{1}{2}n)/(\frac{1}{2}\sqrt{n})$. The larger is $|x|$, the more the sample differs from the most probable result (given the null hypothesis H).

In conversation I have emphasized to other statisticians, starting in 1950, that, in virtue of the "law of the iterated logarithm," by optional stopping an arbitrarily high sigmage, and therefore an arbitrarily small tail-area probability, can be attained even when the null hypothesis is true. In other words if a Fisherian is prepared to use optional stopping (which usually he is not) he can be sure of rejecting a true null hypothesis provided that he is prepared to go on sampling for a long time. The way I usually express this "paradox" is that a Fisherian [but not a Bayesian] can cheat by pretending he has a train to catch like a gambler who leaves the table when he is ahead. Feller (1950, p. 140) discusses optional stopping in a "fair" gambling game and points out that, in virtue of the law of the iterated logarithm, an infinitely rich gambler can be sure of winning if he has the right to stop playing when he chooses (much good it would do him). Surprisingly, Feller [1950] does not mention the effect that optional stopping would have on Fisherian tail-areas [but Feller (1940) does, as does Greenwood (1938), also]. The point is implicit in Jeffreys (1939, Appendix I) and explicit in [Robbins (1952), Anscombe (1954),] Good (1956) and in Lindley (1957). [Jeffreys does not mention optional stopping, and Greenwood and Feller do not mention that optional stopping is acceptable to Bayesians.]

It is intuitively obvious in the binomial sampling experiment that the evidence is summarized by the values of r and n and that there is no point in investigating whether there really was a train to catch except as a criterion regarding the honesty of the statistician.[4] The use of tail-area probabilities is therefore logically shaky but I believe it is useful all the same, and can often be given a rough Doogian justification, at least when samples are not very large. [The point is discussed later.]

It also follows from the likelihood principle alone, without the Bayesian theory, that optional stopping is irrelevant when judging whether $p = 1/2$.[5]

What Is "More Extreme"?

The idea that one outcome is more extreme than another one depends on whether it seems to be "further away" from the null hypothesis. What is meant by "further away" depends on some ordering, precise or vague, of the components of $\overline{H}$. (For a discussion of such orderings see, for example, Kempthorne and Folks [1971, pp. 226-230)].) The statistic chosen for testing the null hypothesis is chosen to reflect this distance. Thus the statistic is a "discrepancy measure" to use the terminology of, for example, Kalbfleisch & Sprott (1976, p. 264). Sometimes more than one statistic is used for the same data because the statistician or his client has more than one non-null hypothesis in mind, each non-null hypothesis being composite. For example a distribution might be tested as a fully specified normal distribution within the class of all normal distributions or within the class of all possible distributions, and different test criteria would be used in these two cases.

The representation of the idea that one outcome is more extreme than another depends on the statistic (= function of the observations) that is used. For example, suppose that the frequencies in the cells of a multinomial distribution having t categories are $n_1, n_2, \ldots, n_t$ (where $\Sigma n_i = N$, the sample size) and that the null hypothesis H specifies that the cell "physical probabilities" (= propensities) are $p_1, p_2, \ldots, p_t$, where $\Sigma p_i = 1$. (A mnemonic for the symbol t, as used here, is that it is often the initial letter of the number of categories such as two, three, ten, twenty, twenty-six, thirty or a thousand.) A statistic that is often used for testing H is X^2 or χ^2 defined by

$$X^2 = \Sigma \frac{(n_i - Np_i)^2}{Np_i}.$$

(X^2 is often called χ^2 [chi-squared] but χ^2 is also used for the asymptotic distribution of X^2 when $N \to \infty$ and when H is true, so the more modern notation for the statistic is X^2.) There are other statistics used for testing H: see, for example, ##547, 862. One special appeal of X^2 to the non-Bayesian is that it resembles Euclidean distance, so it has geometrical simplicity, and the resemblance is especially marked when the p's are all equal to $1/t$ (the "equiprobable case"). Moreover the asymptotic distribution of X^2 (given H), when N is large, is known and tabulated. It is the chi-squared distribution, as we mentioned before, with $t - 1$ "degrees of freedom."

It is intuitively to be expected, and also provable, that if the true probabilities differ from $p_1, p_2, \ldots, p_t$, then X^2 is likely to be larger than if H is true, and roughly speaking the more they differ the larger X^2 is likely to be. Therefore it makes at least some sense to say that one sample is "more extreme" than another one if the former has a larger value of X^2. If some other statistic is used then "more extreme" will have a (somewhat) different meaning. In other words the choice of a "criterion" (= statistic) for testing a hypothesis H always involves some concept of what alternatives to H are under consideration, however vague or precise those alternatives might be, and the deviation of the statistic from its null value should correspond roughly to greater distances from H of the components of $\overline{\text{H}}$, the distances being defined in "parameter space" (or more generally in hypothesis space). Of course "distance" is here to be interpreted in a generalized sense, and it need not be a quadratic function of the parameters. One appealing distance function for multinomials is the "expected weight of evidence per observation," $\Sigma p_j \log(p_j/q_j)$, for discriminating the null hypothesis from some other hypothesis. (The continuous analogue is obvious.) This suggests the use of the statistic $\Sigma p_j \log(tp_j/n_j)$ or of $\Sigma n_j \log[n_j/(tp_j)]$. The latter statistic comes to essentially the same thing as the familiar Neyman-Pearson-Wilks likelihood-ratio statistic. The likelihood-ratio statistic for testing a hypothesis H "within" a hypothesis H′ (where H′ at least is composite) is defined as $\max_i P(\text{E}|\text{H}_i)/\max_j P(\text{E}|\text{H}'_j)$, where (i) E is the observed event, (ii) H is the disjunction of all H_i, (iii) H′ is the disjunction of all H'_j, and (iv) H_i and H'_j are simple statistical hypotheses for all i and j. By saying that H is "within" H′ we mean that each H_i is a component of H′ (besides being a component of H), though the converse is of course not true. Clearly the likelihood ratio tends to be small if H is false but H′ is true, and tends to be smaller the farther away is the true H'_j from any of the H_i's. In this way Neyman and Pearson produced a technique of some generality for generating reasonable test criteria, framed in terms of the non-null hypothesis of interest ot the statistician. Moreover Wilks found, in terms of χ^2, the asymptotic distribution of the likelihood ratio for large sample sizes, under somewhat general conditions, and this made the likelihood ratio convenient to use in the Fisherian manner.

Notice how the likelihood ratio is analogous to a Bayes factor which would be defined as a ratio of weighted averages of $P(\text{E}|\text{H}_i)$ and of $P(\text{E}|\text{H}'_j)$. The weights would be $P(\text{H}_i|\text{H})$ and $P(\text{H}'_j|\text{H}')$ but these probabilities are *verboten* by the non-Bayesian. Just as the value of an integral is often very roughly monotonically related to the maximum value of the integral so one hopes that the Bayesian averages are roughly related to the maximum likelihoods. It would not be entirely misleading to say that the use of the Neyman-Pearson-Wilks likelihood ratio is the non-Bayesian's way of paying homage to a Bayesian procedure. (I have expressed this argument incompletely to save space.)

Although the use of tail-area probabilities may reasonably be called "Fisherian tests of significance" because Fisher made so much use of them, and developed their theory to such an extent, they had a long previous history. One early use was by Daniel Bernoulli (1734). (This example is described by Todhunter (1865, p. 222) and was used by Barnard (1979), in a Bayesian conference in Spain, as a challenge to Bayesians to explain why it was reasonable in Bayesian terms. I was able to reply only briefly at that conference, mainly by giving references to my own work, as usual, where this kind of question had been discussed.)

In Bernoulli's example the null hypothesis is that the normals to the planes of rotation of the various planets are uniformly distributed in direction. One of the statistics he used was the maximum angle between the planes of two planetary orbits. (Todhunter points out some improvements to Bernoulli's approach, but these improvements are barely relevant to the present philosophical discussion.) Clearly this statistic will tend to be small if all the planets rotate approximately in a single plane. Thus the appropriate meaning for "more extreme" in this case is "smaller." Note that the statistic selected by Bernoulli was presumably chosen by him because he had both a null hypothesis and a vague non-null hypothesis in mind. The null hypothesis was that the normals were uniformly distributed in direction, whereas the vague non-null hypothesis was that the normals had a tendency to be close together. He wouldn't have tested the null hypothesis by looking at the maximum angle between planes of orbits unless he had in mind the (vague) non-null hypothesis just described. The smaller the maximum angle, the further away is the specific form of non-null hypothesis likely to be (away, that is, from the null hypothesis) and, at the same time, the less probable is the null hypothesis. I believe it is a feature of all sensible significance-test criteria that they are chosen with either a precise or, more often, a vague non-null hypothesis, in mind. In this respect "non-Bayesians" act somewhat like Bayesians. *If a tail-area probability is small enough then it is worth while to try to make the non-null hypothesis less vague* or even to make it precise, and the smaller the tail-area probability, the more worth while it is to make this attempt.

The "tail-area probability" that Bernoulli obtained would have been somewhat less impressive if he had been able to allow for Pluto because its orbit makes a large angle with the other orbits. To allow for the possibility of such an "outlier" a different statistic could be used, such as the average angle between all pairs of orbits instead of the maximum angle. Or, a little more artificially, one might use the maximum angle, allowing the deletion of one planet, but then some "payment" would be required to allow for the artificiality. It is intuitively obvious that if enough of the normals to the orbits are close enough together, then there must be a physical reason for it even if some of the normals have a large deviation. The method of deleting one of the planets provides an example of selecting a significance test after looking at the data. Many textbooks forbid this. Personally I think that Rule 1 in the analysis of data is "look at the data."

The question of whether the Bode-Titius Law is "causal," a subject that was treated in **#603B** and by Efron (1971), is similar to Bernoulli's problem. It is noteworthy that one of the discussants of the former of these papers used the exceptional planets, combined with a t-test, to argue that "we are in the classic situation which Herbert Spencer once called a 'scientific tragedy'—a theory killed by a fact." Having failed to understand that Spencer's remark does not apply to the question of whether the Bode-Titius Law *needs an explanation*, he repeated the same fallacy in the discussion of Efron's paper and was again refuted. That a reputable statistician could make such an error, and then to persist in it, shows the dangers of being misled by standard cookbook recipes when the conditions for their application are not present. In other words it is an example of the "tyranny of words." A misconception that sometimes arises from the same tyranny is the belief that a P-value less than 5% means that the null hypothesis has probability less than 5%! An example of this blunder is mentioned in a book review in *Applied Statistics 28* (1979, p. 179).

Another example where a standard technique is inadequate was mentioned on p. 60. . . . This example again shows that it is sometimes sensible to decide on a significance test after looking at a sample. As I've said elsewhere this practice is dangerous, useful, and often done. It is especially useful in cryptanalysis, but one needs good detached judgment to estimate the initial probability of a hypothesis that is suggested by the data. Cryptanalysts even invented a special name for a very far-fetched hypothesis formulated after looking at the data, namely a "kinkus" (plural: "kinkera"). It is not easy to judge the prior probability of a kinkus after it has been observed. I agree with Herman Rubin's remark, at the Waterloo conference on scientific inference in 1970, that a "good Bayesian does better than a non-Bayesian but a bad Bayesian gets clobbered." Fisher once said privately that many of his clients were not especially intelligent, and this might have been part of his reason for avoiding Bayesian methods.

A very common example, to support the view that Fisherians allow for the non-null hypothesis, is that they often have to choose whether a tail-area probability should be single-tailed or double-tailed. If we are considering two treatments, and the null hypothesis is that there is no, or a negligible, difference in their effects (the null hypothesis) then it will usually be appropriate to use a double-tail; but, if the question (= the non-null hypothesis) is only whether the second treatment is *better* than the first, then a single-tail is usually appropriate. Clearly a sensible criterion should embody what we want to mean by "more extreme" in terms of the components of the non-null hypothesis. We shall have more to say later regarding the choice of a significance-test criterion, in relation to the "Bayes/non-Bayes synthesis" and to "surprise indexes." It is curious that Fisher introduced general features of statistics for estimation purposes, but not for significance tests. He seemed to select his significance-test criteria by common sense unaided by explicit general principles. Some such general principles were later introduced by Neyman and Pearson who made use of the "power function'' of a significance test and of the likelihood-ratio test with its attractive amount of generality.

It is sometimes stated by Fisherians that the only hypothesis under consideration is the null hypothesis, but I am convinced that this is only a way of saying that the non-null hypothesis is vague, not that it does not exist at all. As so often, in other contexts, the controversy (of whether there is always a non-null hypothesis) is resolved by saying "It depends what you mean." It is a common fallacy that if a concept is not precise then it does not exist at all. I call this the "precision fallacy." If it were true then it is doubtful whether any concepts would have any validity because langauge is not entirely precise though it is often clear enough. It is largely because of the precision fallacy that many statisticians are not Doogians, or do not know that they are. They think that Bayesians all use precise probabilities and do not recognize that the Doogian, who is in many respects a Bayesian, in principle uses "upper and lower probabilities."

To show that tail-area probabilities do not contain all that is needed we ask what if there had been a billion planets and the tail-area of Bernoulli's statistic had been 0.001? Would we then have rejected the null hypothesis? Surely we would have been surprised that the tail-area probability was not much smaller than 0.001, given the non-null hypothesis. Similarly, in the multinomial example, if N were exceedingly large, you would expect to get a very small value for $P(\tilde{X}^2 \geqslant X^2)$ if the null hypothesis were false (where the tilde converts X^2 into a random variable). In other words, there are values of X^2 that are surprising whether the null hypothesis or the non-null hypothesis is true. Jeffreys (1939, p. 324) says that in such circumstances both he and Fisher should be very doubtful. The following Bayesian approach makes the matter clearer.

Bayes Factors and Tail-Area Probabilities

If a criterion S (a statistic) is sensibly chosen it might exhaust most of the weight of evidence against (or for) a null hypothesis H in the sense that $W(\text{H:S}) \approx W(\text{H:E})$. For reasons to be discussed later, X^2 is such a statistic when testing the multinomial null hypothesis mentioned before. Now the asymptotic distribution of X^2 (given H) when $N \to \infty$, is known. As I said before, it happens to be the χ^2 distribution with $t - 1$ degrees of freedom, though it is not essential that the reader should know what this means. This asymptotic distribution is fairly accurate, up to tail-areas as small as 1/100 at any rate, even if N is as small as t. (Some textbooks say that N needs to be as large as $5t$ but it has been known for some time that this is unnecessary.) Suppose we can make some guess at the distribution of X^2 given the (vague) non-null hypothesis, for a given value of N. It seems reasonable to me to assume that it will often resemble a Pearson Type 3 distribution, but with an extra thick tail on the right, and one way of guessing it is to specify its median, quartiles and other quantiles, depending on the application and on "your" judgment. (The quantiles can be judged by imagining bets at various appropriate odds.) The density curves given the null and (vague) non-null [composite] hypothesis might have the general appearance shown in Figure 1.

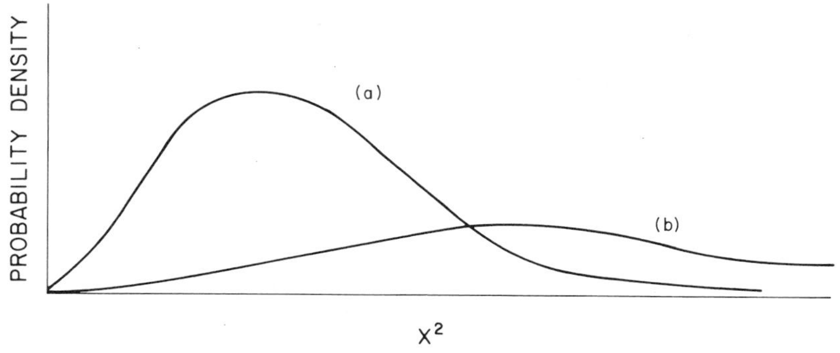

Figure 1. Probability densities of X^2 given H (curve a) and given $\overline{\mathrm{H}}$ (curve b).

When N increases, curve (a), which represents the density of X^2 given H, hardly changes, but the mode of the curve for $\overline{\mathrm{H}}$ (curve [b]) moves to the right. Now the Bayes factor against H, provided by X^2, is the ratio of the ordinate of curve (b) to that of curve (a) at the value of X^2 that occurs. Thus, as N increases, a given value of X^2 provides less and less evidence against H, indeed it ultimately provides an increasing weight of evidence in favor of H; and the same is true of a given tail-area probability $P(\widetilde{X}^2 \geqslant X^2)$.

The above argument, showing that a fixed tail-area probability has diminishing evidential value against H as the sample size N increases, can also be expressed in terms of "errors of the first and second kinds" (as defined below when we discuss the Neyman-Pearson approach). For we can concentrate our attention on any one, say H'_j, of the simple statistical hypotheses of which the non-null hypothesis is a logical disjunction. We can then replace the curve (b) by the probability density curve of H'_j. For any fixed α, the value of β corresponding to H'_j increases as N increases and tends to 1 as $N \to \infty$. Moreover the simple likelihood ratio, for comparing H with H'_j, which is a ratio of ordinates (instead of being a ratio of areas), tends to infinity as $N \to \infty$ if α is fixed. Thus, for any fixed α, however small, the evidence *in favor* of the null hypothesis tends to infinity as $N \to \infty$. Since this is true for each H'_j it must also be true when H is contrasted with $\overline{\mathrm{H}}$. This result hardly requires a Bayesian argument to support it, but only the concept that the simple likelihood ratio is monotonically related to the strength of the evidence.

Another way of looking at the above argument is that, for a fixed α, as N increases, more and more of the components of $\overline{\mathrm{H}}$ become effectively refuted, and we are left with a contracting set of components in a smaller and smaller neighborhood of H. Thus H becomes more and more probable and this result does not even depend on our regarding H as itself only an approximation. Of course the argument is stronger still if we do regard H in this light.

We see then that a given tail-area probability P for X^2 (or for any other statistic) has a fixed meaning neither for a Bayesian nor for a Neyman-Pearsonian, and that the interpretation of P depends on the sample size. In this respect the Bayesian and the Neyman-Pearsonian agree to differ from a Fisherian. The difference is, however, not as large as it might seem because, when the sample size is very large, the tail-area probability will usually either be moderate (e.g. in the range 0.05 to 1) or exceedingly small (e.g. less than 1/10,000), depending on whether the null hypothesis is true or false. [But the difference is large if optional stopping is permitted.]

These comments make one uneasy about certain experiments in parapsychology where the experiments are large and the "proportional bulges" are small. For example, the best subject in Helmut Schmidt's experiments, in a binomial experiment, had a success rate of 52.5% in 6400 trials. (See Frazier, 1979.) As good a result would occur about 1/30,000 of the time in a sample of this size, if $p = 1/2$. But, if we assume that, under the non-null hypothesis, p is uniformly distributed from 0.5 to 0.55 then the Bayes factor against the null hypothesis from this sample is only about $(80\pi)^{1/2}e^8 \div 320 \approx 150$. The factor could be larger if the prior density, given $\bar{H}$, tended to infinity near $p = 1/2$, a possibility suggested on pp. 74 and 78 of Good (1962a) where the matter is discussed in somewhat more detail.[6]

In spite of my logical criticisms of the use of tail-area probabilities, I have found that, for sample sizes that are not extremely large, there is usually an approximate relationship between a Bayes factor F and a tail-area probability P. I shall now discuss this relationship.

Pp. 93-95 of #13 made use of the idea of basing a Bayes factor on the values of a statistic (not necessarily a sufficient statistic), such as X^2. This requires a judgment of the distribution of the statistic given $\bar{H}$. Given H, the distribution of X^2 is often uncontroversially known approximately. I said that

> it would often happen that the factor in favour of H [the null hypothesis] obtained in some such way [by assuming a distribution for a statistic both given H and given $\bar{H}$] would be in the region of three or four times $P(\chi_0{}^2)$ [the tail-area probability of the observed value of the statistic given H]. . . . [In a footnote I remarked that] There are two independent reasons why the factor in favour of H exceeds $P(\chi_0{}^2)$. The first is that to pretend that the result is $\chi \geqslant \chi_0$ when it is really $\chi = \chi_0$ is unfair to H. The second is that $P(\chi \geqslant \chi_0|H) < 1$, so that the factor from the evidence "$\chi \geqslant \chi_0$" is
>
> $$P(\chi \geqslant \chi_0|H)/P(\chi \geqslant \chi_0|\bar{H}) \geqslant P(\chi \geqslant \chi_0|H) = P(\chi_0{}^2).$$

#127, based on lectures of 1955 in Princeton and Chicago, took the matter somewhat further in the following words:

> The Bayes/non-Bayes synthesis is the following technique for synthesizing subjective and objective methods in statistics. (i) We use the neo/Bayes-Laplace philosophy in order to arrive at a factor, F, in favour

of the non-null hypothesis. For the particular case of discrimination between two simple statistical hypotheses, the factor in favour of a hypothesis is equal to the likelihood ratio, but not in general. The neo/Bayes-Laplace philosophy usually works with inequalities between probabilities, but for definiteness we here assume that the initial distributions are taken as precise, though not necessarily uniform. (ii) We then use F as a statistic and try to obtain its distribution on the null hypothesis, and work with its tail-area probability, P. (iii) Finally we look to see if F lies in the range

$$\left(\frac{1}{30P}, \frac{3}{10P}\right).$$

If it does not lie in this range we think again.

(Note that F is here the factor *against* H.)

#547 examined numerical evidence relating a Bayes factor F against H to an orthodox tail-area probability. The application was to significance tests for equiprobability in a multinomial distribution and the Bayes factor was based on all the evidence. It turned out that, in 18 samples,

$$\frac{1}{6P\sqrt{(2\pi N)}} < F < \frac{6}{P\sqrt{(2\pi N)}},$$

where N is the sample size, and in 14 of the samples the 6 could be replaced by 3. The reason for introducing $N^{1/2}$ into the formula was its occurrence in Jeffreys (1939, Appendix I). Jeffreys showed in effect that, in a number of examples, the Bayes factor against the null hypothesis is roughly inversely proportional to $N^{1/2}$ for a given tail-area probability.

For the same multinomial problem, but with a much larger collection of samples, #862 found that $P(F)$, the tail-area probability of F itself, was proportional to $(FN^{1/2})^{-1}$ within a factor of 5. For most pairs (t, N), where t is the number of multinomial categories, we found that $P(F)$ is almost equal to a mathematical function of t, N, and F and does not otherwise depend much on the frequency count $(n_1, n_2, \ldots, n_t)$. This confirmed a basic idea of the Bayes-Fisher compromise, that a *Bayes factor is roughly some function of its own tail-area probability.*

Some similar numerical results were obtained for contingency tables in ##929, 1199.

In virtue of these results, and in spite of my propaganda for the Jeffreys-Good[-Robbins]-Lindley paradox, I personally am in favor of a Bayes/non-Bayes compromise or synthesis. Partly for the sake of communication with other statisticians who are in the habit of using tail-area probabilities, I believe it is often convenient to use them especially when it is difficult to estimate a Bayes factor. But caution should be expressed when the samples are very large if the tail-area probability is not extremely small.

The Neyman-Pearson Approach

The basic idea in the Neyman-Pearson approach to significance tests (Neyman and Pearson, 1933) is to consider the probabilities of errors of the first and second kinds. An error of the first kind is defined as the rejection of the "null hypothesis" H when it is true, and an error of the second kind is the acceptance of H when it is false. The probabilities of these two kinds of error depend of course on the signficance test that is to be used, and the two probabilities are usually denoted by α and β. Here β is regarded as a function of the components of $\overline{\mathrm{H}}$ and $1 - \beta$ is called the *power function.* Also α might be a function of the components of H if H is composite, but we shall suppose for simplicity that H is simple. If the significance test is defined in terms of a statistic with a threshold, then α depends on the threshold. Sometimes the threshold is chosen so that α takes some conventional value such as 0.05; otherwise the power function depends on the value of α.

The concept of a power function is in my opinion most useful when $\overline{\mathrm{H}}$ is a disjunction of a single-parameter set of components. When there is more than one parameter the power function is liable to be difficult to apprehend.

It is not by any means always sensible either to accept or to reject a hypothesis in any sharp sense. If a *P*-value is very small this gives more reason to look for sharp alternatives to it, in a Fisherian context, and this important fact is not captured by the Neyman-Pearson technique.

Neyman and Pearson recognized that a hypothesis cannot be tested unless there is some notion of alternatives so they formalized the matter by assuming the alternatives could be specified accurately. In this respect they were going in the direction of the "sharp" Bayesian, but they stopped short of assuming a conditional prior for the components of $\overline{\mathrm{H}}$ (conditional on $\overline{\mathrm{H}}$).

That the notion of errors of the second kind is useful for orthodox statistics, in one-parameter problems, whatever its weaknesses may be, can be seen from Frieman *et al.* (1978). They found that the emphasis on α (or *P*-value) and the neglect of β had led to many ineffective clinical trials. In 71 "negative" randomized control trials, chosen in a sensible manner from published papers, 50 of the trials had a 10% risk of missing a 50% therapeutic improvement (in a well defined sense). This poor performance might have been avoided if the experimenters had allowed for errors of the second kind when planning their experiments. They would have realized that their samples were too small. But the smallness of the samples in these trials was presumably caused also by the ethical consideration, that as soon as one treatment seems better than another it seems unethical to use the apparently less good treatment. (See also #1146, which gives a brief recent discussion of ethical problems in clinical trials. It seems that the ethical difficulty can only be overcome either by social contract or if patients voluntarily accept compensation for entering the trial.[7] [Or perhaps we can take advantage of the fact that the legality of treatments differs from one country to another.]

It is of historical interest that Neyman (1977) says that he and Pearson were inspired to formulate their theories by Borel (1920) who "insisted that: (a) the criterion to test a hypothesis (a 'statistical hypothesis') using some observations must be selected *not after the examination of the results of observation*, but before, and (b) this criterion should be a function of the observations 'en quelque sorte remarquable.' "

When using the Neyman-Pearson theory, at least in its standard form, the precise significance test and the value of α are supposed to be determined in advance of the experiment or observation. I have already argued that one cannot always sensibly determine a significance test in advance, because, heretical though it may be in some quarters, sometimes the data overwhelmingly suggests a sensible theory after the data are examined. On some other occasions the suggestion is not overwhelming and then it is desirable to collect more data if this is not too expensive. In Daniel Bernoulli's example, and in the example of the Titius-Bode Law, and in many cryptanalytic problems, it is difficult or impossible to collect more data, and in some other situations it is very expensive to do so.[8]

On some occasions the non-null hypothesis is of high dimensionality, or is otherwise a very complicated disjunction of simple statistical hypotheses, with the result that the statistician is unable to apprehend the power function intuitively. In such cases it may be possible to average the power function with weights that are proportional to the prior probabilities so as to obtain a convenient summary of the power function (p. 711 of #862). [I call it the *strength* of the test.] To do this can be regarded as an example of a Bayes/Neyman-Pearson compromise. Some numerical examples of this procedure are given by Crook & Good (1981).

Surprise

The evolutionary value of surprise is that it causes us to check our assumptions (p. 1131 of #82). Hence if an experiment gives rise to a surprising result, given some null hypothesis H, it might cause us to wonder whether H is true even in the absence of a vague alternative to H. It is therefore natural to consider whether a statistical test of H might be made to depend upon some "index of surprise."

There is one such index due to Warren Weaver (1948), and a generalization in ##43, 82. Weaver proposed the index $(\Sigma p_i^2)/p$ where p_i runs through the probabilities of mutually exclusive possible outcomes of an experiment (or observation) and p denotes the probability of the event that actually occurred. This index can be written in the form

$$\lambda_1 = \mathcal{E}(\tilde{p})/p$$

where $\tilde{p}$ denotes the random variable whose possible values are the probabilities of outcomes and $\mathcal{E}$ denotes mathematical expectation. (When there is a continuous infinity of possible outcomes one can of course use probability densities in place of probabilities, but the index is not invariant under transformations of the independent variable. It is invariant under linear transformations.) This form for λ_1 shows that it can replace the microscopic probability of the outcome by a number of more reasonable magnitude. For example, if part of the outcome of spinning a coin is that it lands on a certain point on the floor (which we regard as irrelevant to whether the coin is biased) the microscopic probability of this irrelevance will cancel out when the ratio λ_1 is computed. The surprise index depends on how the outcomes are grouped, especially when there is a continuous infinity of possible outcomes.

As a generalization of λ_1, ##43, 82 proposed

$$\lambda_u = (\mathcal{E}p^u)^{1/u}/p \quad (u > 0)$$

and the limiting form ($u \to 0$)

$$\lambda_0 = \exp \mathcal{E}(\log \tilde{p}) - \log p$$

which is the ratio of the "geometrical expectation" of $\tilde{p}$ to p. If we write $\Lambda_u = \log \lambda_u$, then Λ_u is additive if the results of several statistically independent experiments are regarded as a single experiment whereas the indexes λ_u are multiplicative.[9] An additive property seems more natural. In particular

$$\Lambda_0 = \mathcal{E}(\log \tilde{p}) - \log p$$

and this can be interpreted as the amount of information in the event that occurs minus the expected amount (p. 75 of #13). The most natural of the surprise indexes[10] are λ_1, Λ_1, λ_0, and Λ_0. Moreover Λ_0 has the desirable property that its expectation is zero, that is, the expected amount of surprise vanishes if the null hypothesis is true. (Compare Bartlett, 1952.)

If we prefer not to assume anything about the non-null hypothesis, then a surprise index can sometimes reasonably be used for deciding whether the null hypothesis is suspect, or whether it is supported (as when Λ_0 is negative). Surprise indexes are not yet in common use in statistical practice, but perhaps they should be. One could also make a surprise/Fisher compromise by using the tail-area probability of a surprise index as a criterion for testing a null hypothesis. But sometimes a surprising event is regarded as "merely a coincidence" because we cannot think of any reasonable alternative to the null hypothesis.

I have here ignored some problems of definition of surprise that are related to the meaning of "simplicity." The matters are discussed to some extent in #82.

The Bayesian Influence on Significance Testing

The coherence of the Bayesian philosophy, especially in its Doogian form, can shed light upon and can influence non-Bayesian procedures.

For example, Bochner (1955) once asked me "Why use X^2 [for testing a multinomial null hypothesis H]?" My reply was that, under a wide variety of Bayesian models, the weight of evidence against H is approximately a linear function of X^2 when the weight of evidence is not large. This provides a rough justification for combining independent X^2's by taking a linear combination of them (because independent weights of evidence are additive).

Another example is a proposed procedure for combining several tail-area probabilities P_1, P_2, P_3, . . . that are obtained by distinct significance tests based upon the same data. An informal Doogian argument leads to the rule of thumb of regarding the harmonic mean of P_1, P_2, P_3, . . . as a resultant tail-area probability provided that nothing better comes to mind (#174).

4. EXPLICATIVITY

So far in this article the emphasis has been on whether a hypothesis is probable, but the selection of a hypothesis depends also on its utility or on a quasiutility such as its power to predict or to explain. If this were not so we would always prefer a tautology such as 2 = 2 to any more informative hypothesis. An analysis that allows especially for explanatory power is given in ##599 and **1000**. The analysis introduces a concept called "explicativity" which leads to a sharpened form of the Ockham-Duns razor and which is found to shed light both on significance tests and on problems of estimation.

NOTES

1. I am in the habit of calling non-Bayesian methods "orthodox" because for some years after World War II I was almost the only person at meetings of the Royal Statistical Society to defend the use of Bayesian ideas. Since such ideas are now much more popular it might be better to call non-Bayesian methods "sampling theory" methods and this is often done in current statistical publications. This name covers the use of tail-area probabilities ("Fisherian") and probabilities of errors of the first and second kinds (Neyman-Pearsonian).

2. Intensities of conviction are often called "degrees of belief" but in 1946 or 1947 Gilbert Ryle rejected an article of mine, for *Mind* (containing an outline of #13), partly on the grounds that you either believe something or you do not (and also because the article was somewhat mathematical). He also described the interpretation of $P(A|B)$ in terms of degrees of belief, when B does not obtain, as "make belief." If I had used the expression "intensity of conviction" this joke would have been unavailable to him.

3. By the time a pure likelihood man recognizes that initial probabilities must be allowed for, he calls them initial likelihoods or initial supports instead so as to avoid admitting that he has become a Bayesian!

4. It is intuitively obvious to a layman of not to a Lehmann. Lehmann (1959) made a notable contribution to the sampling-theory approach to significance tests. My wisecrack is not really fair to Lehmann because he recognizes the value of Bayesian decision theory, but there is no reference to optional stopping nor to the law of the iterated logarithm in the index of his book, so presumably his book does not mention the danger of optional stopping in a Fisherian context.

5. It might be thought that binomial sampling, with fixed sample size n, leading to r "successes," would lead to a different tail-area probability, for testing the hypothesis of a specified value for p, than if "inverse binomial sampling" were used (in which r is fixed and sampling is done until r successes are achieved). It turns out, curiously enough, that, if r and n are unchanged, then the tail-area probabilities are equal for the two methods of sampling. This can be inferred from Patil (1960, p. 502). But the unbiased estimates for p are unequal, being r/n for binomial sampling and $(r - 1)/(n - 1)$ for inverse binomial sampling. (See, for example, Kendall and Stuart, 1960, p. 593.)

6. Perhaps an acceptable prior, given $\overline{H}$, would be a mixture of beta distributions scaled to lie in the interval (½, 1). This prior would be similar to the one used in the hierarchical Bayes technique (#862). The implications for Schmidt's data could be worked out when the data are available.

7. Gordon Tullock (1979) suggests that the volunteer might receive a conditional reward in a "double-blind" experiment; that is, the reward would depend on whether the treatment was later revealed to be the more favorable or the less favorable one. It would also sometimes be possible to determine the compensation depending on the effectiveness of the treatment for the particular patient.

8. In both the astronomical examples there is collateral information from the satellites of the planets. But in Bernoulli's problem the evidence was already overwhelmingly strong.

9. The expression $\Lambda_u + \log p$ is sometimes called Rényi's generalized entropy (Rényi, 1961) but he was anticipated in 1954, as explained in the text.

10. The indexes Λ_1 and Λ_0 are closely related to the quadratic and logarithmic fees that encourage honesty and "objectivity of judgments." These were defined respectively and independently by Brier (1950) and in #**26** and there is now a large literature on the topic. See, for example, #690A, Savage (1971), and Hendrickson & Buehler (1972).

CHAPTER 15

Explicativity, Corroboration, and the Relative Odds of Hypotheses (#846)

INTRODUCTION

"Momma do you think it's proper; how
did you react to Poppa?"
Popular song.

In this paper I shall discuss probability, rationality, induction and the relative odds of theories, weight of evidence and corroboration, complexity and simplicity (with a partial recantation), explicativity, predictivity, the sharpened razor, testability and metaphysicality, and gruesomeness.

I agree with some of the things that Popper has said about several of these topics, in his stimulating writings, but I by no means have a Popper fixation.

I. THE PHILOSOPHY OF PROBABILITY AND RATIONALITY THAT I ACCEPT

A. The Shifting Meaning of "Bayesian"

I used to call myself a Bayesian when it was not misleading to do so. I have not changed my position, but the meanings of words change. As Winston Churchill said, his political opinions had not changed when he changed his political alliance: it was the political parties that were marching out of step, or words to that effect. If one could imagine a space of statisticians, in 1950 there were so few Bayesians that they clearly formed a cluster including myself, as in Figure 1a (which, however, is not of course complete). But in 1973 the picture is more like that shown in Figure 1b, where it would be less misleading to say that I represent a Bayes/non-Bayes compromise. The meanings of words are often determined by clusters in abstract spaces.

At last count there were 46656 varieties of Bayesians (#765) so I shall have

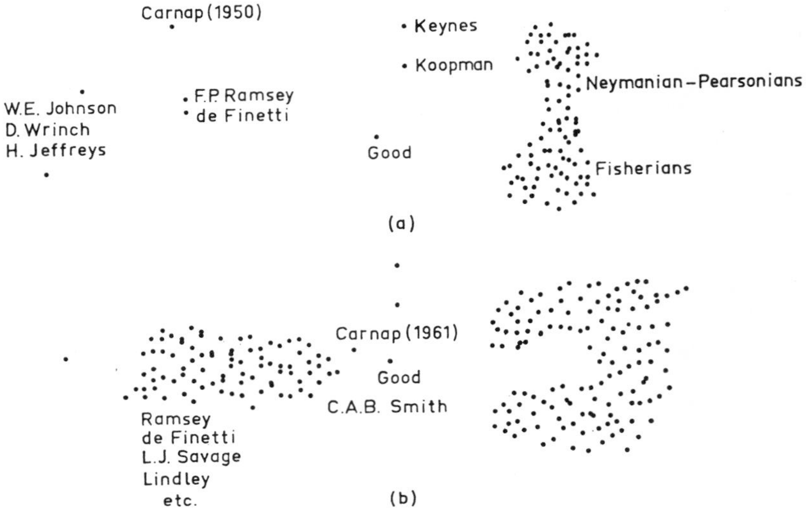

Figure 1. Statistician's space: (a) 1950, (b) 1973.

to describe to some extent where I stand. I shall try to be succinct because there are several other topics to discuss. (Previous descriptions of my philosophy are given, for example, in ##13, **26**, **230**, 290, 398, 522, 547, 679, **838**.)

"Bayesian" is not the only ambiguous word that might give trouble during this conference. "Probability" has at least five meanings (for example, ##**182**, 398, 522, p. 16 of #750), and the expressions "subjective probability" and "logical probability" or "credibility" both need clarification.

B. Subjective and Logical Probabilities

I use the expression "subjective probability" in the sense of "personal probability" if an attempt at coherence or consistency with axioms has been made. For a purely snap judgment of a degree of belief or an intensity of conviction, without such an attempt at coherence, I use the expression "psychological probability," the kind that has been investigated, for example, by John Cohen in children. By "logical probability" or "credibility" (an expression used by F. Y. Edgeworth and by Bertrand Russell) I mean the *unique* rational belief, if it exists. Carnap (in Carnap and Jeffrey, 1971, p. 13) said that what I call "subjective probability" would be better called "rational credibility," but since it depends on the person or group of people making the judgment, whom I call "you," I think "subjective" or "personal" (or multisubjective or multipersonal) is a better term, although *logico-subjective* (or *logico-multisubjective*) would be even better apart from its clumsiness. (The differences between the upper and lower probabilities, that is, the widths of the intervals of estimation, might be much larger for groups than for individuals.) There is a continuous gradation from psychological probability to logical probability. Whether logical probability exists or not it is an ideal to hold in mind like absolute truth. Probabilities might be strictly logical and constructive relative to a precisely formulated and appropriately chosen language and then only if exact rules could be given for computing them, as was Carnap's original program (Carnap, 1950). Between 1950 and 1961 Carnap moved close to my position in that he showed a much increased respect for the practical use of subjective probabilities.

In my opinion both subjective probabilities and logical ones should be based on external reality and not just on an arbitrary language (cf. p. 48 of #13). In so far as they are based on a language, the choice of this language is likely to be largely subjective or multisubjective in practice. In this choice of a language a large measure of rationality is possible in principle, because, as a guiding principle in its selection, we could try to make it as economical as possible for expressing what we would like to express (cf. p. 24 of #617). There are reasons for believing that languages have some measure of efficiency; for example, this assumption leads to an economical economic explanation for the approximate validity of the Zipf law of distribution of word frequencies (Mandelbrot, 1953; #130; p. 575 of #524, where it is mentioned that Zipf's law "is unreliable but is often enough a good approximation to demand an explanation"). It should be held in mind that part of the description of a language is its statistical properties which unfortunately dictionaries do not adequately supply. Carnap ignored this point as far as I know. If a word is used with frequency 10^{-12} it might be difficult to decide that it definitely belongs or definitely does not belong to the language. It is simpler to say that its frequency of use in the language (or in some dialect at some place and time and in some context) is 10^{-12} and leave it at that.

Languages and sublanguages have evolved during the course of billions of man-hours of usage, and have adaptively achieved some measure of efficiency. So it is not surprising that they are helpful for your judgments of initial probabilities of theories although they are not yet as helpful as we should like them to be. Note that the probability that the next book you see will be blue is not equal to the reciprocal of the number of names of colors, but is closer to the frequency of use of the word "blue" in English, when referring to a color. Credibilities based on artificial langauges can be much worse. Certainly artificial languages can be designed for the special purpose of being misleading. George Orwell (1949) emphasized this point and Goodman's paradox of grue and bleen demonstrates it. I shall return to the gruesome paradox later.

In one of his attacks on subjective probability, Popper argued that it was impossible to generate knowledge from ignorance (Popper, 1957). But it seems to me that this is precisely what mammals have been doing for the last billion years, and I even once heard Popper state that science is based on a swamp. Just as language develops adaptively, both in men and in man, in order to cope with reality, there is no reason why probability judgments should not develop in much the same way. So it seems to me that Popper has been inconsistent.

One cannot infer probabilities by using the axioms of probability unless one starts with some probabilities or functions of probabilities as input, and these have to be based on judgments. One example of a judgment that the frequentist will be prepared to make is that of a probability based on a large sample. Then we can use Bayes's theorem in reverse so as to obtain inequalities relating his implicit initial probabilities, the existence of which he would like to whistle away. No amount of whistling will remove them. This use of Bayes's theorem in reverse is a legitimate method of inference and is especially effective when combined with the Device of Imaginary Results (##13, 398, 547).

There is no need of magic here. Even an intelligent robot would be advised to adopt a subjectivistic theory of probability and rationality, in the interests of self-preservation. More precisely, the robot, like a man, should behave *as if* it adopted these theories as an approximation. The reason magic might seem to be present is that we do not know how judgments are made; if we did we would not call them judgments but inferences. Judgments depend on complicated neural circuits and half remembered information and not on known simple algorithms (#183; p. 115 of #411). If we knew the structure of these circuits the engineers would call them "logic circuits" and judgments would have been reduced to logic. Until such time it seems better to accept subjectivism as a practical necessity.

Subjectivism is bridled. Various sets of desiderata or constraints have been shown to lead to essentially the usual axiomatics of the theories of probability and rationality. (See, for example, F. P. Ramsey, 1931; C. A. B. Smith, 1961.) We all have to make subjective probability judgments but the person who recognizes this clearly enough is prepared to constrain his judgments so that they tend to satisfy a certain set of axioms. In short his judgments are to that extent more likely to have a measure of objectivity than those of the objectivist who refuses to allow his subjective judgments to be so constrained. Honest objectivism leads inevitably to subjectivism. Contrary to Agassi's thesis, it is the *denial* of the need of subjectivism, not its *acceptance* that is the chronic illness.

C. Brief Outline of My Philosophy

I believe that, of the various interpretations of probability, the most operational, the one closest to action, is subjective or personal probability, for it enables us to extend ordinary logic into a useful general-purpose system of reasoning and decision-making. Moreover I believe that a theory of subjective probability is necessary as a reasonably consistent basis even for the measurement of physical probabilities. (The distinction between physical and subjective probability was emphasized, for example, by Poisson, 1837, and Carnap, 1950.) (See also ##**659, 838**.)

I believe that subjective probabilities and utilities are only partially ordered. This means that a comparison can sometimes be made between a pair of probabilities or a pair of utilities but not necessarily always. This comes to the same thing as ascribing an interval of values to each probability or utility, and thus an upper and a lower probability, an upper and a lower utility, and an upper and a lower expected utility. This theory differs from the currently more popular one of sharp probabilities and utilities. I reserve historical comments until later.

To be more precise it is better to describe my philosophy as an adherence to the "black box theory of probability and rationality." The black box is supposed to contain the axioms of the subject; it has an input that consists of inequalities between probabilities, probability ratios, utilities, expected utilities, etc. These inequalities constitute a "body of beliefs." The output consists of *discernments.* By a "discernment" I mean a judgment that becomes compulsory once it has been deduced. The inside of the black box is supposed to operate with sharp probabilities and utilities although the judgments and discernments are only partially ordered. The advantage of this theory is that it is dead realistic yet as simple as possible because it is based on a more or less classical set of axioms while not being committed to the notion of sharp judgments of probabilities and utilities. The purpose of the theory is to enlarge the body of beliefs and to detect inconsistencies in it, whereupon the judgments need revision (##13, 26, 230). In this respect it resembles Aristotelean logic.

The principle of rationality of Type I is the recommendation to maximize expected utility. The principle of rationality of Type II is the same except that the costs of theorizing and calculation are allowed for. The Type II principle of rationality is not consistent but at any given moment it should appear reasonably consistent to you. The notion of an inconsistent logic is not one that logicians support as yet. But if we imagine a futuristic robot making use of a subjectivistic theory it is obvious that all it can do at any given moment is to use the judgments that it has made up to that moment. I can see no reason to suppose that a person could do any better without divine guidance. This is a sufficient reason for calling my theory subjectivistic rather than credibilistic. But I regard it as mentally healthy to think of subjective probabilities as estimates of credibilities although I am not sure that credibilities exist. It's like half-believing in God. One might say that a belief in credibilities constitutes a religion, whereas subjective probabilities emerge more directly from ordinary experience.

D. Historical Comments

J. M. Keynes put forward a theory of partially ordered logical probabilities whereas F. P. Ramsey dealt with sharp subjective probabilities. B. O. Koopman proposed a theory of partially ordered so-called "intuitive probabilities," which could be either logical or subjective, but I think they were intended to be logical. Koopman was not concerned with utilities. The theory of Dorothy Wrinch and Harold Jeffreys was one of sharp credibilities, and again utilities did not come into the argument. My philosophy can be regarded as based on a combination of those of J. M. Keynes and F. P. Ramsey. In the interests of simplicity I often talk about sharp probabilities and utilities, and then I justify it by means of the Type II principle of rationality. That is, I make a high order judgment that in some circumstances this use of sharp probabilities and utilities is *good enough* for the application, in fact that it roughly maximizes expected utility because it avoids complexities with little loss of realism.

de Finetti showed that a consistent set of sharp probabilities imply, in certain circumstances, that the probabilities could be expressed as if they were subjective expectations of apparent physical probabilities. It is impossible to disprove solipsism. Therefore de Finetti's important theorem *had* to be true! He never says that physical probabilities exist but that a consistent subjectivist would act in certain respects as if they did. In my opinion one might just as well assume that the physical probabilities exist as well as the subjective ones. After all, Ramsey's approach to the axioms, as elaborated by L. J. Savage and others, also shows that a rational person will act *as if* he had subjective probabilities and utilities. To quote from #617, "Thus physical probabilities can be described as metaphysical in the same sense that the postulate of the existence of the external world is metaphysical. All kinds of probability are metaphysical but some are more metaphysical than others. Subjective probabilities involve a single "as if" whereas physical probabilities involve two "as if's." . . . Physical probabilities are metametaphysical."

Perhaps *all* our concepts are no more than "as if" in which case we might just as well drop the as if's in discussion (**#838**). If a concept can *always* be used in an "as if" way then one might as well say that the concept is "real." Perhaps this is an adequate definition of reality. The reason it sounds like an inadequate definition is that in ordinary usage when we use the expression "as if" we mean than there *are* exceptions. In other words, an object or a concept can be said to be real or to exist if *for all purposes* it is as if it were there.

Popper's main criticism of theories of subjective probability is that they are incapable of expressing physical independence (Popper, 1957). I think he overlooked de Finetti's theorem.

My criticism of the permutability assumption is that it would be abandoned if the evidence showed it should be, and this means that it was not held in the first place except as an approximation. But I have little objection to approximations. [For the meaning and history of "permutability" see #398, p. 13.]

E. Evolving or Dynamic Probabilities

[See #938.]

II. COMPLEXITY: A RECANTATION

In #599 I defined the complexity of a proposition H as the amount of information in it, namely $I(\mathrm{H}) = -\log P(\mathrm{H})$. I wish to withdraw this definition of complexity, although I still accept the definition of information (p. 75 of #13; #505). It disturbed me at the time that the proposition 0 = 1, which looks simple, should be said to be infinitely complex, although it does imply any other proposition however complex. My definition was further weakened when

K. S. Friedman (1973) pointed out an objection to my postulate that the complexity of H · K is greater than that of H. He showed that it leads quickly to the conclusion that A v B is simpler than A, which is counterintuitive. What I should have said is that H · K is more complex than H when H and K are entirely independent propositions. In fact it seems very reasonable to assume that in this case

$$\kappa(H \cdot K) = \kappa(H) + \kappa(K).$$

But if $\kappa(H)$ is some function f of the probability of H it follows that identically

$$f(xy) = f(x) + f(y)$$

from which it follows that $f(p)$ must be proportional to log p, and we are back to the previous unsatisfactory definition in which 0 = 1 is infinitely complex. I conclude that the *complexity of a proposition cannot be defined in terms of its prior probability alone.* Fortunately this has very little effect on my paper on the sharpened razor (#599).

Valéry (1921, and p. 109 of the 1968 printing) called a figure "geometric" if it can be traced by motions which can be expressed in a few words. (I wonder whether he got this idea from Lemoine, 1902, referenced by Coolidge, 1916, p. 170.) He is here explicating the concept of simplicity. It must have been obvious for centuries that the complexity of a statement had *something* to do with the length of that statement and clearly this definition can be sharpened by describing the complexity in terms of the smallest number of words that could be used for making an equivalent statement. This definition depends on the language used but often we want to talk about the complexity *in the real world* of a hypothesis, rather than in some specific language. So, just as before, the quantitative definition of complexity depends on the choice of an economical language for describing the world or some field of interest.

One consequence of this approach, even before we have identified the appropriate language, is that if we have two theories that are totally unrelated placed end to end we are forced to say that the complexity of the conjunction of the two theories is the sum of the separate complexities as I assumed before.

Perhaps the best plan is to define the complexity of a theory, not as minus the logarithm of the prior probability that the theory is true, but as minus the logarithm of the probability that the linguistic expression that describes the theory would occur in the language when the statistical properties of the language are specified up to say di-word frequencies. Moreover we must take the largest possible value of this probability by an appropriate linguistic transformation. Since the word "and," or the corresponding symbol in a formal logic, say an ampersand or period, naturally has a high probability, it will still be approximately true that the complexity of a conjunction of two theories that have nothing to do with one another is approximately the sum of the complexities, or slightly greater because of the symbol for conjunction. Likewise, a proposition and its negation have approximately equal complexities, for example $0 \neq 1$ and $0 = 1$ are about equally simple.

But if we adopt this definition of complexity is it still true that we like a theory to be simple? (More precisely, of two theories that explain the same facts, the simpler is preferable.) The answer is that it *is* still true but it is more important that its probability should be high *(but see Section V)* rather than that the maximum probability of its linguistic expression should be high. Although there is obviously a rough relationship between these two probabilities they are not equal. The probability of its linguistic expression is more closely related to brevity of the statement than is the probability of the theory itself. A very easy way to see this is by considering a theory and its negation, H and $\bar{H}$. They take about the same number of words to assert yet one of them might have a probability close to 0 and another close to 1. We would not wish to regard both of these as equally good theories though they have nearly equal complexities (or simplicities).

Apparently then there is no clear-cut relationship between probability and complexity when negations or disjunctions are allowed. As far as I can see the best prospect of establishing a relationship would be to restrict our attention to *conjunctions* of propositions that are in some sense atomic propositions. (For a definition of atomic propositions, see, for example, Good, 1952.) This raises the question of whether scientific theories can be expressed without using the words "not" or "or" or their equivalents. (This is reminiscent of an "affirmative logic" of D. van Dantzig.)

A relationship between the probability and the complexity of a theory or hypothesis H is that if H is very complex then its initial probability is low and the probability tends to zero when the complexity tends to infinity. But it is also possible for a theory to be simple and improbable, the theory that 0 = 1 being an extreme example. Brevity is the soul both of wit and of high probability but is not a sufficient condition for either.

What does this do to Ockham's razor? Ockham's razor is sometimes expressed in the form that of two hypotheses, both of which explain the same facts, the simpler one is to be preferred. In accordance with what I have just said it would be more appropriate to say that the more *probable* hypothesis is to be preferred in these circumstances. In this form the razor would contradict Popper, who likes improbable theories, but a compromise can be reached as we shall see later.

III. THE PROBABILITIES AND THE RELATIVE ODDS OF THEORIES

Perhaps the true fundamental laws of physics are infinitely complex, but it would be dogmatic to regard this as certain. (In the discussion David Miller acutely pointed out that Gödel's work shows that even arithmetic is infinitely complex in that an unlimited number of independent axioms can be introduced. My reply was that, with non-zero probability, the fundamental laws of physics could be formulated without reference to Gödel. Moreover one could argue that Gödel's proof describes, in finite terms, the recipe for constructing the transfinite sequence of new axioms. In effect, each new axiom can be expressed "the axioms that have been mentioned previously form a self-consistent system" so that the "infinite complexity" can be collapsed into the phrase "etcetera, etcetera, etcetera.") Let us denote the logical or subjective probability that they are infinitely complex by c. Then $1 - c$, which exceeds 0, is the total probability of all the mutually exclusive theories that are not infinitely complex. These can each be expressed in a finite number of words, therefore there are not more than a countable infinity of them. Thus their probabilities cannot all be zero, because the sum of a countable (enumerable) number of zeros is zero (compare Wrinch and Jeffreys, 1921; #191). This refutes Popper's claim that, in an infinite universe, the probability of a fundamental law is always zero (Popper, 1959, p. 363). Popper apparently believes that the universe is an unbreakable cypher. It might be or it might not be: there is hardly any evidence either way.

Similarly each fundamental dimensionless constant of nature, if it *is* a constant, has *a priori* a non-zero probability of being any given computable number (p. 55n of #13). It also has a non-zero probability of being non-computable, in which case its numerical value will never be precisely specifiable.

Once it is conceded that universal laws cannot all have zero initial probabilities, there is no reason for picking on any self-consistent one of them and saying that it does have zero probability. The only sensible conclusion is that all those laws that are not self-contradictory have positive probability.

Against this thesis it might be argued that languages evolve and the meanings of words change, so that there is no clear-cut language for expressing all the laws. But I don't think it affects the thesis. The new words and forms of speech can either be defined in terms of the old language or can be incorporated into an axiomatic scheme together with rules of application. I think the English of the fifteenth century would be more than sufficient, when combined with definitions and axioms, to describe the whole of modern science. The concepts of science, even when new, are not so new as to be indescribable, as if they were mystical experiences.

On the other hand, since all science and all language is based on a swamp it may never be possible to express any idea so that the meaning is completely unambiguous. Otherwise we'd say that Adam was the son of a monkey, and that either the egg or the chicken came first. From this point of view there are obvious difficulties in pinning down what is meant by the probability of a theory. We cannot expect to get a sharp value for the probability if we cannot obtain a sharp meaning for the statement of the theory. But as far as I know it is not this semantic problem that Popper regards as underlying his thesis that general theories have zero probabilities. It would not be appropriate to say that a proposition had zero probability merely because its meaning was not completely clear.

Even apart from the semantic difficulty we do not yet have adequate formal procedures for fixing sharp initial probabilities for theories. The probabilities are small and are difficult to judge (compare **#603B**), but I claim most of us have implicit judgments about lower bounds for the initial probabilities that some theories are true, for example that chromosomes exist (in our world). Otherwise we would not accept them on the basis of the experimental evidence. I shall return to this point when discussing induction. (See also de Finetti, 1971.)

Some vague rules can be given. The initial probability of a theory will tend to be small if its statement is long or if it has many probabilistically independent adjustable parameters (Jeffreys, 1961, p. 246). Also the initial probability depends on what competitive theories there are. But it is impracticable to consider all the theories, even the reasonably simple ones, in advance of experiments, owing to the "combinatorial explosion." By this I mean that the number of statements of length n is asymptotically of the form e^{kn} and so is exponentially large for large n. We cannot even list all statements ten words long. So we often look at the experimental results first, and then formulate hypotheses. We usually test them by further experiments, but farfetched hypotheses would need extra strong corroboration because *to say that they are farfetched is to say that their initial probabilities are low.*

One can make sharper judgments about the relative odds of two hypotheses concerned with the same subject matter than about the separate probabilities. This helps to explain why we sometimes consider hypotheses or theories in pairs (pp. 66, 83-84 of #13). Suppose that H_1 and H_2 are two mutually exclusive hypotheses and let

$$H_3 = \overline{H_1 \vee H_2}.$$

Then H_1, H_2, and H_3 form a set of mutually exclusive and exhaustive hypotheses. Let their initial probabilities, given some background assumptions, which I take for granted, be p_1, p_2, and p_3 $(p_1 + p_2 + p_3 = 1)$. Now it easily proved that

$$P(H_1 | H_1 \vee H_2)/P(H_2 | H_1 \vee H_2) = p_1/p_2.$$

Therefore, if $H_1 \vee H_2$ is denoted by G, we have,

$$O(H_1 | G) = p_1/p_2$$

where O denotes odds. (The odds corresponding to a probability p are defined as $p/(1-p)$.)

Now de Finetti (1968/70) has claimed that it is easier to judge the probability of an event E than of a statistical hypothesis (indeed that the latter judgment cannot be made). This of course is consistent with his disbelief in the reality of physical probabilities, but it may often be good advice in any case. Let us consider how we can use a judgment of $P(E)$ to infer a value for p_1/p_2. We must assume that $P(E)$ is well approximated by $P(E|G)$ otherwise we could not take G for granted. So

$$P(E) \approx P(E|G) = P(E|H_1)P(H_1|G) + P(E|H_2)P(H_2|G).$$

Now $P(E|H_1)$ and $P(E|H_2)$ are known tautologically, at least when H_1 and H_2 are simple statistical hypotheses, and in any case are often easier to judge than

p_1/p_2. So we now have a linear equation for $P(\mathrm{H}_2)$, which is $p_1/(p_1+p_2)$. Thus we arrive at a value for the *relative odds*

$$p_1/p_2 = O(\mathrm{H}_1/\mathrm{H}_2) = O(\mathrm{H}_1/\mathrm{H}_2 | \mathrm{H}_1 \vee \mathrm{H}_2).$$

This one way of judging p_1/p_2 without judging p_1 itself.

But my own view is that is sometimes less difficult to make a direct judgment of p_1/p_2 than of $P(\mathrm{E})$. I shall return to this matter when answering one of Agassi's arguments of 1960.

IV. WEIGHT OF EVIDENCE EXPLICATES CORROBORATION

When Peirce (1878) used the expression "weight of evidence" in a technical sense, though in a popular article, he was obviously talking about corroboration. He did not express it symbolically and made rather heavy weather of it. [In fact, he got it slightly wrong. See #1382.] In succinct modern notation, the weight of evidence in favor of H provided by E given G is

$$W(\mathrm{H}:\mathrm{E}|\mathrm{G}) = \log O(\mathrm{H}|\mathrm{E}\cdot\mathrm{G})/O(\mathrm{H}|\mathrm{G}),$$

the logarithm of the factor by which the odds of H are multiplied when E is observed, given G as background information. Turing (1941) called this factor "the factor in favor of H." Since, by two applications of Bayes's theorem, this factor is seen to equal $P(\mathrm{E}|\mathrm{H}\cdot\mathrm{G})/P(\mathrm{E}|\overline{\mathrm{H}}\cdot\mathrm{G})$ it may also be called the *Bayes factor* in favor of H (provided by E given G), or perhaps the Bayes-Jeffreys-Turing factor. The theorem that this factor was equal to the probability ratio, or simple likelihood ratio, was mentioned by Wrinch and Jeffreys (1921, p. 387). Weight of evidence was called "support" by Jeffreys (1936), but in his book (Jeffreys, 1939/61) he dropped this term because he always assumed that the initial probability of H was 1/2, that is, that the initial odds were 1, so that the weight of evidence was always equal to the final log-odds of H. Turing used the term *deciban*, by analogy with the acoustic term decibel, for a unit of weight of evidence, and we called weight of evidence "decibannage," a clumsy term. We also called it the *log-factor*. The concept can be generalized (#599) to the weight of evidence in favor of H_1 as compared with H_2,

$$W(\mathrm{H}_1/\mathrm{H}_2:\mathrm{E}|\mathrm{G}) = \log\frac{O(\mathrm{H}_1/\mathrm{H}_2|\mathrm{E}\cdot\mathrm{G})}{O(\mathrm{H}_1/\mathrm{H}_2|\mathrm{G})} = \log\frac{P(\mathrm{E}|\mathrm{H}_1\cdot\mathrm{G})}{P(\mathrm{E}|\mathrm{H}_2\cdot\mathrm{G})},$$

an equation that is true even if H_1 and H_2 are not mutually exclusive.

The concept of weight of evidence was central to my first book (#13) and occurred also in at least 32 other publications (Good 1954, 1965; ##174, 191, 210, 211, 221, **223B**, 245, 315, 397, 398, **508**, **518**, 524, 541A, 547, 570, 574, 599, **603B**, 618, 622, **659**, 690A, 699, 701, 705, 755, 795, 798, 810). What I say thirty-three times is true.

Kemeny and Oppenheim (1952), in a paper I overlooked, introduced the expression *factual support for a hypothesis* (provided by evidence). They laid down a number of desiderata and from these desiderata arrived uniquely at the explicatum

$$\frac{P(\mathrm{E}|\mathrm{H}) - P(\mathrm{E}|\overline{\mathrm{H}})}{P(\mathrm{E}|\mathrm{H}) + P(\mathrm{E}|\overline{\mathrm{H}})}.$$

In my notation this is $\sinh\{W(\mathrm{H:E})/2\}$.

Popper (1954; 1959) also assumed a variety of desiderata for corroboration and showed his desiderata were self-consistent by exhibiting at least two formulae, in terms of logical probabilities, that would satisfy them. He says (1959, p. 394) "I regard the doctrine that the *degree of corroboration or acceptability cannot be a probability* as one of the most interesting findings of the philosophy of knowledge." This finding had been taken for granted by some earlier writers.

Minsky and Selfridge (1961) again independently used the expression "weight of evidence" in the same sense that it had been used by . . . myself. . . . The expression was used by J. M. Keynes (1921, p. 71) in a less satisfactory sense, to apply to the total bulk of evidence whether any part of it supports or undermines a hypothesis, almost as if he had the weight of the documents in mind.

Without knowing of the work of Kemeny and Oppenheim, I took Popper's desiderata, modified them a little and was able to determine all the formulae that would satisfy the modified set (#211). The most convenient of these is weight of evidence. Although my arguments were quite different from those of Kemeny and Oppenheim it is interesting that the weight of evidence is a monotonic function of their explicatum, namely $\log[(1+x)/(1-x)]$. It has the additional convenience of having the additive property,

$$W(\mathrm{H:E\cdot F}) = W(\mathrm{H:E}) + W(\mathrm{H:F|E}).$$

In particular if E and F are independent pieces of evidence then we get ordinary additivity. Somewhat more compelling desiderata are given in #599. The additive property makes it reasonable to regard weight of evidence as a quasiutility, whose expectation might reasonably be maximized when true utilities are difficult to judge. But we shall meet other quasiutilities.

It would take too long to describe the methods of arriving at weight of evidence as the best explicatum of corroboration along the lines of #211 and some of #599. There is, however, a short argument that can be given (p. 127 of #599). Assume that corroboration is a real function of $P(\mathrm{E}|\mathrm{H})$ and $P(\mathrm{E}|\overline{\mathrm{H}})$ that, together with $P(\mathrm{H})$, mathematically determines $P(\mathrm{H}|\mathrm{E})$. Then it must be a monotonic function of weight of evidence. This would give a wide variety of possible explicata but for convenience we might as well use the weight of evidence itself because of its additive property.

Weight of evidence can be expressed in a third way, namely in terms of "amount of information" $I(\mathrm{H:E|G})$ (for example, p. 126 of #599). This is defined by

$$I(\mathrm{H}:\mathrm{E}|\mathrm{G}) = \log \frac{P(\mathrm{E}|\mathrm{H}\cdot\mathrm{G})}{P(\mathrm{E}|\mathrm{G})},$$

and we have

$$W(\mathrm{H}:\mathrm{E}|\mathrm{G}) = I(\mathrm{H}:\mathrm{E}|\mathrm{G}) - I(\bar{\mathrm{H}}:\mathrm{E}|\mathrm{G})$$
$$W(\mathrm{H}_1/\mathrm{H}_2:\mathrm{E}|\mathrm{G}) = I(\mathrm{H}_1:\mathrm{E}|\mathrm{G}) - I(\mathrm{H}_2:\mathrm{E}|\mathrm{G})$$
$$I(\mathrm{H}:\mathrm{E}\cdot\mathrm{F}|\mathrm{G}) = I(\mathrm{H}:\mathrm{E}|\mathrm{G}) + I(\mathrm{H}:\mathrm{F}|\mathrm{E}\cdot\mathrm{G}).$$

Various other properties of weight of evidence are mentioned in the listed references.

As six examples of its use I mention here

(i) For evaluating the hypothesis of gravitational frequency shift by using the Mössbauer effect (Good, 1960).

(ii) For evaluating the evidence in favor of General Relativity as compared with Newtonian physics (in two forms) by the observations of the deflection of light by the gravitational field of a star. The calculations could be done without difficulty if they have not yet been done, provided that the "law of error" is known. The calculations of weights of evidence from (i) and (ii) do not of course determine the final odds of General Relativity as compared with Newtonian physics. We need to take into account all other evidence and also the relative odds of the two theories. In my opinion the relative initial odds in favor of Newtonian physics do not exceed 10000 whereas the factor against it is far greater than 10000. (I believe that the relative initial odds are much less than 10000 but I have used an extravagantly large value to indicate how little accuracy is needed. I suspect that much of the opposition to inductivism arises through overlooking this point, and for this perhaps some of the blame should be ascribed to [those] inductivists who insist on the use of sharp probabilities.) Therefore, in my opinion, General Relativity has very heavy odds on *as compared with* Newtonian physics. I believe most physicists would agree with me once they understood what I am saying.

(iii) For medical diagnosis (##755, 798).

(iv) For computing the probability that a person will contract lung cancer; for example, a heavy smoker with a morning cough scores 70 centibans, and if he lives in a highly urban area he scores another 32 cb, a total of 102 cb, or roughly a factor 10 on the initial odds (#570).

(v) For weighing the evidence concerning authorship from the frequency of use of vocabulary (Mosteller and Wallace, 1964).

(vi) For an appealing proof and improvement of one of Shannon's coding theorems in communication theory (#574).

Now consider Nelson Goodman's hypothesis H_1 that emeralds are grue, meaning green until 1990 and blue thereafter. Compare this with the hypothesis H_2 that emeralds are green until *at least* the year 2000. Clearly,

$$W(\mathrm{H}_1/\mathrm{H}_2:\mathrm{E}|\mathrm{G}) = 0,$$

where G is our background knowledge and E asserts that all emeralds seen to date have been green. (I assume there is some other test for deciding whether a stone is an emerald.) Therefore,

$$O(H_1/H_2:F|G) = O(H_1/H_2|G),$$

the final (or posterior) odds are equal to the initial (or prior) odds. Now in Goodman's artificial language H_1 is about as easily stated as H_2. But (a) it is *physically* more complicated, and (b) it loses an enormous factor on its initial probability because of the arbitrary choice of the parameter 1990. So here we have an example where the weight of evidence is zero and a judgment of initial probabilities *must* be made, and moreover the example shows that simplicity of expression as a linguistic string does not necessarily provide high initial probability, especially in a language deliberately designed to mislead. (Compare Good, 1970, p. 113; and p. 24 of #617.)

V. EXPLICATIVITY AND PREDICTIVITY

There are at least three kinds of explanation:

(i) *Semantic explanation* or *elucidation.*

(ii) *Informative explanation* which increases the probability of an explicandum E.

(iii) *Purely theoretical explanation* which increases the *evolving* probability of an explicandum. [For more on evolving or dynamic probability see **#938.**]

All that needs saying about semantic explanation is that it is an aim of dictionary makers and of philosophical conferences. Scientific explanation is more concerned with informative and theoretical explanation, and I should like to try to elucidate the distinction between these two.

Suppose that we wished to explain cell division (mitosis) E and that we discovered it depended on some electrostatic field. This discovery H would provide a partial *informative* explanation, because it would be new relevant information that would increase the conditional probability of E, $P(E|H) > P(E)$. To give an example of a purely *theoretical* explanation I shall refer to a discussion I once had with Dr. Agassi in about 1960. What I shall say here is essentially the same as in p. 129 of #599, except that there I brought in notions of complexity or simplicity which are unnecessary.

Popper (1959) argued that useful theories are improbable ones. I disagreed with this and Agassi argued along the following lines: At one time the motions E of the planets seemed very improbable, yet these motions follow from the hypothesis H of the inverse square law of gravitation. (There were some discrepancies but I am ignoring them for the moment.) Therefore H must be even more improbable than E. To this I gave the following reply about seven years later (#599).

> $P(E)$ suddenly increased greatly, as judged by the astronomers of the time, as a consequence of Kepler's and Newton's calculations. I do

> not mean merely that $P(E|H) \gg P(E)$, but that $P(E)$, *now* judged to satisfy
>
> $$P(E) = P(H)P(E|H) + P(\bar{H})P(E|\bar{H}) > P(H)P(E|H) \approx P(H),$$
>
> is judged to be much larger, in ratio, than it had been judged to be before it was noticed that $P(E|H) \approx 1$ . . . . We are *forced* to the view that explanations depend on *evolving* probabilities when the explanation does not involve any new empirical observations. It is surprising in retrospect that this argument was ever overlooked.

That example should elucidate "theoretical explanation." (Agassi used the slightly better example of Fresnel's laws following from Maxwell's equations but I've used the gravitational example because of its greater familiarity. The discrepancies I mentioned would at first be ascribed to the influence of heavenly bodies not yet detected rather than to incorrectness of the inverse square law. For the implications of the inverse square law were extremely close to the observed orbits, even though some of the discrepancies were statistically significant. Allowing for these discrepancies, $P(E)$ was much smaller than $P(H)$, even after Newton, but still far far greater in ratio than $P(E)$ was before.)

Popper (1959, p. 403) recognized that it would be interesting to have some measure for the explanatory power of a hypothesis, and I developed the subject further in #599. By means of the desideratum-explicatum approach I was able to show that a satisfactory explication for the explanatory power of H of evidence E could be defined as the mutual information between E and H, namely

$$I(H:E) = \log[P(E|H)/P(E)].$$

More precisely I called this the explanatory power in the weak sense and I found it necessary also to define explanatory power in a strong sense. To explain this let us suppose that we add an arbitrary hypothesis K to H that has nothing to do with the evidence. Then the conjunction H · K would have the same weak explanatory power for E as does H but it does not deserve to be regarded as having the same explanatory power if we object to the "clutter," namely the irrelevant hypothesis K.

Instead of the clumsy expression "strong explanatory power" let me use the single word *explicativity*. . . . [For more on explicativity, see #**1000**.]

VI. INDUCTION AND THE TRUTH OF THEORIES

When reading Popper's work I assume, by the principle of induction, that the words he uses have much the same meaning as they seem to have had in the past in other writings. Sometimes there might be an exception, such as his use of the term "simplicity," but even when I detect Humpty-Dumptyism it is because, by scientific induction, I have an understanding of normal English usage. ("When *I* use a word . . . it means just what I choose it to mean . . . " —*Through the Looking Glass*, Chapter 6.) Then again, when he says that induction is not used for the acceptance of scientific theories but by their "proving their mettle" in virtue of our honest attempts to refute them, I assume that he has noticed this happening in the past and, by scientific induction, he expects this to continue in the future. He has formulated a scientific hypothesis here, belonging to the area of the sociology of scientists, a hypothesis that, if it can be accepted at all, will either have to be accepted by scientific induction, or because it proves its mettle by surviving our honest attempts to refute it.

Hume argued that induction cannot be logically justified because induction is needed to justify it. Equally we could argue that mettle-proving cannot be logically justified except either by induction or by another mettle-proving operation, so we are once again in an infinite regress (#191), or we hit metaphysics.

Similarly Popper often makes statements in the present tense of the form that "we can learn from experience" (Popper, 1962, p. 291). I think what he means is that in the past we have learned from experience but there is presumably an implication that we shall go on doing so. If he means that then he seems to have accepted a principle of induction as applied here to a hypothesis in psychology.

I conclude (i) that Popper really relies on induction in spite of his disavowals, and (ii) that even if induction could be replaced by mettle-proving, Hume's claim of the impossibility of a complete justification of induction would not be effectively by-passed.

As far as I know, the best mathematical treatment of pure induction was given by Keynes (1921, Chap. XX), Wrinch and Jeffreys (1921), apparently independently, and Huzurbazar (1955). Since the kudology is difficult I shall call the results the First and Second Induction Theorems. The First Induction Theorem states that, *if* $P(H|G) > 0$, *and if* $E_1, E_2, E_3, \ldots$ *are all implied by* $H \cdot G$, *then*

$$P(E_{n+1}E_{n+2} \ldots E_{n+m}|E_1E_2 \ldots E_nG) \to 1$$

when m *and* n *tend to infinity in any manner.* Note that this first theorem says nothing about the final probability of H, in fact the conclusion does not mention H at all. But the Second Induction Theorem, slightly improved here, states that *if* $P(H|G) > 0$, *and if (again)* $E_1, E_2, E_3, \ldots$ *are all implied by* $H \cdot G$, *and if moreover* $P(E_1 \ldots E_n|\bar{H} \cdot G) \to 0$ *as* $n \to \infty$, *then*

$$P(H|E_1E_2 \ldots E_nG) \to 1.$$

Proof of the Second Theorem. After n observations, the Bayes factor in favor of H is equal to $P(E_1 \ldots E_n|H \cdot G)/P(E_1 \ldots E_n|\bar{H} \cdot G)$ which tends to infinity. Therefore the odds of H tend to infinity, that is, its probability tends to 1 as asserted. Q.E.D.

A sufficient condition for $P(E_1 \ldots E_n|\bar{H} \cdot G) \to 0$ is of course that the infinite product

$$\Pi P(E_{n+1}|E_1E_2 \ldots E_n\bar{H}G)$$

diverges to zero. For example, it would be sufficient to have, for large n,

$P(E_{n+1}|E_1E_2 \ldots E_n\overline{H}G) < 1 - 1/n$. Keynes uses the somewhat stronger condition with $1 - \epsilon$ instead of $1 - 1/n$.

Note how difficult it is to establish high probability for a hypothesis unless one is able to state other hypotheses clearly enough to be able to judge $P(E_{n+1}| E_1 \ldots E_n\overline{H}G)$.

As an example of the First Induction Theorem let me give a proof that a printed language is not a Markov process of any finite order. (The proof of this in #524 made unnecessarily heavy weather of it.) If we have two texts that agree to n letters, then the probability that the next letter will be the same in both tends to 1 as n tends to infinity. This is impossible for a Markov chain of finite order unless it runs into a periodic sequence which I shall assume is not the case for real languages. The explanation might be that the two texts had to have come from the same source, or any other hypothesis that implies that the two texts are identical.

Both induction theorems depend on the assumption that the probability of H is not zero, whereas Popper claimed (for example, 1962, p. 281) that they typically are zero when H is a general law of Nature. He supports his thesis by reference to Carnap (1950, p. 571), but the argument had already been answered by Wrinch and Jeffreys and I have I believe strengthened the argument in Section III of the present paper. It should be pointed out that part of Carnap's theory is contained in Perks (1947) and in Johnson (1932).

At first sight the Second Induction Theorem might seem impossible. For suppose our hypothesis is of the form $H \cdot K$ where K has, up to the present, had nothing to do with the experimental evidence. In fact K might even be a metaphysical statement incapable of either refutation or verification. Then how can the probability of $H \cdot K$ tend to 1? The answer is of course that the condition that $P(E_1 \ldots E_n|G \cdot \overline{H \cdot K})$ is small is not satisfied in this case, for $H \cdot K$ can be false although H is true.

It can easily happen that a scientific theory is, for practical purposes, in terms of observable deductions, equivalent to other theories. For example, quantum mechanics can be expressed in at least two forms, Schrödinger's and Heisenberg's (which are at least nearly equivalent) and perhaps many other forms. In such cases the question of the meaning of the "truth" of the theory needs discussion. If the "as if" meaning of truth is accepted, then any one of a set of equivalent theories is just as true as any other. [My usage of the expression "as if" is somewhat different from Hans Vaihinger's.]

If you ask whether I think quantum mechanics has a high probability of being true, I would say that it depends on a judgment of whether there are likely to be entirely new kinds of tests that could be devised, possibly related to rare elementary particles, or to conditions of exceptionally high complexity (as in a living system) or of exceptionally high density (as in a white dwarf or perhaps in a space-time singularity). To that extent I agree with Popper: so long as one can think of qualitatively new kinds of test of a theory that have not yet been applied, the theory has not, in accordance with your subjective judgment, been properly tested. Note that this follows clearly enough from the theory of subjective or logical probability and from the induction theorems that belong to these theories. In the example just given it is a question of testing K, however well H has been tested, if you wish to make H · K probable. Thus Popper's emphasis on "honest attempts at refutation" is a *consequence* of inductivism although the inductivists seem not to have emphasized it enough.

To summarize my position on induction from a practical point of view: Provided that we are prepared to estimate probabilities of theories *conditional* on a disjunction of reasonably well specified theories, then the Second Theorem of Induction can be used to achieve near certainty. But we should not forget that this *is* conditional on one or other of these theories being true. In fundamental physics we cannot be sure that we have not overlooked some other theories, so that the First Theorem of Induction must be used if we want a non-conditional and non-evolving statement and it refers to the near certainty of observational results deduced from theories and not to the theories themselves. (Both induction theorems also have the weakness of being only limit theorems.) But, in other subjects, near certainty is achievable for theories; for example, the theory that chromosomes exist seems firmly established.

If Newtonian mechanics were stated with appropriate limitations, including upper bounds on all relative velocities and lower bounds on the accuracies of the observations, I would expect its discrepancies from the Special Theory of Relativity to be negligible. In this modified form, Newtonian mechanics would not be refuted as compared with Special Relativity, but would merely explain a smaller collection of observations and would have less explicativity and less predictivity. Such a modified form of Newtonian mechanics would perhaps be strictly true, and a subjectivist might be able to say that its probable truth had been established inductively. Such an approximative form of Newtonian mechanics is of course still extremely useful and cannot reasonably be said to have been refuted.

Incidentally Popper has beliefs about the truth of metaphysical statements (Popper, 1962, p. 195), so why not about some scientific ones also?

I have one further isolated point to make before going on to the seventh and last section, namely that the problem of induction is a *special case* of that of estimating probabilities in multinomial distributions, the case where all the entries are in one cell. This probability estimation problem has some history much of which can be found in ##398, 547 where it is also related to the problem of testing the "null hypothesis" of equiprobability. (There are $2^t - 1$ possible null hypotheses of the form "all p's [are] equal except those that are zero," where t is the number of cells. These could all be tested by the same method.)

VII. TESTABILITY, SIGNIFICANCE TESTS, AND CHECKABILITY

Popper (1962, p. 36) defines the testability of a theory as the degree to which it is exposed to refutation and he uses this as the demarcation between science and non-science. He says (1962, p. 256):

> . . . a system is to be considered scientific only if it makes assertions which may clash with observations; and a system is, in fact, tested by attempts to refute it. Thus testability is the same as refutability, and can therefore likewise be taken as a criterion of demarcation.

And again on p. 197:

> Every serious test of a theory is an attempt to refute it. Testability is therefore the same as refutability, or falsifiability.

Since I regard refutation and corroboration as both valid criteria for this demarcation it is convenient to use another term, *checkability*, to embrace both processes. I regard checkability as a measure to which a theory is scientific, where "checking" is to be taken in both its positive and negative senses, confirming and disconfirming. (I was unable to find a better word in Roget's *Thesaurus.*)

As an example I would say that the two propositions $0 = 1$ and $1 = 1$ are equally checkable, namely not at all, because their probabilities are respectively 0 and 1 and ever more shall be so. But $0 = 1$ is easily falsifiable whereas $1 = 1$ is not, so Popper should call the former scientific and the latter metaphysical. Perhaps he excludes purely mathematical propositions from his scheme in an *ad hoc* manner, whereas "checkability" needs no *ad hoc* appendage.

Again consider the proposition that there is consciousness after death. If false it cannot be refuted, but its opposite could be refuted if false, though not necessarily by living men. So Popper should regard the hypothesis as metaphysical but not its opposite. I think they are equally scientific or equally metaphysical (#243).

Or consider the proposition that chromosomes exist, or even that horses still exist on earth (fortunately,—we might need them again). One of the best pieces of evidence is that they can be observed. Are we then to say that the nonexistence of chromosomes and horses is more scientific than their existence? I think it would be a strange use of language.

When Eddington organized the expedition to detect whether light was deflected by the gravitational field of a star, during an eclipse, I believe he was excited by the General Theory of Relativity and was anxious to prove it right not wrong. On theories available, there were three possible "expected" deflections, one being zero, and all three would be combined with experimental error. The degree to which any of these theories was scientific should surely not depend on whether the experimenter was trying to refute any one of them. Thus Popper's remark, which I just quoted (1962, p. 256), does not hold water. His remark was scientific in the sense of being refutable, in fact I have just refuted it to my own satisfaction.

That Popper was not entirely happy with this remark is made clear by the following quotation from p. 36 of the same book: "Confirming evidence should not count *except when it is the result of a genuine test of the theory*; and this means that it can be presented as a serious but unsuccessful attempt to falsify the theory." (His italics.) So apparently seeing a horse doesn't count as support for the hypothesis that horses still exist.

It has always struck me as surprising that Popper does not discuss statistical significance tests in his books (1959, 1962). At least the expression is not in the index of either of them. In statistics a so-called "simple statistical hypothesis" is a hypothesis H such that, for a certain set of propositions E, the probabilities $P(E|H)$ are known tautologically, that is, by the definition of H. It is not usually pointed out that many scientific theories satisfy this definition. For example, provided a precise assumption is made for the law of error of observation, the following theories are simple statistical hypotheses: Newtonian Mechanics, Special Relativity, General Relativity, Quantum Mechanics, and Classical Statistical Mechanics (p. 912 of #322; p. 40 of #13).

Often a particular simple statistical hypothesis of interest is called a "null hypothesis," and, more often than not, the negation of a simple statistical hypothesis H is a so-called "composite hypothesis" which is a logical disjunction of one or more simple statistical hypotheses, usually a continuous infinity of them (in the mathematical model). For example, a null hypothesis might assert that some real parameter is equal to zero, and the non-null hypothesis might assert that it is not. In this case the non-null hypothesis "abuts" the null hypothesis, in fact we might as well say that it *includes* it. Then a test of the null hypothesis can be regarded as a test *within* a larger class of hypotheses. I shall say that such a null hypothesis is *immersed.*

It is part of the lore of orthodox statistics that an immersed null hypothesis cannot be made more probable, but that all one can do is to test it and see if it can be rejected at various levels of significance. This seems to be very close to Popper's attitude to scientific theories. Many statisticians *are* prepared to accept immersed null hypotheses, at least provisionally and as approximations. Jeffreys (1939/61) adopts a Laplace/Bayesian approach with the modification of assuming non-zero initial probabilities for immersed null hypotheses (usually 1/2) so acceptance of them became possible for him in a stronger sense, namely that they could gradually become more probable, and could approach certainty. My own position is slightly different (p. 90 of #13) in that I regard most null hypotheses in statistics as only approximate, but my conclusions are similar to those of Jeffreys. In physics some immersed null hypotheses *might* be exactly true.

But even in this Bayesian approach corroborating evidence for an immersed null hypothesis can arrive only slowly whereas refutation can arrive quickly. So it makes sense to say that immersed null hypotheses, including the scientific theories I just mentioned, are readily refutable if false but could be corroborated only slowly. I think it is this phenomenon, a deduction from a modern Bayesian philosophy, that can provoke people into assuming that a theory is scientific only to the extent that it is refutable.

But there are simple statistical hypotheses where the number of alternatives under consideration in some experiment is small, or even only one. I have already mentioned some examples of scientific theories of this kind in Section IV: the Mössbauer effect as tested in the laboratory, General Relativity as tested by the deflection of light, and some medical diagnosis problems, and a null result for the Michelsson-Morley experiment and various hypotheses in genetics. In these cases corroboration can be just as useful as refutation, or in other words one can obtain a large positive or a large negative weight of evidence. In fact when there are just two simple statistical hypotheses under consideration their logical relationship is a symmetrical one and they are equally scientific.

Thus, a modern Bayesian philosophy shows why refutability is usually more important than corroborability but not always. In other words the Bayesian philosophy explains how it is possible for a philosopher of science to fall into a dogmatic position and put *all* the emphasis on refutation: it explains the existence of a Popperian philosophy and at the same time improves it.

The concept of checkability can be put on a more quantitative basis. Consider the case where there are just two simple statistical hypotheses under consideration, H and $\overline{\text{H}}$. If, for a given cost, we move the probability of H from p to q, the revealed checkability should be a function of p and q, and this function should be unchanged when p is replaced by $1 - p$ and at the same time q is replaced by $1 - q$. To choose between such functions is not easy, but perhaps some guidance can be obtained by generalizing the problem to n hypotheses and also to a continuous infinity of them. The somewhat analogous work on the measurement of decisions, the "change of mind with respect to a class of acts" (#315), gives a plausible clue, namely that the "divergence" could be used as a measure of the *revealed* checkability for a set of hypotheses in the light of a definite experimental result. The divergence (Jeffreys, 1946; or 1961, p. 179; Kullback, 1959) is

$$\Sigma(p_i - q_i)\log\frac{p_i}{q_i}$$

where the initial and final probabilities of the hypotheses are respectively $p_1, p_2, \ldots, p_n$ and $q_1, q_2, \ldots, q_n$. (If $p_i \approx q_i$ for all i, the divergence $\approx 2\Sigma(p_i - q_i)^2/(p_i + q_i)$.) This reduces to

$$(p - q)\log\left(\frac{p}{1-p}\bigg/\frac{q}{1-q}\right)$$

when $n = 2$. (Compare #243 where the formula did not include the factor $(p - q)$.)

For a given cost we imagine the expected value of the divergence maximized by choice and design of an experiment, and the result may be called the checkability (or scientificality) of the set of theories for that cost. (Granting Foundations please note!) The expectation will depend on the relative initial probabilities of the n hypotheses, so it will depend on your judgment. Clearly a hypothesis of negligible initial probability should not contribute much to the checkability of the set of n hypotheses.

In high-energy physics qualitatively new kinds of experiments seem to have a propensity to become successively more expensive. Expense certainly has something to do with testability. To quote,

> Until recently it might have been regarded as metaphysical to conjecture that the other side of the moon is populated by animals hundreds of miles high that occasionally sling a surreptitious flying saucer at the earth. This hypothesis has now been refuted [at considerable expense] and was therefore not metaphysical. It used to be safe to maintain the even more far-fetched theory that the earth rested on an elephant standing on a tortoise [even larger animals] : it cost nothing to accept it when it was unverifiable, and it comforted people who suffered from a fear of falling (p. 493 of #243).

I can see no way to avoid arbitrariness in the magnitude of the cost unless we equate it to the total value of the observable universe. At one cent per kilowatt-hour, this comes to about $\$10^{61}$. If a theory in high-energy physics costs more than this to be tested, then it is metaphysical.

Part IV. Information and Surprise

The Appropriate Mathematical Tools for Describing and Measuring Uncertainty (#43)

1. INTRODUCTION

In this paper I shall be concerned less with decisions that are made than with those that are rational. But the paper will have some relevance to economics since the decisions of *Homo Sapiens* are not entirely irrational. Moreover the "theory of rational decisions" or rational behavior includes a completely general theory of probability and is therefore applicable to economics, whether human decisions are rational or not. The economist himself should attempt to be rational.

2. SCIENTIFIC THEORIES

. . . The function of the theory is to introduce a certain amount of objectivity into your subjective body of judgments, to act as shackles on it, to detect inconsistencies in it, and to increase its size by the addition of discernments. It is not misleading to describe the discernments as implied judgments. We do not yet know precisely how the mind makes judgments: if we did we could build a machine to do all our thinking for us. Until we can do this it will be necessary to describe scientific techniques with the help of suggestions as well as axioms and rules. . . .

3. DEGREES OF BELIEF

I shall now make thirteen remarks about degrees of belief. . . .

(ii) Following Shackle (1949) we could order our degrees of belief by the potential degrees of surprise associated with them. Personally, if I used this method I think I would use, in reverse, the definition of the surprise index given by Weaver (1948), namely $\left(\sum_i p_i^2\right)/p_1$, the expected value of a probability divided by the probability of the event that actually occurs. Here $p_1, p_2, \ldots$ are the probabilities of mutually exclusive propositions that you consider appropriate to regard as separate. The meaning of the qualification "in reverse" is that you can first write down a partial ordering of (a set of inequalities concerning) your subjective estimates of the (potential) surprise indexes and only then convert these inequalities into inequalities circumscribing the values of $p_1, p_2, p_3 \ldots$ to some extent. If you apply this method you could cultivate a judgment about the numerical values of surprise indexes by first considering some examples in which the values of $p_1, p_2, p_3 \ldots$ were known in advance with considerable precision, as in some games of chance.

It is possible to modify Weaver's definition of a surprise index without affecting its main properties. It may, for example, be preferable to use $\prod_i (p_i/p_r)^{p_i}$ or its logarithm

$$\Sigma p_i(\log p_i - \log p_r).$$

(This suggestion may be compared with Bartlett, 1952.) Both Weaver's surprise index and this modified one have the property that the index for the combined event E & F, where E and F are independent, is a function, namely the product or sum, of the two separate indexes. A general class of indexes with the multiplicative property is

$$p_r^{-1}\,[\mathcal{E}(p_i^{\,n})]^{\frac{1}{n}} = p_r^{-1}\,[\textstyle\sum_i p_i^{\,n\ +1}]^{\frac{1}{n}}.$$

Weaver's index is the special case $n = 1$. If the logarithmic form of any of these indexes is used, the index of surprise associated with turning up any card in a well-shuffled pack of playing cards would be zero. The index would, however, become positive if the card happened to be predicted in advance. Negative values of the logarithmic surprise indexes correspond to events that are in a sense "more probable than the average." (For some further related comments see Good, 1953b.)

Consider a bet on a "double event" at a race-course. The logarithmic surprise index (in any of the senses) for the double event is the sum of the separate logarithmic indexes. Hence these surprise indexes cannot be functions of Shackle's degree of surprise, which is equal to the greater of the two degrees of surprise assigned to the separate events. [I regard this as a defect in Shackle's theory.] . . .

(x) It is convenient to use the word "you" for the person to whom the degrees of belief belong. The word can be interpreted in the singular or plural. The subjective theory thus includes what de Finetti (1951) calls "multisubjective problems."

(xi) The postulate that credibilities exist is useful in that it enables other people to do some of our thinking for us. Naturally we pay more attention to some people's judgment than to others'. . . .

4. UTILITIES

. . . People's judgments of utilities are, I think, [usually] liable to disagree more than their probability judgments. Utilities can be judged with a fair amount of agreement when the commodity is money but not when deciding between, say, universal education and universal rowing.

It is possible that infinite utilities could occur in questions of salvation and damnation, as suggested by Pascal (1670), and expressions like $\infty - \infty$ may occur when deciding between rival religions. To have to argue about such matters as a necessary preliminary to laying down the axioms of probability would weaken the foundations of that subject.

Some philosophers regard the phrase "degree of belief" as metaphysical. They would presumably prefer to use the theory with a body of decisions rather than a body of beliefs. It is more convenient however to work with a joint body of beliefs and decisions. There is not much harm in introducing a little metaphysics provided it is simple, convenient, free from contradictions and expressed axiomatically.

JUSTIFICATION OF THE PRINCIPLE OF RATIONAL BEHAVIOR

The simplest attempt to justify the principle of maximizing expected utilities is in terms of some form of the law of large numbers. This law is inapplicable if some of the utilities are very large. For example, if the possible positive and negative utilities of marriage are large then the law of large numbers is inapplicable in monogamous societies. Nevertheless the law of large numbers does show that the principle of rational behavior is consistent with the theory of probability. This justification is analogous to the frequency (long-run) definition of probability. It shows that the theory is self-consistent, and then the theory can be regarded as containing an implicit definition of probability and utility, even when there is no question of long runs.

Other types of justification of the principle of rational behavior have been given by von Neumann and Morgenstern (1947), Savage (1951), Marschak (1951), Lindley (1953) and perhaps Samuelson. Marschak shows that the principle is equivalent to a more convincing one that he calls the "rule of substitution between indifferent prospects." This rule seems to be essentially the same as some assumptions made by von Neumann and Morgenstern. Savage uses a similar rule called the "sure-thing principle." . . .

6. AXIOMS AND RULES

The theory of probability that I accept is based on six axioms. [See #13.] The origins of these axioms will not be discussed here. . . .

Typical axioms are

A1. $P(E|F)$ is a non-negative number (E and F being propositions).

A4. If E is logically equivalent to F then $P(E|G) = P(F|G), P(G|E) = P(G|F)$.

There are two less "obvious" axioms called the sum and product axioms. They have been shown to be at least partly conventional by Jeffreys (1948), Schrödinger (1947), Barnard (1951) and Good (App. III of #13). [See also R. T. Cox (1946).] Any strictly increasing continuous function of probability can be used instead of probability, without any effect on the partial ordering, although the axioms would be transformed. It is mainly in this sense that the axioms are to some extent conventional.

Finally there are two axioms that say that there exists a zero probability but that not all probabilities are zero. I am also prepared to use an axiom of complete additivity as a mathematical convenience but do not regard it as essential to any application. It is purely metaphysical. It is concerned with the probability of the logical disjunction of a countable infinity of propositions. (*Note:* By a *purely metaphysical statement* I mean one which is incapable of observable logical consequences. The actual enunciation of the statement has, however, material consequences. A metaphysical statement is *logically useful* if it lends simplicity to thought. The enunciation of logically useless metaphysical statements can also be useful [or harmful], roughly in the same sense that music is useful. [Such statements do not correspond to propositions.])

It is possible to replace A4 by

A4′. If you have *proved* that E is logically equivalent to F then $P(E|G) = P(F|G)$, etc.

The adoption of A4′ amounts to a weakening of the emphasis on consistency and enables you to talk about the probability of purely mathematical propositions. . . . [See #938.]

7. EXAMPLES OF SUGGESTIONS

[This section has been omitted.]

8. RATIONAL BEHAVIOR

. . . In the applications of the principle of rational behavior some complications arise, such as—(i) . . . My impression of Shackle's method (1949), of focussing attention on special outcomes, is that it is often an excellent time-saver and could therefore be incorporated as a suggestion in the technique of rational behavior. I suspect, however, that it would occasionally contradict the principle of rational behavior and I would then classify the use of it as unreasonable. If most people use the principle it must be relevant for economists. . . .

9. FAIR FEES

. . . When we engage a professional expert to make probability estimates $p_1, p_2, \ldots, p_n$ for the n mutually exclusive events, we may have already formed our own "amateur" estimates $a_1, a_2, \ldots, a_n$ of the probabilities. By regarding ourselves as one of the two experts we see that the fair fee to pay will be of the form $k.\log(p_r/a_r)$ if E_r occurs, where k is a factor depending on the utility of the forecast. This fee can be seen to have the desirable property that its expectation is maximized if the estimates p_i are all equal to the true probabilities or credibilities, t_i say. It is therefore in the expert's interest to give objective estimates.

In #26 I assumed that the payment should be of the form $k.\log(p_r b)$ where b is independent [of] r. By insisting that the expected payment should be zero when the professional's probability estimates are the same as the amateur's, I found that $\log b$ had to be of the form of Shannon's "entropy," $-\Sigma a_i \log a_i$. But I cannot now see why b should be independent of r.

The objective expectation of the payment to the professional, if his judgment is objective, and if the a_r's are all equal, is expressible in terms of entropy; otherwise the expression is of the more general "cross-entropy" form. (Cf. Good, 1950/53, 180.)

If Daniel Bernoulli's formula for the utility of money is assumed, and if we wish to pay $k.\log(p_r/a_r)$ utility units, then the money can be written $c\{(p_r/a_r)^k - 1\}$, where c is the professional's initial capital.

These fair fees could be sued as a method of introducing piecework into the Meteorological Office.

When making probability estimates it may be help to *imagine* that you are to be paid in accordance with the above scheme. . . .

10. MINIMAX SOLUTIONS

[This section has been omitted.]

APPENDIX. AN APPROXIMATELY INVARIANT FORM OF THE FOCUS-OUTCOME

The conditions under which the focus-outcome method is rational can be roughly expressed in the following manner. . . .

CHAPTER 17

On the Principle of Total Evidence (#508)

Ayer (1957) raised the question of why, in the theory of logical probability (credibility), we should bother to make new observations. His question was not adequately answered in the interesting discussion that followed. . . . The question raised by Ayer is related by him to a principle called by Carnap (1947), "the principle of total evidence," which is the recommendation to use all the available evidence when estimating a probability. Ayer's problem is equally relevant to the theory of subjective probability, although, as he points out, it is hardly relevant to the theory of probability in the frequency sense.

In this note, Ayer's problem will be resolved in terms of the principle of rationality, the recommendation to maximize expected utility. . . .

Our conclusion is that, in expectation, it pays to take into account further evidence, provided that the cost of collecting and using this evidence, although positive, can be ignored. In particular, we should use all the evidence *already* available, provided that the cost of doing so is negligible. With this proviso then, the principle of total evidence follows from the principle of rationality.

Suppose that we have r mutually exclusive and exhaustive hypotheses, H_1, $H_2, \ldots, H_r$, and a choice of s acts, or classes of acts, $A_1, A_2, \ldots, A_s$. It will be assumed that none of these classes of acts consists of a perpetual examination of the results of experiments, without ever deciding which of $A_1, A_2, \ldots, A_s$ to perform. Let the (expected) utility of act A_j if H_i is true be $U(A_j|H_i) = u_{ij}$. Suppose that, on some evidence, E, we have initial (prior) probabilities, $p_i = P(H_i|E)$. If just this evidence is taken into account, then the (expected) utility of act A_j is $\Sigma_i p_i u_{ij}$ and the principle of rationality recommends the choice $j = j_0$, the value of j that maximizes this expression; and therefore the (expected) utility in the rational use of E is

$$\max_j(\Sigma_i p_i u_{ij}) = \Sigma_i p_i u_{ij_0}.$$

We now consider making an observation whose possible outcomes are E_1, E_2, . . . , E_t, where $P(E_k|H_i) = p_{ik}$ $(i = 1, 2, \ldots, r; k = 1, 2, \ldots, t)$. Let

$$q_{ik} = P(H_i|E.E_k) = p_i p_{ik}/\Sigma_i p_i p_{ik},$$

the final (posterior) probability of H_i if E_k occurs. (We denote logical conjunction by a full stop or period.) If in fact E_k occurs, then the expected utility of the use of E_k combined with E becomes

$$\max_j(\Sigma_i q_{ik} u_{ij}).$$

Now the initial probability of E_k is $\Sigma_i p_i p_{ik}$, so that the expected utility, in deciding both to make the new observation and to use it, is

$$\Sigma_k(\Sigma_i p_i p_{ik})\max_j(\Sigma_i q_{ik} u_{ij}) = \Sigma_k \max_j \Sigma_i p_i p_{ik} u_{ij}.$$

Accordingly we should like to prove that

$$\Sigma_k \max_j \Sigma_i p_i p_{ik} u_{ij} \geqslant \max_j \Sigma_i p_i u_{ij},$$

with strict inequality unless the act recommended by the principle of rationality is the same irrespective of which of the events E_k *occurs;* in other words unless there is a value of j, mathematically independent of k, that maximizes $U(A_j|E.E_k) = \Sigma_i q_{ik} u_{ij} = \Sigma_i p_i p_{ik} u_{ij}/\Sigma_i p_i p_{ik}$, or equivalently that maximizes $\Sigma_i p_i p_{ik} u_{ij}$.

Since $\Sigma_k p_{ik} = 1$, the above proposition follows from the following Lemma by putting $f(j,k) = \Sigma_i p_i p_{ik} u_{ij}$.

LEMMA. *Let* f(j,k) *be any real function of* j *and* k. *Then*

$$\Sigma_k \max_j f(j,k) \geqslant \max_j \Sigma_k f(j,k)$$

with strict inequality unless the matrix $\{f(j,k)\}$ *has a "dominating row."* (By a "dominating row" of a matrix we mean a row in which each element is at least as large as any element in its own column.)

Proof of Lemma. Let a value of j that maximizes $\Sigma_k f(j,k)$ be j_0. Clearly $\max_j f(j,k) \geqslant f(j_0,k)$, since this would be true *however* j_0 were defined. The inequality is strict, when the definition of j_0 *is* used, unless $f(j,k)$ and $\Sigma_k f(j,k)$ are maximized by the same value of j. Therefore

$$\Sigma_k \max_j f(j,k) \geqslant \Sigma_k f(i_0,k) = \max_j \Sigma_k f(j,k).$$

This inequality is strict unless, for all k, $f(j,k)$ and $\Sigma_k f(j,k)$ are maximized at the same value of j. This establishes the Lemma and hence completes the resolution of Ayer's problem in terms of the principle of rationality.

At this point an opponent might say "You have justified the decision to make new observations and to use them for the choice of the act A_j, but you have not justified the use of all observations that have already been made." To this we can reply, "The observations already made can be regarded as constituting a record. The process of consulting this record is itself a special kind of observation. We have justified the decision to make this observation and to use it, provided that the cost is negligible. In other words we have justified the use of all the observations that have been made, and this is the principle of total evidence."

Our opponent might then say, "What you have shown is that, when faced with the two following possibilities, it is rational to select the second one:

(i) Not make an observation;
(ii) To make the observation and to use it for the choice of A_j;
but you have ignored a third possibility, namely
(iii) Make the observation and then not use it."

My reply would be "if we make an observation and then do not use it, this is equivalent to putting it back into the record. We have shown that it would then be irrational to decide to leave the observation in the record and not to use it, since there is a better course of action, namely to take it out (observe the record) and use it. You will now suggest other possibilities, such as the making of an observation, putting it on record, taking it out, putting it back, and so on, several times, and finally not using it. Our previous argument, with an obvious modification, shows that any such procedure is irrational, and it remains for you to suggest that your vacillating procedure should be continued for ever. But this would be a perpetual examination of the results of experiments, without a decision, and we have ruled this out by an explicit assumption."

The simple mathematical theorem of the present note is not entirely new. Raiffa and Schlaifer (1961), p. 89, refer to the expected value of sample information, and seem implicitly to take for granted that it is positive. Lindley (1965), p. 66, explicitly states part of the theorem without proof. Perhaps the main value of the present note is that it makes explicit the connection between Carnap's principle of total evidence and the principle of rationality, a connection that was overlooked by seventeen distinguished philosophers of science.

CHAPTER 18

A Little Learning Can Be Dangerous (#855)

It has been proved that, under certain assumptions, it pays you "in expectation" to acquire new information, when it is free. A precise formulation of this thesis, together with a proof, was given in #**508** with historical references to Carnap, Ayer, Raiffa and Schlaifer, and Lindley. The "expectation" in this result is *your own* expectation. In the present note it will be pointed out that the result can break down when the expectation is computed by someone else. It is of course familiar that an experiment can be misleading by bad luck, but this is not by itself a justification for ignoring free information. It is perhaps surprising at first sight that, "in expectation," an experiment can be of negative value to a subject, in the opinion of *another* person who may or may not be better informed or have better judgment.

The argument supporting the main thesis of this note is very short, but I should like to take the opportunity of replying to a criticism of #**508**. Let us first recall the result of that note in precise form.

We refer to "you," the subject, by the symbol S. Suppose that there are r mutually exclusive and exhaustive hypotheses $H_1, H_2, \ldots, H_r$, and a choice of s acts, or classes of acts, $A_1, A_2, \ldots, A_s$. Let your expected utility of act A_j if hypothesis H_i is true be $U_S(A_j|H_i) = u_{ij}$ $(i = 1, 2, \ldots, r; j = 1, 2, \ldots, s)$ and let your initial (prior) probability of H_i be $P_S(H_i) = p_i$. (There will always be some previous evidence, but we take it for granted and omit it from the notation.) You now have the option of performing an experiment, or making an observation, whose possible outcomes are $E_1, E_2, \ldots, E_t$, where $P(E_k|H_i) = p_{ik}$ $(i = 1, 2, \ldots, r; k = 1, 2, \ldots, t)$. Here there is no need for a suffix S if we regard the hypotheses as being simple statistical hypotheses, for in this case the conditional probabilities p_{ik} are "tautological" (having values known by the definitions of the hypotheses and not subject to dispute nor to empirical verification). Suppose further that the cost of the experiment is zero (or negligible), and that the expected utilities u_{ij} and the probabilities p_i have sharp (precise) values. Then the theorem proved in #**508** is that, before the experiment is performed, the expected utility available to you would not be decreased in the light of the experiment, and would be increased unless the act recommended by the principle of rationality (maximizing expected utility) is the same irrespective of the outcome of the experiment. So it is always rational for you to perform a costless experiment.

It has been pointed out to me (by Isaac Levi and Teddy Seidenfeld) that, in the light of my repeated emphasis on the need for non-sharp subjective probabilities and utilities, I should have dealt also with the case where the u_{ij}'s and p_i's are not sharp. The use of "comparative" utilities and probabilities leads one to the conclusion that there will be situations where there is no rational choice between two options. As Dr. Levi said in effect (private communication) the result of an observation might lead to a vaguer state of information than one had before, and so to a state of confusion. I think one way of coping with this criticism is to invoke the "black box" theory of rationality (see ##13, **26**, **230**). In this theory, inequality judgments are plugged into the black box, but the calculations within the box are assumed to be performed with sharp values. The box will imply a discernment that we cannot be worse off by performing the experiment, and this result can therefore be regarded as proved if it is assumed that the black box theory is self-consistent. It was proved self-consistent by C. A. B. Smith (1961) [if ordinary mathematics is consistent]. Although this argument seems a little too glib, I cannot see anything wrong with it. At any rate in what follows I shall assume sharp values.

Suppose that some other person, T, who might be a demiurge (T for *Teufel*), has some other set of initial probabilities $P_T(H_i)(i = 1, 2, \ldots, r)$ although he agrees with the values of p_{ik} and u_{ij} (for all i, j, and k), and he also knows S's initial probabilities. It might be conjectured that, in T's opinion, S is again better off in expectation if he performs the zero-cost experiment. But the following example shows that this conjecture is false.

To disprove the conjecture we need very little notation because it can be done with a very simple example. We suppose that there are just two hypotheses, either a certain coin is "fair" or it is double-headed, and that the acts available to S are to bet on one or other of these two hypotheses in a level bet. S must make the bet without examining the coin, but he is given the option of being told the outcome of a single toss of the coin.

It so happens that initially S regards it as just "odds on" (probability greater than a half) that the coin is fair, whereas T *knows* that the coin is indeed fair. (Or T's estimate of the probability is close to 1.) S might be entirely rational in his opinion, in fact we can imagine that his subjective probabilities are "credibilities" if such things exist; but T happens to have more information.

Then T knows that with chance one half, the coin will come up heads and that this would cause S rationally to regard the wrong hypothesis as "odds on." In this case S will lose his bet. Thus T knows that S has nothing to gain and something to lose if he opts for the experimental information, and therefore T's expectation of S's utility is decreased if S opts for the experiment. T would do S a service by not allowing him to acquire the new information. This would be true even if the experiment consisted of several tosses of the coin, because we have considered a rather extreme situation where T *knows* the truth about the coin. In less extreme situations the result would be true only for experiments involving a small amount of information. When ignorance is bliss, a little learning can be poison according to another man.

CHAPTER 19

The Probabilistic Explication of Information, Evidence, Surprise, Causality, Explanation, and Utility (#659)

My purpose in this paper is to review some of my life's work in the mathematics of philosophy, meaning the application of mathematics in the philosophy of science. Apart from the clarification that the mathematics of philosophy gives to philosophy, I have high hopes for its application in machine intelligence research, just as Boolean logic, a hundred years after its invention, became important in the design of computers. I think philosophy and technique are both important, but I do not intend to argue the case for the use of subjective probability on this occasion. I would just like to quote a remark by Henry E. Daniels (c. 1956) that *each statistician wants his own methods to be adopted by everybody.*

I shall cover a variety of topics and will have to be too succinct for complete clarity. But I hope to give some impression of the results attained, and I shall refer to the original sources for fuller details.

Some unifying themes for this work are the simple concepts of weight of evidence and amount of information, and also what Carnap calls the *desideratum-explicatum* approach to the analysis of linguistic terms. The desiderata are extracted from normal linguistic usage of a term, and an explicatum that satisfies the desiderata constitutes a sharpened form of the term, likely to be of more use in scientific contexts than the original vaguely defined term. Of course, the explicatum is not necessarily uniquely determined by the desiderata, and this ambiguity gives us an opportunity of enriching the language.

The desideratum-explicatum approach has been used by several writers for arriving at the usual axioms of subjective probability and utility. In the present paper, I shall take the usual axioms of probability and utility for granted. By doing so, I do not imply that judgments are precisely representable by numbers; on the other hand, I think that all judgments are judgments of inequalities: see for example, Keynes (1921), #13. I hope that this point is well taken since I do not intend to repeat it; and I hope what I shall say later will not give the impression of being overprecise merely because symbols are used.

In order to clarify my presuppositions, it is necessary to make one or two comments concerning the principle of rationality, the recommendation to maximize expected utility. I call this *rationality of type 1* and, when allowance is also made for the cost of theorizing, *rationality of type 2* (##290, **679**). Rationality of type 1 implies complete logical consistency with the axioms of rationality; but rationality of type 2 should be adopted in practice. For example, you should not normally and knowingly allow any blatant contradictions in your judgments and discernments when the axioms of rationality are taken into account; but an exception is reasonable if a decision is extremely urgent. In particular, apparently non-Bayesian methods are often acceptable to me. I think this compromise resolves all the important fundamental controversies in statistics, but we shall go on arguing because, being mortal, we are anxious to justify our existence, and for other reasons not mentioned in polite society.

The notion of rationality of type 2 is closely related to that of evolving probabilities. An evolving probability is one that changes in the light of reasoning alone, *without the intervention of new empirical information.* I shall return to this matter later.

People will say that the principle of rationality is inapplicable in a situation of conflict, and that then the theory of games and minimax solutions are more fundamental. I do not agree. I think the principle of rationality is still the overriding principle; but you should, of course, take into account your opinions concerning your opponent's (randomized) strategy (which opinions might be largely based on his past behavior) and whether his past behavior appears to have been based on yours. In particular, if you are convinced that he is playing a minimax strategy then, by von Neumann's theorem, you maximize your expected utility by also adopting a minimax strategy in a zero-sum game.

I mentioned that point in order to emphasize that the principle of rationality has no real exceptions, although pseudo-utilities can be used, as discussed later. [I now call them quasiutilities.]

It is convenient to make a distinction between information and evidence, and I shall discuss information first. My approach is fairly closely related to that of Shannon, the main difference being that he was concerned with the average amount of information transmitted by a communication channel, whereas my approach is in terms of the amount of information concerning one proposition that is provided by another one.

Let E, F, G, and H be propositions, and let $I(H{:}E|G)$ denote the amount of information concerning H provided by E when G is given throughout. Let us consider the explication of $I(H{:}E|G)$ as a real number in terms of probabilities.

We write E.F for the logical conjunction of E and F.

From six reasonable axioms (#505) we can deduce that $I(H{:}E|G)$ must be a continuous increasing function of the *association factor* (Keynes's term)

$$P(H.E|G)/[P(H|G)P(E|G)] \text{ or } P(E|H.G)/P(E|G).$$

If we are concerned only with preserving the ordinal relations (inequalities) between amounts of information, then we might as well, by convention, select the logarithm of the association factor, to some base exceeding unity, since this choice leads to simple additive properties. I shall do this.

The analysis applies whether the probabilities are physical, logical, or subjective, in which case we are talking about physical, logical, or subjective information respectively.

The main axiom assumed was that $I(\text{H:E.F}|\text{G})$ is some function of $I(\text{H:E}|\text{G})$ and $I(\text{H:F}|\text{E.G})$.

In terms of communications, H can be interpreted as the hypothesis that a particular message was transmitted on a particular occasion, and E as the event that a particular message was received. Then, upon taking expectations with respect to both H and E we obtain the rate of transmission of information, as in Shannon's theory of communication.

A philosophical approach is not necessary for the theory of communication since the main theorems of that theory deal with efficient coding for transmission through a given communication channel; and these theorems can be proved without even giving a definition for the rate of transmission of information. Nevertheless, I think the theorems are easier to understand when such a definition is given.

I should now like to discuss *evidence.* This differs from information in the following respect. In legal and other circumstances, we talk about the evidence *for or against* some hypothesis, but we talk about the information *relevant to or concerning* a hypothesis. Thus the notion of evidence, as ordinarily used, makes almost explicit reference both to a hypothesis H and to its negation. On the other hand, *information* concerning a hypothesis H does not seem to refer primarily to the question of discriminating H from its negation. It seems linguistically appropriate to regard the *weight of evidence,* in favor of a hypothesis H, provided by evidence E, as identifiable with the degree to which the evidence corroborates H as against its negation $\overline{\text{H}}$. Note that I am distinguishing between evidence E and weight of evidence, just as in ordinary English.

Now Popper (1959) laid down nine compelling desiderata that corroboration C should satisfy. To these I (##211, 599) made some minor modifications and also added the assumption that $C(\text{H:E.F}|\text{G})$ is some function of $C(\text{H:E}|\text{G})$ and $C(\text{H:F}|\text{E.G})$. This assumption is reasonable although it is not as compelling as the other axioms. But, since it is possible to assume it and still find an *explicatum* that satisfies all the conditions, this explicatum can be expected to be the most convenient one to accept. Any explicatum that does not satisfy this assumption would come under suspicion. It seems reasonable to say that if an explicatum satisfies all the compelling axioms and also the reasonable ones, then it is better than an explicatum that satisfies only the compelling axioms.

The assumptions made in #211 did not lead to a unique explicatum so in #599 I added the further assumption that corroboration is objective in some circumstances, i.e. that if $C(\text{H:E})$ depends only on $P(\text{H})$, $P(\text{E}|\text{H})$, and $P(\text{E}|\overline{\text{H}})$, then it depends only on $P(\mathrm{E}|\mathrm{H})$ and $P(\mathrm{E}|\bar{\mathrm{H}})$. This approach led to the conclusion that the best explicatum for corroboration is some monotonic increasing function of weight of evidence W, defined (#13) as the logarithm of the Bayes factor (or Bayes-Jeffreys-Turing factor) in favor of a hypothesis provided by evidence. Formally, the weight of evidence is

$$W(\mathrm{H}:\mathrm{E}|\mathrm{G}) = \log F(\mathrm{H}:\mathrm{E}|\mathrm{G}),$$

where the base of the logarithms merely determines the unit of measurement of weight of evidence, and the factor F is defined by

$$\begin{aligned} F(\mathrm{H}:\mathrm{E}|\mathrm{G}) &= P(\mathrm{E}|\mathrm{H}.\mathrm{G})/P(\mathrm{E}|\bar{\mathrm{H}}.\mathrm{G}) \\ &= O(\mathrm{H}|\mathrm{E}.\mathrm{G})/O(\mathrm{H}|\mathrm{G}), \end{aligned}$$

where O denotes odds. (The odds corresponding to a probability p are defined as $p/(1-p)$.) It is because of the expression in terms of odds that Turing (1941) suggested the excellent name *factor in favor of a hypothesis.* Note that

$$W(\mathrm{H}:\mathrm{E}|\mathrm{G}) = I(\mathrm{H}:\mathrm{E}|\mathrm{G}) - I(\bar{\mathrm{H}}:\mathrm{E}|\mathrm{G})$$

and this gives especially clear expression to the previous remark that evidence is concerned with a comparison of a hypothesis with its negation.

Sometimes we wish to talk about the weight of evidence in favor of H as against, or as compared with, some other hypothesis H′. A convenient notation for this is $W(\mathrm{H}/\mathrm{H}':\mathrm{E})$ and it is equal to $W(\mathrm{H}:\mathrm{E}|\mathrm{H} \vee \mathrm{H}')$, where the sans serif (v) denotes logical disjunction. A similar notation can be used for Bayes factors F and corroborations C. The colon can be read *provided by.*

The expression W for weight of evidence is more clearly appropriate in terms of ordinary English than is I for amount of information since (Bartlett, c. 1951) it is not linguistically clear whether information relevant to H should ever be allowed to take negative values. It is not unreasonable of course, since information *can* be misinformation.

The calculations of the expressions $I(\mathrm{H}:\mathrm{E}|\mathrm{G})$ and $W(\mathrm{H}:\mathrm{E}|\mathrm{G})$ both usually depend on a Bayesian philosophy, especially the former. But W is non-Bayesian when both H and $\bar{\mathrm{H}}$ are simple statistical hypotheses, since in this case the factor in favor of H is equal to the simple likelihood ratio. Note though that even in this case, the interpretation as a factor on the odds has more intuitive appeal to the non-statistician. It would be more appropriate to say that the name *likelihood ratio* is jargon than to say it of *weight of evidence* or *factor in favor of a hypothesis.*

Proofs of some coding theorems more general than Shannon's Fundamental Theorem can be expressed with much intuitive appeal in terms of weight of evidence, and it turns out that amount of information occurs in this theory merely as an approximation to weight of evidence (#574).

An entertaining philosophical application of weight of evidence occurs in the discussion of Hempel's paradox of confirmation. In a nutshell, the paradox is that since a case of a hypothesis supports it, and since the assertion that all crows are black is logically equivalent to the assertion that all non-black objects are non-crows, it follows that the observation of a white shoe supports the hypothesis that all crows are black. This paradox might seem trivial but if left unresolved it would undermine the whole of statistical inference. I shall here merely mention some of the relevant references (##199, 245, **518**, **600**, and Hempel, 1967).

Since I have just said that a paradox might undermine statistical inference, I should like to take this opportunity to correct an error I made (p. 382 of #522) when discussing Miller's paradox concerning the axioms of probability. Succinctly, his paradox can be expressed thus: for all propositions E we have

$$P(\overline{E}) = P(E|P(E) = P(\overline{E})) = P(E|P(E) = \tfrac{1}{2}) = \tfrac{1}{2}.$$

My attempted resolution was that $P(E) = P(\overline{E})$ is impossible unless $P(E) = \frac{1}{2}$ and no sensible theory of probability permits an impossible proposition to be *given*. (This is permitted in Popper's theory, which is sensible, but the argument is unaffected.) But, as pointed out by Miller (1970), this resolution is wrong since $P(E) = P(\overline{E})$ is not impossible unless it is *known* that $P(E) \neq \frac{1}{2}$. One moral is that it is dangerous to allow the propositions to the right of the vertical stroke to make explicit reference to probabilities, as pointed out on p. 41 of #13 and by Koopman (1940a, p. 275). This paradox is more or less of the self-referring kind familiar in logic. But this resolution of the paradox will not work if the given information is interpreted in terms of long-run frequencies. Then (#599) I think we can resolve the trouble by quoting more precisely the theorem in the theory of subjective or logical probability that lies behind the intuitive feeling that $x = P(E|P(E) = x)$. This theorem, which is a form of the law of large numbers, is that, provided that the initial density of $P(E)$ is positive at x, then $x = P(E|s = x)$, where s denotes the limiting relative proportion of *successes* in an infinite sequence of *trials*, and where x is an assigned real number. Since $P(\overline{E})$ is not an assigned number, Miller's paradox is not a threat to the axioms of subjective or logical probability, as least as I understand them.

An algebraic analogue of Miller's paradox was pointed out by Mackie (1966). It can be expressed succinctly thus: $\frac{1}{2}$ = (The value of a if $a = \frac{1}{2}$) = (The value of a if $a = 1 - a$) $= 1 - a$. Here the resolution is that an expression involving a cannot legitimately be said to be a *value* of a, so that the last step in the argument is illogical. [See also #1159.]

Weight of evidence is especially appropriate for medical diagnosis and one example occurs in connection with the analysis of the relationship between lung cancer on the one hand and smoking, morning cough, and degrees of rurality or urbanity on the other.

It turned out in this example that degree of urbanity (rural, urban, highly urban) gave a weight of evidence that could be added to that derived from the evidence concerning coughing and smoking, but that coughing and smoking could not be treated as additive. In other words there was evidential interaction between coughing and smoking in relation to lung cancer, but there was not important interaction between urbanity on the one hand and coughing and smoking on the other. In this statement, the interaction can be defined as

$$W_1(\mathrm{H{:}E,F}) = W(\mathrm{H{:}E.F}) - W(\mathrm{H{:}E}) - W(\mathrm{H{:}F}) = I(\mathrm{E{:}F|H}) - I(\mathrm{E{:}F|\overline{H}}).$$

Second-order interactions are naturally defined by the expression

$$W_2(\mathrm{H{:}E,F,G}) = W(\mathrm{H{:}E.F.G}) - W(\mathrm{H{:}F.G}) - W(\mathrm{H{:}G.E}) - W(\mathrm{H{:}E.F}) + W(\mathrm{H{:}E}) + W(\mathrm{H{:}F}) + W(\mathrm{H{:}G})$$

as in ##210 and 755. It will be useful for diagnostic purposes if it turns out in many instances that interactions of the second order can be ignored.

In my work, I put much stress on the notion of weight of evidence. This enthusiasm is shared in a book by Myron Tribus (1969), and has also been much used, though is somewhat different manners, by Jeffreys (1939/61) and Kullback (1959). Entropy enthusiasts . . . include E. T. Jaynes, Jerome Rothstein, and S. Watanabe. . . .

The next topic I should like to discuss is the concept of *surprise*. . . . [See ##**43**, 82.]

My next topic is *probabilistic causality*. . . . [See #**223B**.]

I believe that the philosophical analysis of *strict* causality, itself by no means trivial, can be inferred as a limiting case of this treatment of *probabilistic* causality.

Next consider *explanation*. . . . [See #**1000**.]

My final topic is *utility*, especially in relation to the utility of a distribution (##198, 211, 618). Statisticians have been giving their customers estimates of distributions of random variables for a long time, and they hardly ever stop to consider what the loss in utility is if the estimate is not accurate. In order to find out something about this, let us denote by $U(G|F)$ the utility of asserting that a random variable $\mathbf{x}$ has the distribution G when the true distribution is F. Let us suppose that $U(G|F)$ is some form of generalized expectation of $\nu(\mathbf{x}, \mathbf{y})$, where $\nu(\mathbf{x}, \mathbf{y})$ denotes the utility of asserting that the value of the random variable is $\mathbf{y}$ when it is really $\mathbf{x}$. We naturally assume that $\nu(\mathbf{x}, \mathbf{y}) < \nu(\mathbf{x}, \mathbf{x})$. Compelling desiderata are (a) if a constant is added to ν then the same constant is added to U; (b) additivity for mutually irrelevant vectors

$$U(GG^*|FF^*) = U(G|F) + U(G^*|F^*);$$

(c) invariance under non-singular transformations of $\mathbf{x}$: $U(\mathrm{G|F})$ is unchanged if a non-singular transformation $\mathbf{x} = \psi(\mathbf{x}')$, $\mathbf{y} = \psi(\mathbf{y}')$ is made, subject to the obvious desideratum that the transformed form of ν is

$$\nu(\psi(\mathbf{x}'), \psi(\mathbf{y}')).$$

A doubly infinite system of functionals satisfying these desiderata is

$$U_a^{\beta}(\mathrm{G|F}) = \frac{1}{\beta}\log\int dF(\mathbf{x})\left[\int e^{a\nu(\mathbf{x},\mathbf{y})}dG(\mathbf{y})\right]^{\beta/a}$$

where $0 \leqslant a \leqslant \infty$, $0 \leqslant \beta \leqslant \infty$. (Another system of solutions can be obtained by interchanging F and G.) When $\beta \to 0$ we obtain

$$U_a(\mathrm{G|F}) = \frac{1}{a}\int dF(\mathbf{x})\log\int e^{a\nu(\mathbf{x},\mathbf{y})}dG(\mathbf{y}).$$

If it required further that (d) $U(F|F) \geqslant U(G|F)$ for all F and G we must let $a \to \infty$ and when G has a density function g we obtain *up to a linear transformation* (irrelevant for utility measures)

$$U_{\infty}^{\beta}(G|F) = U_{\infty}(G|F) = \int \log\{g(\mathbf{x})|\Delta(\mathbf{x})|^{-\frac{1}{2}}\}\, dF(\mathbf{x}) \tag{1}$$

where

$$\Delta(\mathbf{x}) = \left\{ \left.\frac{\partial^2 \nu(x,y)}{\partial y_j \partial y_k}\right|_{\mathbf{y}=\mathbf{x}} \right\} j,k = 1, 2, \ldots .$$

(The solutions with F and G interchanged cannot satisfy condition [*d*].)

If also F has a density function *f*,

$$U_{\infty}(F|F) = \int f(\mathbf{x}) \log\{f(\mathbf{x})|\Delta(\mathbf{x})|^{-\frac{1}{2}}\}\, d\mathbf{x}. \tag{2}$$

This can be regarded as minus an *invariantized* entropy and equation (1) as an invariantized *cross-entropy.*

When there is *quadratic loss,* that is, when ν is a quadratic form, the factor involving $\Delta(\mathbf{x})$ is constant and can be ignored and equation (1) reduces to an ordinary cross-entropy which was called by Kerridge (1961) the *inaccuracy* of G when F is true. This reduction occurs also if ν is any twice differentiable function of a quadratic form.

The formulae could be used in the design of experiments and in the summarizations of their results. Another interpretation can be given if there is a density function $f_0(\mathbf{x})$ proportional to $|\Delta(\mathbf{x})|^{\frac{1}{2}}$, that is, if $\int|\Delta(\mathbf{x})|^{\frac{1}{2}} d\mathbf{x}$ converges. Then $U_{\infty}(F|F)$ can be minimized by taking F to have the density function f_0, so that, with this utility measure, f_0 is the *least favorable* initial density. It is thus the minimax initial density (Wald, 1950). The principle of selecting the least favorable initial distribution may be called *the principle of least utility.* For discrete $\mathbf{x}$ it leads to the principle of maximum entropy (Jaynes, 1957) if $\nu(\mathbf{x}, \mathbf{x})$ is constant, and for continuous $\mathbf{x}$ if $|\Delta(\mathbf{x})|$ is constant. Otherwise, for continuous $\mathbf{x}$ it leads to a principle of maximum invariantized entropy. I think it is very interesting to see that the principle of maximum entropy can be regarded as a minimax procedure; and also to see how it should apparently be applied for continuous distributions: that U_{∞} reduces to negentropy when the loss function is quadratic or a function of a quadratic.

Suppose that $\mathbf{x}$ is the parameter in the distribution $T(\mathbf{z}|\mathbf{x})$ of another random vector $\mathbf{z}$, the density function being $t(\mathbf{z}|\mathbf{x})$. In the Jeffreys-Perks invariance theory (Jeffreys, 1946; Perks, 1947) the initial density is taken as the square root of the determinant of the information matrix

$$\left\{ -\int t(\mathbf{z}|\mathbf{x}) \frac{\partial^2 \log t(\mathbf{z}|\mathbf{x})}{\partial \mathrm{x}_j \partial \mathrm{x}_k} d\mathbf{z} \right\} j,k = 1, 2, \ldots$$

(strictly Perks was concerned only with the one-dimensional case). This comes to the same as using the principle of least utility with ν defined in terms of the expected weight of evidence

$$\nu(\mathbf{x},\mathbf{x}) - \nu(\mathbf{x},\mathbf{y}) = \int \log \frac{dT(\mathbf{z}|\mathbf{x})}{dT(\mathbf{z}|\mathbf{y})} dT(\mathbf{z}|\mathbf{x})$$

for distinguishing the true from the assumed distribution of **z**. This formula follows at once from equation (1), quite *irrespective of the minimax interpretation,* for we must have by definition

$$\nu(\mathbf{x},\mathbf{y}) = U_\infty(T(\mathbf{z}|\mathbf{y})|T(\mathbf{z}|\mathbf{x}))$$

if the only use of **x** is to serve as a parameter in the distribution of **z**. Hence $\Delta(\mathbf{x})$ is equal to Fisher's information matrix. Here ν and Δ are defined for the *random variable* **x** and should not be confused with the corresponding functions for **z**.

Thus we see that Harold Jeffreys's invariant density can be derived from a minimax procedure provided that utility differences are identified with weight of evidence. The disadvantage of Jeffreys's brilliant suggestion can therefore be attributed to that of minimax procedures in general.

The maximization of entropy is a very reasonable method for the estimation of probabilities in contingency tables, and in Markov chains, since it leads to hypotheses of generalized independence that are satisfactory to the intuition of statisticians. But in a problem such as medical diagnosis it is reasonable, when acquiring information, to try to *minimize* the entropy of a set of mutually exclusive diseases. In a medical diagnostic search tree, one is involved both with the estimation of probabilities and with the acquisition of new information. Hence a reasonable procedure is to try to minimax the entropy in the sense of the theory of games (##592, 755: rival formulae are given in these references).

Very closely related to and somewhat more general than the principle of maximum entropy is a principle of minimum discriminability in which expected weights of evidence are used in place of the entropy (Kullback, 1959). The formula for U_∞ gives some support for this. A satisfying property of minimum discriminability was shown in #522: if we have a chain of hypotheses H_1, H_2, . . . , H_n, concerning various probabilities, where the hypotheses satisfy increasing sets of linear constraints, and if we introduce additional constraints, and determine the next hypothesis H_{n+1} by minimum discriminability from any one of the earlier hypotheses, then we always arrive at the same hypothesis H_{n+1}. This is by no means obvious, but the proof is not difficult.

When we aim to maximize the expectation of any expression, that expression can be regarded as a pseudoutility or quasiutility, whether it be entropy, weight of evidence, or something else. As a historical matter, my first introduction to the use of expected weight of evidence as a measure of the value of a statistical investigation was in 1941 when working with Turing on a war-time project in which he invented sequential analysis [independently of Wald and of Barnard]. Expected weight of evidence also occurs in my definition of a decision (#315), but I shall not discuss that here.

It is useful to use quasiutilities when true utilities are difficult to estimate, as they often are, especially in problems of inference. The use of weight of evidence as a quasiutility is especially appropriate when we are trying to decide whether a hypothesis or its negation is true. If the negation of the hypothesis is sufficiently vague, or if we are not sure which of several hypotheses we are really interested in, then the entropy serves as a reasonable quasiutility. Moreover, owing to an additive property of entropy, the principles of maximizing and minimizing entropy are consistent when applied to a pair of completely independent problems. For a further discussion of this point see #592.

There seems to be a constant interplay between the ideas of entropy and expected weight of evidence, or *dientropy* as it might be called since it refers to two distributions. Even in statistical mechanics, it seems that expected weight of evidence is a useful concept, as conjectured by Good (1950/53) and demonstrated by Koopman (1969) for non-equilibrium problems. [See also Gibbs in my index.]

My conclusion is that the mathematics of the philosophy of inference is a useful and interesting pursuit.

[The discussion of this paper is omitted.]

CHAPTER 20

Is the Size of Our Galaxy Surprising? (#814)

Eddington (1933/52, p. 5) after pointing out that the earth and the Sun are of middling size, *qua* planet and star respectively, says, "So it seems surprising that we should happen to belong to an altogether exceptional galaxy." It is not quite as exceptional as Eddington thought; thus van de Kamp (1965, p. 331) said that the photometric studies by Stebbins and Whiteford in 1934 "did much to do away with the notion that our galaxy was a 'continent' and that the others were 'islands.' " But still our galaxy is a large one, and Eddington's remark raises an interesting logical question: Is the expected size of *our* galaxy larger than the average size of a galaxy? The answer is yes, for the following reason.

Let p_n be the probability that a galaxy is of "size" n, that is, it contains just n planets inhabitable by human beings. Then the average size of a galaxy is

$$\mu = \sum_{n=1}^{\infty} np_n.$$

But if we know that *we* are in a certain galaxy, the likelihood of the hypothesis that it contains n inhabitable planets is proportional to n. Therefore, by Bayes's theorem, the final (posterior) probability that this galaxy contains just n inhabitable planets is proportional to np_n; so the expected size of a galaxy, conditional on our being in it, divided by the average size μ of a galaxy, is

$$\frac{\Sigma n^2 p_n}{(\Sigma np_n)^2} = 1 + \frac{\sigma^2}{\mu^2} \qquad (1)$$

which exceeds 1, as asserted. (Here σ^2 is the variance of the size of a galaxy.) The argument can be applied in other contexts; for example, in my next reincarnation, assumed to be on this planet, I might very well be Chinese but would be surprised to be born in Liechtenstein. . . .

[The remainder of this paper consisted of an attempt to evaluate formula (1) by fitting a formula to the size of galaxies.]

Part V. Causality and Explanation

CHAPTER 21

A Causal Calculus (#223B)

1. INTRODUCTION

This paper contains a suggested quantitative explication of probabilistic causality in terms of physical probability. (Cf. Reichenbach, 1959, Chap. 3; Wiener, 1956, pp. 165-190.) The main result is to show that, starting from very reasonable desiderata, there is a unique meaning, up to a continuous increasing transformation, that can be attached to "the tendency of one event to cause another one." A reasonable explicatum will also be suggested for the degree to which one event caused another one. It may be possible to find other reasonable explicata for tendency to cause, but, if so, the assumptions made here will have to be changed.

I believe that the first clear-cut application in science will be to the foundations of statistics, such as to an improved understanding of the function of randomization, but I am content for the present to regard the work as contributing to the philosophy of science, and especially to what may be called the "mathematics of philosophy." Light may also be shed on problems of allocating blame and credit. I hope to consider applications to statistics on another occasion. [See #1317.]

In #180 I tried to give an interpretation of "an event F caused another event E" without making reference to time. It was presumably clear from the last three paragraphs, which were added in proof [the words "added in proof" were omitted in error, with a peculiar effect], that I was not satisfied with my attempt [see Appendix I]. I shall describe the note as the "previous paper" but it will not be necessary for the reader to refer back.

The present paper is more ambitious in that it is quantitative, but less so in that it assumes, at least at first, that F is earlier than E. (It may be possible to interpret the explicatum more generally.) As in the previous paper I shall take for granted the notion of physical (= material) probability. In order to avoid misunderstanding I must mention my opinion that in so far as physical probability can be measured it can be done only in terms of subjective probability, but this opinion will not affect the arguments below. Likewise the notion of an "event" will be taken for granted. In some earlier drafts I included material dealing with the meanings of "event," "probability," and "definition," and with the modifications of the analysis required in order to cope with the possibility that the future may affect the past. I have omitted this material here for the sake of brevity.

2. NOTATION AND GENERAL OUTLINE

Propositions and events will be understood in a very general sense, and will be denoted by the symbols E, F, G, H, and U. These will be combined by means of the logical connectives "." meaning "and," "—" meaning "not," and "v" meaning "or." A vertical stroke, "|," will mean "given," as in the expression $P(\mathrm{E}|\mathrm{H})$, the probability of E given H. Similarly $O(\mathrm{E}|\mathrm{H})$ will mean that the odds of E given H, i.e. $P(\mathrm{E}|\mathrm{H})/P(\bar{\mathrm{E}}|\mathrm{H})$. Sometimes some or all of what is "given" is omitted from the notation for the sake of brevity. A colon will be used to mean "provided by" or "by" or "from," as in

$$I(\mathrm{E}:\mathrm{F}|\mathrm{G}) = \log(P(\mathrm{E}|\mathrm{F.G})/P(\mathrm{E}|\mathrm{G})),$$

which can be read from left to right as the amount of information concerning E provided by F given G. Another example of the colon notation is

$$\begin{aligned} W(\mathrm{H}:\mathrm{E}|\mathrm{G}) &= \log(P(\mathrm{E}|\mathrm{H.G})/P(\mathrm{E}|\bar{\mathrm{H}}.\mathrm{G})) \\ &= \log(O(\mathrm{H}|\mathrm{E.G})/O(\mathrm{H}|\mathrm{G})) \\ &= I(\mathrm{H}:\mathrm{E}|\mathrm{G}) - I(\bar{\mathrm{H}}:\mathrm{E}|\mathrm{G}), \end{aligned}$$

the "weight of evidence concerning H provided by E given G." (See, for example, #211.)

The general plan of the paper is to suggest explicata for:

(i) $Q(\mathrm{E}:\mathrm{F})$, or Q for short, the "causal support for E provided by F, or the tendency of F to cause E." The explicatum that the argument forces upon us is the weight of evidence against F if E does not happen, $W(\bar{\mathrm{F}}:\bar{\mathrm{E}}|\mathrm{U.H})$, where U and H are defined below. In order to formulate enough desiderata it is necessary to introduce some auxiliary notions.

(ii) The strength of a causal chain joining F to E.

(iii) The strength of a causal net joining F to E. Causal chains and nets will be defined in Sections 8 and 11.

(iv) $\chi(\mathrm{E}:\mathrm{F})$, or χ for short, the contribution to the causation of E provided by F, i.e. the degree to which F caused E. This will be defined as the strength of a causal net joining F to E, when the details of the net are completely filled in, so that there are no relevant events omitted. I avoid the use of the letter C for either Q or χ, because it has been used to mean corroboration: see Popper (1959). An example is given in Appendix I to show that Q and χ cannot be identified.

It would not be appropriate to define χ as the limit of strengths of more and more detailed nets; for, if space and time are continuous, the limiting operation could be done in a biased manner so as to get entirely the wrong result; like a lawyer making a case by special selection of the evidence. We must have the whole truth in order to define χ in principle. (Compare the first Appendix.) If, however, the events fill the relevant parts of space and time, and we let the events become smaller and smaller, then the limit should be unique.

In practical uses of the notion of causality, judgments of approximate irrelevance are always made in order to reduce the complication of the causal net.

It is possible to draw an analogy between a causal net and an electrical resistance network, with a resistance in each link. In this analogy it is necessary to imagine a rectifier placed in each link in order to prevent a flow of causal influence backward in time. It is then tempting to define the degree of causality between the input and output of the causal net as the effective resistance of the corresponding causal network. This analogy suggests that the causal resistances should be defined in such a manner that they are additive for chains in "series," and such that their reciprocals are additive for chains in "parallel." It turns out that the analogy cannot be pressed as far as this, but it is one of the themes of the paper.

The main part of the paper consists of a list of assumptions, and deductions from them, leading up to the explicatum for *Q*. Afterwards a general explicatum will be suggested for χ, but this will not be deduced in the same formal manner. It is fairly convincingly unique for causal nets of the "series-parallel" type, and has a certain cogency in the general case. [I regard *Q* as more operational.]

3. SMALL EVENTS

Until near the end of the paper all events will be assumed to occupy small volumes of space (more precisely: have small diameters) and occupy small epochs of time. For the most part "space" could be interpreted in a more general sense than as ordinary three-dimensional space; for example, it could be phase space or Hilbert space. On the other hand time will be assumed to be well-ordered and one-dimensional. There must be some sort of metric in both space and time, in order that smallness and continuity should have a meaning. If the metrics of space and time have been mixed up, as in the theory of relativity, then they will be assumed to be unmixed by the use of a fixed frame of reference. (Dr. O. Penrose has pointed out that the present work is consistent with the theory of relativity provided that causal influence does not travel faster than light.)

4. CAUSAL SUPPORT, OR TENDENCY TO CAUSE

Let H denote all true laws of nature, whether known or unknown, and let U denote the "essential physical circumstances" just before F started. When we talk about "essential physical circumstances" we imply that the exact state has a probability distribution. An equivalent description is to say that a system is one of an "ensemble." (I must admit that there is more than meets the eye in this description, since in quantum mechanics the word "state" can be given at least eight interpretations, seven of which may be relevant here. See Appendix III.)

In order to arrive at explicata for Q and χ I have found it necessary to discuss them in an interconnected manner; i.e. there do not appear to be enough desiderata for Q, considered by itself, to circumscribe its possible explicata to a satisfactory extent.

In the present section the ground will be cleared by discussing what Q and χ depend upon. It is convenient to think of this dependence in terms of notation, which seems to bring out the main points better than a purely verbal discussion. For example, the symbols Q and χ are abbreviations for $Q(\text{E:F})$, and $\chi(\text{E:F})$, and these expressions are themselves abbreviations for $Q(\text{E:F}|\text{U.H})$ and $\chi(\text{E:F}|\text{U.H})$. To take U and H for granted, and omit them from the notation, is parallel to linguistic usage. If we say that it is bad for eggs to throw them in the air, we take it for granted that there is a law of gravitation, and that there is a large gravitational body nearby.

Events later than F and earlier than E may be relevant to χ but not to Q. Accordingly I shall assume that $Q(\text{E:F})$ depends only on $P(\text{E}|\text{F})$, $P(\text{E}|\overline{\text{F}})$, $P(\text{E})$, and $P(\text{F})$. It is natural to define $Q(\text{E:F}|\text{G})$ as the same function of these four probabilities, but made conditional on G.

Even $Q(\text{E:F}|\text{U.H})$ is an incomplete notation. If the subjective element is to be removed from the expression "F caused E," then it must be expanded to "F, as against $\overline{\text{F}}_\text{D}$, caused E rather than E′," where the suffix, D, to $\overline{\text{F}}$ (the negation of F), represents a complete specification of the relative probabilities of the mutually exclusive events whose disjunction is $\overline{\text{F}}$. (D represents a probability distribution.) We could use a notation like

$$Q(\text{E/E}':\text{F}/\overline{\text{F}}_\text{D}|\text{U.H})$$

or

$$Q(\text{E:F/F}_\text{D}|\text{U.H.(E v E}')),$$

the degree of causation of E rather than E′ by F rather than $\overline{\text{F}}_\text{D}$.

The failure to recognise all the variables on which tendency to cause is based was for me one of the stumbling blocks in capturing the notion of probabilistic causality, if indeed I have fully succeeded even now.

It should be held in mind that E v E′ is regarded as taken for granted in the four probabilities on which Q is assumed to depend, when we are concerned with the causation of E rather than E′. When we take E v E′ for granted we may write $\overline{\text{E}}$ instead of E′.

5. ASSUMPTIONS AND DEDUCTIONS LEADING TO THE EXPLICATUM FOR Q

In order to make my assumptions clear I shall list them in the form of axioms, A1, A2, . . . ; and the deductions from them will be called theorems T1, T2, . . . , for ease of reference. On a first quick reading the justifications and proofs should *quite definitely* be skipped, but I have not postponed them to a later section. (I did so in an earlier draft, but the cross-referencing made the paper more difficult to read.) The justifications of the most easily acceptable axioms, and the proofs of the easily proved theorems will be omitted.

$$\text{Let } P(\mathrm{F}) = x,\ P(\mathrm{E}|\mathrm{F}) = p,\ P(\mathrm{E}|\overline{\mathrm{F}}) = q,\ P(\mathrm{E}) = r, \text{ so that}$$
$$r = xp + (1 - x)q,\ x = (r - q)/(p - q).$$

Unless $p = q$ (in which case $r = p = q$), x is a function of p, q, and r. Therefore by an assumption of the previous section we have:

A1. *Q(E:F|G) is a function of* p, q, r, *unless perhaps* p = q = r. *We call this function* Q(p, q, r) *so that the symbol* Q *has two slightly different meanings. The symbol G will usually be taken for granted and omitted.* Here $p = P(\mathrm{E}|\mathrm{F.G})$, $q = P(\mathrm{E}|\overline{\mathrm{F}}.\mathrm{G})$, $r = P(\mathrm{E}|\mathrm{G})$.

A2. Q *is a real number or* ∞ *or* $-\infty$; *but it may be indeterminate for special values of* p, q, *and* r, *such as when two of them are equal or one of them is 0 or 1.* (It seems sensible to look for a scalar explicatum rather than a "vector.")

The next two axioms deal with monotonicity and continuity.

A3. (i) Q *increases with* p *if* q *and* r *are held constant.*

(ii) Q *decreases when* q *increases if* p *and* r *are held constant.*

A4. Q *is continuous except when it becomes infinite or indeterminate, if there are such points.*

A5. *If* P(*F*) $\neq$ *1, meaning, as usual,* P(*F*|*U.H'*) $\neq$ *1, then* Q *has the same sign as* p − r, *and therefore also the same sign as* p − q, *and as* r − q; *and if these expressions vanish we may say that F has no tendency to cause E, and we put* Q = *0.* (This axiom removes a gloss from A1.)

A6. *Any causal net joining F to E, as defined below in Section 11, has a causal strength,* S, *and a causal resistance,* R. *These are positive numbers, except that if* p = q = r, *or if* p *or* q *is 0 or 1, we may get zero or infinite resistance or strength.* (An important part of the definition of a causal net is that it consists only of events that actually occurred or will have occurred.)

A7. *There is a functional relationship between* R *and* S, S = f(R), R = g(S), *where* f *and* g *are absolute functions inverse to one another.*

A8. f *and* g *are continuous decreasing functions.*

A9. *$\chi(E{:}F)$ is the strength of the complete causal net joining F to E. More precisely, it is the limit, as the sizes of the events tend uniformly to zero, of the strengths of nets; where each net of the sequence joins F to E, consists of a finite number of events, and omits no events temporarily between F and E.* It is not claimed that this axiom is formulated with complete rigor, but it is used in only a weak form for the explication of Q (in the proof of T14). It is introduced at this early stage in order to supplement the outline in Section 2. If we assume that the degree to which F caused E has an objective meaning, with a precise numerical value, we are committed to the idea that there is a complete world, uninfluenced from outside. Outside influence could be allowed for by assuming that the numerical values are not absolutely precise.

A10. *The strength of the causal net consisting of F and E alone is equal to Q(E:F) when this is positive, and is otherwise zero.* We shall clearly get nowhere unless we assume some relationship between Q and S, and A10 is the simplest reasonable relationship that could be assumed.

The strength and resistance of a net, $\mathfrak{n}$, joining F to E, are denoted by $S(\mathrm{E{:}F}|\mathfrak{n})$ and $R(\mathrm{E{:}F}|\mathfrak{n})$.

A11. *Let $\mathfrak{n}$ be a "chain," $F = F_0 \rightarrow F_1 \rightarrow \ldots \rightarrow F_n = E$. Then $S(E{:}F|\mathfrak{n})$ is some function, $\phi(s_0, s_1, \ldots, s_{n-1})$, of the strengths, $s_0, s_1, \ldots, s_{n-1}$, of the links.* Here

$$s_i = S(\mathrm{F}_{i+1}{:}\mathrm{F}_i|\mathrm{U}_i.\mathrm{H}),$$

where U_i represents the essential physical circumstances just before F_i began. (Causal chains are formally defined in Section 8.)

T1. $\phi(s) = s$. (Proof from A10 and A11.)

A12. *ϕ is a symmetrical function, i.e. the arguments of the function can be permuted without changing its value.*

A13. *ϕ is a non-decreasing function of each of its arguments.* That is, a chain cannot be weakened by strengthening any of its links without changing the strength of any of the others.

T2. *ϕ vanishes if the chain is cut, i.e. if any of the links is of zero strength.* (A cut chain can be uncut by filling it in in more detail.) We may alternatively say in this case that there was no causal chain.

A14. *If two consecutive links are replaced by a single link of equivalent strength, then the strength of the chain is unchanged. Formally,*

$$\begin{aligned}&\phi(s_0, s_1, \ldots, s_{n-1})\\ &\qquad = \phi(s_0, s_1, \ldots, s_{i-1}, \phi(s_i, s_{i+1}), s_{i+2}, \ldots, s_{n-1}).\end{aligned}$$

A15. *A chain is not weakened by "omitting" one of its links, i.e.*

$$\begin{aligned}&\phi(s_0, s_1, \ldots, s_{i-1}, s_{i+1}, \ldots, s_{n-1})\\ &\qquad \geqslant \phi(s_0, s_1, \ldots, s_{n-1}).\end{aligned}$$

T3. *A chain is no stronger than its weakest link.* (From A10 and A15.)

Definition. Let the maximum possible causal strength be σ. This is either a positive finite number or $+\infty$.

A16. $S(\mathrm{E{:}F}|\mathfrak{n}) \leqslant \sigma$, *for any net*, $\mathfrak{n}$. (This axiom is a mere restatement of the definition.)

T4. $Q(\mathrm{p, q, r}) \leqslant \sigma$. (From A10 and A15.)

A17. *If any of the links of a chain is of strength* σ, *then it can be "omitted" in the sense of A15, without strengthening the chain.*

T5.
$$\phi(s_0, s_1, \ldots, s_{i-1}\sigma, s_{i+1}, \ldots, s_{n-1}) = \phi(s_0, s_1, \ldots, s_{i-1}, s_{i+1}, \ldots, s_{n-1}).$$

(From A15 and A17.)

T6. *If every link of a chain is "cast-iron," then the chain is cast-iron, i.e.* $\phi(\sigma, \sigma, \ldots, \sigma) = \sigma$. (From T1 and T5.)

A18. *A chain of* n *links all of the same fixed strength*, τ, *where* $\tau < \sigma$, *is as weak as you like if* n *is large enough. Formally* $\phi(\tau, \tau, \ldots, \tau) \to 0$ *as the number of arguments tends to infinity.*

A19. ϕ *is a continuous function of all its arguments when they are all less than* σ; *and, if* $s_i \to \sigma$, *then the value of the function tends to the value it would have with* $s_i = \sigma$. The reason for the clumsy expression of this axiom is that σ may be $+\infty$.

T7. *If a chain has* n *links, all of the same strength*, τ, *where* $\tau < \sigma$, *then the chain is as strong as you like if* n *is fixed and* τ *is close enough to* σ. *Formally, if* n *is fixed, then*

$$\phi(\tau, \tau, \ldots, \tau) \to \sigma \text{ when } \tau \to \sigma.$$

(From T6 and A19.)

T8. *There is a function,* g, *such that, identically,*

$$\phi(s_0, s_1, \ldots, s_{n-1}) = g^{-1}(g(s_0) + g(s_1) + \ldots + g(s_{n-1})).$$

The function, g, *is defined for all non-negative arguments not exceeding* σ, *and is itself non-negative, continuous and strictly decreasing, and* $g(0) = \infty$, $g(\sigma) = 0$. *We may define* g *as* $+\infty$ *when its argument is negative.*

Proof. Consider the function $\phi(s,t)$ of just two variables. By A19, A13, A12, and A14, this function may be said to be continuous, monotonic, commutative, and associative. It follows that it is of the form $g^{-1}(g(s) + g(t))$, where g is a strictly monotonic continuous function. The use of the symbol g is justified since A7 and A8 can be satisfied with this function.

The mathematical theorem just invoked was apparently first published by Abel (1826, . . .). It was rediscovered several times, such as by me in 1940 (unpublished) and by Aczél (1948, 1955) who improved it. See also Jànossy (1955) or *Math. Rev. 10*, 685, and *16*, 1127. What it amounts to is that ϕ can be calculated on a suitably calibrated slide-rule.

The results $g(0) = 0$, $g(\sigma) = \infty$, follow from A18 and A19. Q.E.D.

We may satisfy A6, A7, and A8, which are the only axioms that involve R, by identifying $g(S)$ with R. This identification is no restriction on the explication of Q. As a consequence of this identification we have the following theorem.

T9. *The resistance of a chain is equal to the sum of the resistances of its links.*

A20. *Consider the causal net shown in the diagram below, in which* $P(\mathrm{F}) = x$, $P(\mathrm{G}_j|\mathrm{F}) = p_j$, $P(\mathrm{G}_j|\overline{\mathrm{F}}) = q_j$, $P(\mathrm{G}_j) = r_j = xp_j + (1-x)q_j$, $P(\mathrm{E}|\mathrm{G}_1 \vee \mathrm{G}_2 \vee \mathrm{G}_3) = 1$, $P(\mathrm{E}|\overline{\mathrm{G}}_1 \,.\, \overline{\mathrm{G}}_2 \,.\, \overline{\mathrm{G}}_3) = 0$, $P(\mathrm{E}) = r$, $j = 1, 2, 3$, *and where* G_1, G_2, G_3 *are independent given* F *and also given* $\overline{F}$. *Then the strength of the net is a function of the strengths of the three separate chains, and this function is continuous, monotonic increasing in each variable, commutative (cf. A12), and associative (cf. A14).*

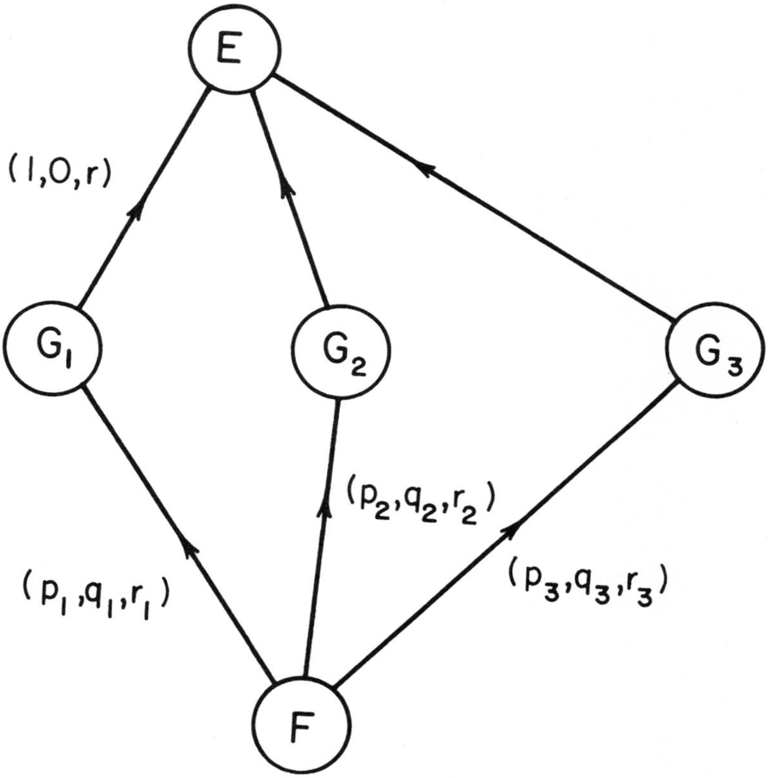

Figure 1.

T10. *The strength of the net of A20, generalized to* m *chains in parallel, is of the form*

$$h^{-1}(h(s_1) + h(s_2) + \ .\ .\ .\ + h(s_m)),$$

where $s_1, s_2, \ldots, s_m$ *are the strengths of the individual chains. The function* h *is defined for all non-negative arguments not exceeding* σ, *and is itself nonnegative, continuous and strictly increasing, and* $h(0) = 0$, $h(\sigma) = \infty$. (The theorem of the generalized slide-rule implies this theorem, just as it implied T8 above.)

We are now at liberty to call $h(S)$ the strength of a causal net, in place of S, provided we are content to determine the explicatum of S and Q only up to a continuous increasing transformation. It might be thought for a moment that this change of notation would invalidate T9. But since T8 is now true with $g(x)$ replaced by $g(h^{-1}(x))$, we can simply rename this function "$g(x)$" in order to validate T9. With these conventions we have:

T11. *The strength of the net of A20, generalized to* m *chains in parallel is the sum of the strengths of the individual chains.* When applying this theorem the independence condition mentioned in A20 should not be overlooked.

It appears that the analogy with electric networks is not bad, although the function $f(x)$ turns out later not to be $1/x$, but another self-inverse function.

A21. *For the net of A20 (a "firing squad"), with G3 omitted, i.e. with only two chains in parallel, it will be assumed that the strength of the net is equal to the tendency of F to cause E worked out as if the only events were those in the net. This assumption is an extension of A10.* Note that it would be unreasonable to assume this coalescence property for dependent events, for if we did so we could collapse any net into a single event.

Although I think A21 is eminently reasonable, especially in view of later developments, as in Section 9, I believe it to be the weakest part of my argument, and I conjecture that the replacement of this axiom by other assumptions would be the most fruitful method of finding other explicata of tendency to cause, if they exist.

T12. *Identically, if* $p_1 \geqslant q_1$, $p_2 \geqslant q_2$, $0 < x < 1$, *then*

$$S(p_1 + p_2 - p_1p_2, q_1 + q_2 - q_1q_2, x(p_1 + p_2 - p_1p_2) + (1 - x)(q_1 + q_2 - q_1q_2))$$
$$= S(p_1, q_1, xp_1 + (1 - x)q_1) + S(p_2, q_2, xp_2 + (1 - x)q_2).$$

This follows at once from T11 by making the identification mentioned in A21.

A22. $Q(p, q, r)$ *is an analytic function when* $0 < p < 1$, $0 < q < 1$, $0 < r < 1$, $p \neq q$.

The only purpose of this axiom is to enable us to extend a formula proved for a large set of values of (p, q, r) to all values except those for which Q may be infinite or indeterminate. I think only an extreme purist would object to A22. It could be avoided by assuming instead that Q is anti-symmetric in the sense

$$Q(p, q, xp + (1 - x)q) = -Q(q, p, xq + (1 - x)p).$$

T13.

$$Q(p, q, r) = u(x) \log \frac{1 - q}{1 - p} = u\left\{\frac{r - q}{p - q}\right\} \log \frac{1 - q}{1 - p},$$

where $u(x)$ *is a non-negative analytic function of* x.

This theorem, and the next one, will be superseded by T15.

Proof. By A10 and A22, we may replace S by Q in T12, and drop the inequalities $p_1 \geqslant q_1$, $p_2 \geqslant q_2$. Let

$$\psi(\xi, \eta, x) = Q(1 - e^{\xi}, 1 - e^{\eta}, x(1 - e^{\xi}) + (1 - x)(1 - e^{\eta})),$$
$$p_1 = \exp \xi_1, \; q_1 = \exp \eta_1, \text{ etc.}$$

Then

$$\psi(\xi_1 + \xi_2, \eta_1 + \eta_2, x) = \psi(\xi_1, \eta_1, x) + \psi(\xi_2, \eta_2, x).$$

On putting $\eta_1 = \eta_2 = 0$, and provisionally regarding x as a constant, we get a well known functional equation whose only continuous solution is easily seen to be of the form

$$\psi(\xi, 0, x) = \xi \, . \, u(x),$$

where $u(x)$ is a function of x only. (The only other solutions are in fact non-measurable: see Hamel, 1905, or Hardy *et al.*, 1934, p. 96.) Likewise $\psi(0, \eta, x) = \eta \cdot w(x)$, where $w(x)$ is a function of x only. Therefore

$$\begin{aligned}\psi(\xi, \eta, x) &= \psi(\xi + 0, 0 + \eta, x)\\ &= \psi(\xi, 0, x) + \psi(0, \eta, x) = \xi u(x) + \eta w(x).\end{aligned}$$

Therefore

$$Q(p, q, xp + (1 - x)q) = u(x) \, . \, \log(1 - p) + w(x) \, . \, \log(1 - q).$$

T13 now follows from A5 combined with the equation

$$r = xp + (1 - x)q.$$

Q.E.D.

A23. *Consider a radioactive particle in a certain state, which I shall call the "white" state. In any time interval,* t, *it has probability* e^{-at} *of remaining in the white state throughout the interval if it starts the interval in that state. If it does not remain in the white state, then it proceeds to another state called here the "black" state, from which there is no return. Now let F be the event that the particle is in the white state at the start of an interval of duration T and let E be the event that it is in the white state at the end of this interval. Then we assume that, if F and E both occurred,* $\chi(E:F)$ *does not depend on the unit in terms of which time is measured.*

A24. *If F. E implies G, and* $F \rightarrow G \rightarrow E$ *is a chain, then this chain is of the same strength as* $F \rightarrow E$.

T14. $R(p, 0, r) = v(r/p) - k \, . \, \log p$,
where v(x) *is a non-negative analytic function of* x, *and* k *is a positive constant.*

Proof. Consider the radioactive particle described in A23. Let $P(\mathrm{F}) = x$. The degree to which F caused E is the limit of the strengths of finite chains obtained by breaking up the time interval $(0, T)$ into a "Riemann dissection" (see A9). Since g is a continuous function (A8) the resistances of these finite chains must also tend to a limit, which we may call the causal resistance from F to E. This must be some function of x, a, and T, say $R^*(x, a, T)$. By A23 we see that for any positive constant, k, the resistance must be equal to $R^*(x,\ ka,\ T/k)$. Since this is independent of k it must be of the form of $R^*(x,\ aT)$.

Now, by a continuity argument, we may generalize T9 to continuous chains, and hence deduce that, for any positive T and U we have

$$R^*(x,\ aT) + R^*(1,\ aU) = R^*(x,\ aT + aU).$$

By giving x the value 1 and subtracting from the equation with arbitrary x, we see that $R^*(x,\ aT)$ is of the form

$$R^*(x,\ aT) = \nu(x) + R^*(aT),$$

where, identically,

$$R^*(aT_1 + aT_2) = R^*(aT_1) + R^*(aT_2),$$

so that $R^*(aT)$ is of the form

$$R^*(aT) = k_1 aT.$$

Now, by repeated use of A24, we see that

$$R(p,\ 0,\ xp) = R^*(x,\ aT),$$

where $p = e^{-aT}$. Thus

$$R(p,\ 0,\ r) = \nu(r/p) - k \ .\ \log p.$$

Q.E.D.

T15. $$Q(p,\ q,\ r) = \log(1-q) - \log(1-p),$$
$$R(p,\ 0,\ r) = -\log p,$$

where the base of the logarithms may be taken as e. Q(p, q, r) *is mathematically independent of* r, *and may be abbreviated to* Q(p,q). *It can be written in other ways:*

$$Q(\mathrm{E}:\mathrm{F}|\mathrm{G}) = \log\frac{P(\overline{\mathrm{E}}|\overline{\mathrm{F}}\ .\ \mathrm{G})}{P(\overline{\mathrm{E}}|\mathrm{F}\ .\ \mathrm{G})} = \log\frac{O(\overline{\mathrm{F}}|\overline{\mathrm{E}}\ .\ \mathrm{G})}{O(\mathrm{F}|\mathrm{G})}$$
$$= W(\overline{\mathrm{F}}:\overline{\mathrm{E}}|\mathrm{G}) = -W(\mathrm{F}:\overline{\mathrm{E}}|\mathrm{G}),$$

the weight of evidence against F if E does not happen. More precisely, Q *is uniquely determined only up to a continuous analytic increasing transformation. Among all the explicata there is just one apart from a scale factor (choice of unit), for which theorems T9 and T11 are true.* We lose no real generality, and we gain simplicity, by choosing this explicatum.

Proof. By T13, T14, and A7, we have the identity

$$f(\nu(r/p) - \log p) = -u(r/p)\ .\ \log(1-p).$$

Let $\nu(x) = y$, $-\log p = z$, $\log f(y+z) = \rho(y+z)$. Then $\rho(y+z)$ is of the form

$$\rho(y+z) = \rho_1(y) + \rho_2(z).$$

If $\nu(x)$ is not a constant, we can differentiate and deduce that $\rho(y)$ is a linear function of y, from which we can soon derive that $\log(1-p)$ is a power of p.

Since this is false it follows that $v(x)$ is a constant, and hence also that $u(x)$ is a constant.

The theorem now follows from the remark that the choice of the base of the logarithms is equivalent merely to the choice of units of measurement of strength and resistance. We may call the units "natural," "binary," or "decimal," according as the base is e, 2, or 10. In this paper I shall use natural units. Possible names would be "natural causats" and "natural tasuacs."

Note that the explicatum for Q was by no means obvious in advance, nor was it obvious that all the desiderata could simultaneously be satisfied.

It is interesting to note that, if, contrary to most of the discussion, we assume E to be earlier than F, and if the universe has the "Markov" property, defined below, then the tendency of F to cause E is zero. This result may very well have been taken as a desideratum, but was in fact noticed only after the explicatum was obtained. By the Markov property is meant here that, for prediction, a complete knowledge of the immediate past makes the remote past irrelevant.

T16. *The relationship between* R *and* S *is symmetrical, namely*

$$R \geqslant 0,\ S \geqslant 0,$$
$$e^{-R} + e^{-S} = 1,$$

or equivalently,

$$R = -\log(1 - e^{-S}),\ S = -\log(1 - e^{-R}).$$

Further,

$$R(p, q, r) = \log(1 - q) - \log(p - q).$$

This is an immediate corollary of A7 and T15.

Thus the function f is its own inverse, g. It is tempting to permit negative and imaginary values because some of the formalism is faintly reminiscent of Feynman's formulation of quantum mechanics, but I shall not pursue this matter here.

T17. *If a chain consists of* n *links whose* p's *and* q's *are* (p_i, q_i), *where* $p_i \geqslant q_i$, *then its causal strength is*

$$-\log\left\{1 - \prod_i \frac{p_i - q_i}{1 - q_i}\right\}.$$

This follows from T16 and T9.

Before reading the proofs in the present section the reader will probably prefer to read the next two sections, in which some examples are given.

6. TWO-STATE MARKOV PROCESSES

The radioactive process described in Axiom 23 can be slightly generalized by permitting return from the black to the white state, with a parameter β corresponding to the α of the white-to-black transition. We have a two-state Markov process with continuous time. The parameters α and β are of course both non-negative. In the special case of the radioactive particle we have $\beta = 0$.

It can be shown that

$$Q(\mathrm{E}:\mathrm{F}) = \log[(\alpha + \beta e^{-(\alpha+\beta)T})/(\alpha - \alpha e^{-(\alpha+\beta)T})].$$

If the particle ever entered the black state during the time interval, T, the chain would be cut and the degree of causality would be zero. Assuming that this does not happen, we can calculate $\chi(\mathrm{E}:\mathrm{F})$ by applying a Riemann dissection to the interval, so as to obtain a causal chain consisting of a finite number of events, and then proceed to the limit as the fineness of the dissection tends to zero. By applying T17 and A9 we find that

$$\chi(\mathrm{E}:\mathrm{F}) = -\log(1 - e^{-\alpha T}),$$

which is mathematically independent of β.

For large T, both Q and χ are exponentially small, but Q is smaller than χ and is much smaller if β is large. This is reasonable since, if β is large, the initial state makes little difference to the probability of being in the white state at the end of the interval.

Note that χ is the degree to which being in the white state rather than in the black state at the end of the interval was caused by being in the white state rather than in the black state at the start of the interval. A similar explicit description can of course be given for Q.

7. PARTIALLY SPURIOUS CORRELATION

A well known pitfall in statistics is to imagine that a statistically significant correlation or association is necessarily indicative of a causal relationship. The seeing of lightning is not usually a cause of the hearing of thunder, though the two are strongly associated. Such associations and correlations are often described as "spurious," a better description than "illusory." They may also be *partially spurious*, and the explicata for Q and χ should help with the analysis of such things. Smoke and dust might be a strong cause of lung cancer, but smoking only a weak cause. Even so, the correlation between smoking and lung cancer may be high if there is more smoking per head in smoky districts. I mention this only as an example, and have not made a special study of this problem.

Note that

$$Q(\mathrm{E}:\mathrm{F}.\mathrm{G}/\overline{\mathrm{F}}.\overline{\mathrm{G}}) = Q(\mathrm{E}:\mathrm{G}|\overline{\mathrm{F}}) + Q(\mathrm{E}:\mathrm{F}|\mathrm{G}),$$

so that the tendency to cause can be split into components, somewhat in the manner of an analysis of variance. For example, the tendency for lung cancer to be caused by smoking and living in a smoky district, as against not smoking and living in a clean district, is equal to the tendency through living in a smoky district, given no smoking, plus the tendency through smoking, given that the district is smoky. It is also equal to the causal tendency through living in a smoky district, given that one smokes, plus the tendency through smoking, given that the district is clean. This approach to the analysis of spurious correlation is entirely different from, and more quantitative than, the approach used by Simon (1957).

Let

$$K(E:F) = -I(\overline{E}:F),$$

the *intrinsic* causal tendency of E by F. It is related to Q in essentially the same way that I is related to W, since

$$Q(E:F) = K(E:F) - K(E:\overline{F}),$$
$$Q(E:F/F') = K(E:F) - K(E:F').$$

K does not depend on the negation of F, so its use enables us to avoid the distribution, D, of Section 4. We have

$$K(E:F.G) = K(E:F) + K(E:G|F),$$

so that K can be split up into contributions from various sources in a simpler manner than Q. In my opinion both K and Q will probably have useful applications in statistics and physics.

The remainder of this paper is primarily concerned with the extension of the explication of causal strength to general nets, in order that degree of causality should be generally explicated. The next section however contains a formal definition of a causal chain, which strictly was required in what has already been discussed. I postponed it in order not to interrupt the thread of the argument.

8. CAUSAL CHAINS

Let $F = F_0, F_1, \ldots, F_{n-1}, F_n = E$, be $n + 1$ events such that (for $i = 0, 1, \ldots, n-1$):

(*i*) F_i and F_{i+1} are contiguous in space and time, or approximately so.

(*ii*) No two of the events overlap much in space *and* time.

(*iii*) All the events occurred or will have occurred, i.e. they "obtain" but I prefer to write simply "occurred."

(*iv*) F_{i+1} started later than F_i did.

(*v*) F_i had a positive tendency to cause F_{i+1}.

(*vi*) If F_i is given, then the probability of F_{i+1} is unchanged if one or more of the earlier events did not occur, i.e. we have a Markov chain.

(*vii*) If the chain is embedded in a completely detailed chain containing intermediate events, then condition (*v*) will remain true for the more detailed chain.

Then we say that $F_0, F_1, \ldots, F_n$ or $F_0 \rightarrow F_1 \rightarrow \ldots \rightarrow F_n$ is a causal chain connecting F to E. Perhaps it should be called a "putative causal chain" if condition (*vii*) has not been established. In practice all causal chains are putative, but there are degrees of putativity.

The failure of condition (*v*) may be said to "cut the chain."

A causal net will be formally defined in Section 11. A chain is a special case of a net.

[For more discussion of causal chains see Reichenbach (1956, index reference under "Causal chain") and (1958 index reference under "Casual chain" [sic]).]

9. INDEPENDENT CAUSAL TENDENCIES

Let $G_1, G_2, \ldots, G_m$ be independent given H, and also independent given $H.\bar{E}$. Then it is easily proved, with the help of T15, that the tendencies to cause E are additive in the sense of the theorem below. It therefore seems reasonable to say in these circumstances that the G's have *independent tendencies to cause E given H.* The events G_1, G_2, and G_3 of A20 exemplify this definition, with H = F, and also with $H = \bar{F}$; that they are independent given $\bar{E}$ is trivial since their probabilities are then all zero.

T18. *If $G_1, G_2, \ldots, G_m$ have independent tendencies to cause E given H, then*

$$Q(E{:}G_1G_2 \ldots /\bar{G}_1\bar{G}_2 \ldots |H) = \Sigma_i Q(E{:}G_i|H).$$

Note that

$$Q(E{:}G_1|G_2 \ldots G_mH) = Q(E{:}G_1|H),$$

a natural requirement for independent causal tendencies.

The nets of A20 and T10 also exemplify the following definition:

A bundle of parallel independent causal chains from F to E is a class of chains from F to E such that, apart from F and E, each event on each chain is, given F and given $\bar{F}$, probabilistically independent of any collection of events on other chains, and also such that the penultimate events have independent tendencies to cause E, given their pasts.

10. SERIES-PARALLEL NETWORKS

As an extension of T11 it is natural to define the strength of a bundle of independent causal chains as the sum of the strengths of the individual chains.

For a "chain of bundles," in a self-explanatory sense, we can first calculate the resistance by summing the resistances of the individual bundles, and then obtain the strength from T16. We can extend the process to bundles of chains of bundles and so on. In other words we can construct natural rules for evaluating the causal strength of any "series-parallel" net. Topologically these are the same as the two-terminal series-parallel networks whose enumeration was considered by MacMahon (1892) or Riordan (1958, pp. 139-143). Not all networks are of this type.

11. CAUSAL NETS "HAVING INDEPENDENCE"

Let $\mathfrak{n}$ be a class of events all of which occurred. For each event, G, in $\mathfrak{n}$, there is a subclass of earlier events, $G_1, G_2, \ldots, G_k$, which so to speak, "lead in" to G. By "lead in" is meant that the probability of G, given which of $G_1, G_2, \ldots, G_k$ occurred and which did not, is independent of any further assumptions of which other events in $\mathfrak{n}$, earlier than G, occurred. (Note that not all the events in $\mathfrak{n}$ are regarded as "given" even though they all actually occurred. This should cause neither surprise nor confusion to those who are familiar with the idea of a conditional probability.) We may think of k oriented links joining $G_1, G_2, \ldots, G_k$ to G. If the whole class, $\mathfrak{n}$, is connected together by means of such links we describe $\mathfrak{n}$ as a *causal net.* If E is the latest of the events in the net, and can be reached from each other event by passing through a succession of links in the right direction, then the causal net will be said to *lead to E.* If F is the earliest of the events in $\mathfrak{n}$, and each other event can be reached from F by passing through a succession of links in the right direction, then the causal net will be said to *lead from F.* If both conditions are satisfied, the net will be said to *lead from F to E.* For example, a net leading to E could have the form of a "tree," but a net leading from F to E could be a tree only if it were a chain.

In this definition we may call $G_1, G_2, \ldots, G_k$ the *immediate predecessors* of G. A causal net will be said to *have independence* if, for each G in the net, the immediate predecessors have independent tendencies to cause G given the past.

For each link $G_i \rightarrow G$, having a "p" and a "q,"

$$p = p_i = P(G|G_i),\ q = q_i = P(G|\overline{G}_i),$$

let the *quasiprobability,* π, be defined as

$$\pi = \left[\frac{p-q}{1-q}\right],$$

in which the "square" brackets indicate that $\pi = 0$ if $q \geqslant p$. The quasiprobability reduces to p when $q = 0$. We know from T17 that the quasiprobabilities are multiplicative for a chain, and the strength of the chain is the same as if the quasiprobabilities were ordinary probabilities and the q's were all zero. Also, from T15, we have

$$S(p,q) = -\log(1-\pi),$$

so that for a bundle of the type occurring in T10 the quasiprobabilities again behave like probabilities, in view of the additivity of the strengths of the chains. (The term "pseudoprobabilities" would conveniently refer, by analogy with the pseudo-random numbers that are often used in Monte Carlo calculations, to the apparent probabilities that occur in a deterministic, but pseudo-indeterministic, set-up.)

Let us now consider an arbitrary finite causal net having independence and leading from F to E. We should like a general procedure for defining the strength of such a net that will include the results for the nets already considered, and which is simple, and which does not lead to a contradiction. I believe that the procedure illustrated in the following example satisfies these conditions. It would of course be more satisfactory if some convincing axioms could be laid down that would uniquely determine the procedure.

In the diagram, the quasiprobabilities $\pi_1, \pi_2, \ldots, \pi_8$ are assigned, and pertain to the links of the net. It will be easier to appreciate the example if the π's are at first thought of as ordinary probabilities (with all the q's equal to 0).

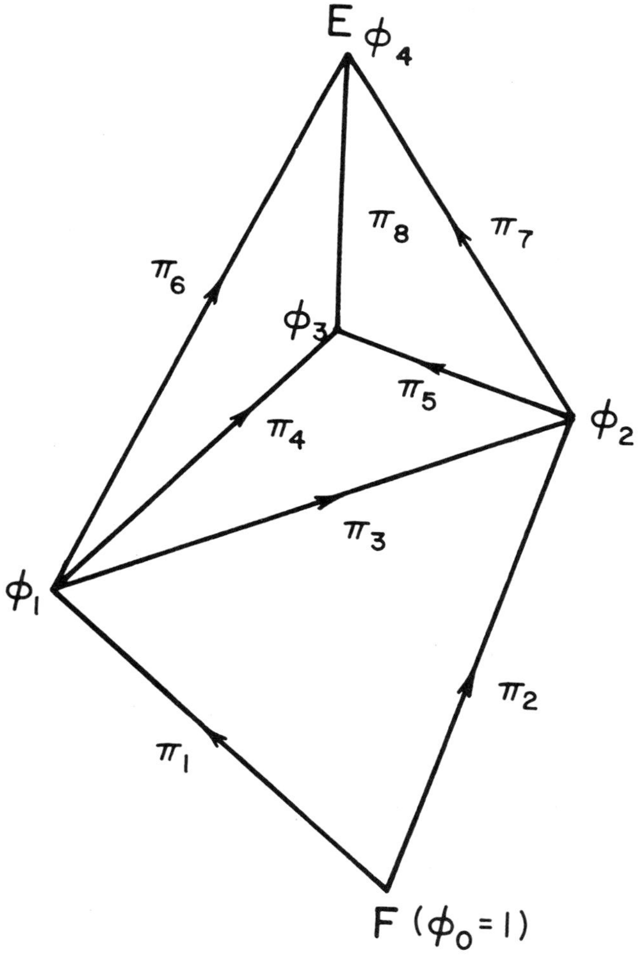

Figure 2.

The ϕ's may be thought of as quasiprobabilities of the *events*. They are defined successively as follows:

$$\phi_0 = 1.\ \phi_1 = \pi_1.$$
$$\phi_2 = 1 - (1 - \pi_2)(1 - \phi_1\pi_3).$$
$$\phi_3 = 1 - (1 - \phi_1\pi_4)(1 - \phi_2\pi_5).$$
$$\phi_4 = 1 - (1 - \phi_1\pi_6)(1 - \phi_2\pi_7)(1 - \phi_3\pi_8).$$
$$S(\mathfrak{n}) = Q(\phi_4, 0) = -\log(1 - \phi_4).$$

The reader should perhaps check that this procedure contains the previous ones as special cases.

12. CAUSAL NETS IN GENERAL

It will often be possible to divide up a time-slice preceding E into non-overlapping events whose causal influences on E are approximately but not absolutely independent. Let such a dissection of the time-slice be $F_1, F_2, \ldots, F_m$. We need a definition of the strength of the causal link $F_1 \to E$ that will reduce to the value given previously in the case where F_1 and $F_2.F_3. \ldots F_m$ are *causally independent* with respect to E, in the same sense as that defined above for nets having independence. A simple definition having the required property is

$$S(E{:}F_1|F_2 \ldots F_m) = \log\frac{P(\overline{E}|\overline{F}_1F_2 \ldots F_m)}{P(\overline{E}|F_1F_2 \ldots F_m)} = W(\overline{F}_1{:}\overline{E}|F_2 \ldots F_m),$$

when this is positive; otherwise $S = 0$.

This definition reduces to the previous use of the expression $S(E{:}F)$ in the case of causal independence. But the strengths of the lead-ins do not add up to $S(E{:}F_1, F_2, \ldots, F_m)$ unless the F's do have independent causal influences on E. We can cope with this difficulty by the introduction of "interaction terms" in a sense analogous to the use of this expression in the literature of the design of statistical experiments. (See, for example, Davies, 1954, index reference under "Interaction.")

We can think of an extra node in the causal net leading to E corresponding to every subset of the events $F_1, F_2, \ldots, F_m$. For example, there will be a node corresponding to the pair $(F_1.F_2)$. The strength of the link from the node $(F_1.F_2)$ to E will then be taken as the "interaction" term

$$t_{12} = s_{12} - s_1 - s_2,$$

where

$$\begin{aligned} s_{12} &= \log\frac{P(\overline{E}|\overline{F}_1\overline{F}_2F_3 \ldots F_m)}{P(\overline{E}|F_1F_2F_3 \ldots F_m)} \\ &= Q(E{:}F_1F_2/\overline{F}_1\overline{F}_2|F_3 \ldots F_m) \\ &= W(\overline{F}_1\overline{F}_2/F_1F_2{:}\overline{E}|F_3 \ldots F_m) \\ &= I(\overline{F}_1\overline{F}_2{:}\overline{E}|F_3 \ldots F_m) - I(F_1F_2{:}\overline{E}|F_3 \ldots F_m). \end{aligned}$$

The strength of the link to $(F_1 F_2)$ from an earlier node $(G_1 G_2)$ is

$$W(\overline{G}_1\overline{G}_2/G_1 G_2 : \overline{F}_1\overline{F}_2/F_1 F_2 | G_3 G_4 \ldots),$$

where $G_3, G_4, \ldots$ are the other immediate predecessors of F_1 and F_2.

The definitions of the s's are forced, if we regard conjunctions of the F's as single events. An example of a third-order interaction is

$$t_{1234} = s_{1234} - s_{234} - s_{134} - s_{124} - s_{123} + s_{12} + s_{13} + \ldots + s_{34} - s_1 - s_2 - s_3 - s_4,$$

where the notation is now self-explanatory. In any piece of causal analysis one would try to choose the dissection of the time-slice so as to make the high-order interactions negligible.

Since

$$s_{123\ldots m} = \Sigma s_i + \Sigma t_{ij} + \Sigma t_{ijk} + \ldots,$$

our enlarged causal net has the property of additivity of strengths of lead-ins that we previously had for causally independent lead-ins. It is therefore now potentially possible to apply the method of Section 11 to define the causal strength of an arbitrary finite net from F to E.

13. DEGREES OF CAUSATION

We may now define $\chi(E{:}F)$ as the limit of the strength of the net joining F to E and containing all intermediate events, when the events are made smaller and smaller. I have not proved that this limit exists. The proof, if possible, would depend on a physical theory, and would be mathematically intricate. Note the implication: whether degrees of causality exist is a matter of physics, even if we take for granted that physical probabilities exist.

In practice one must always over-simplify or simplify in order to be able to judge, estimate, or guess, the value of $\chi(E{:}F)$. (In the past, χ has been given only a few values, such as "small," "moderate," and "large.") There is always the possibility that something has been overlooked. Even in a statistical experiment involving randomization, from which we can apparently deduce that some $\chi(E{:}F)$ is large, in fact E and F may both have been caused by some preceding event. The table of random numbers might have been seen by the famous lady tea-taster (Fisher, 1949, Chap. 2), or there may have been some psychokinesis. We are always thrown back on judgment.

14. BIG EVENTS

So far the analysis has assumed F and E to be small events. If F is big we may imagine it split up into many small events, and imagine all these to be "short-circuited" from an earlier "input node." By "short-circuited" is meant that the resistances of all the imaginary links are taken to be zero. We may apply a similar process to a big E by short-circuiting its small parts to a future output node. The previous methods may then be applied even if F does not end before E begins.

[For references to other work on probabilistic causality see the Introduction of this book.]

Acknowledgements. This paper owes much to correspondence and discussion with Mr E. M. L. Beale, Professor Bruno de Finetti, Professor K. R. Popper, Professor L. J. Savage, Mr Christopher Scott, and especially to Dr Oliver Penrose. The Referees and Editor have also been helpful.

APPENDIX I. CORRECTION OF SOME ERRORS IN PREVIOUS WORK

Reichenbach (1956, p. 204) says that F is causally relevant to E if $P(E|F) > P(E)$ and if there is no set of events earlier than or simultaneous with F that "screens off" E from F. By "screens off" he means that the probability of E given these other events is unchanged if F is also given. The property is analogous to the Markov property.

It seems to me that this definition is not acceptable as it stands for much the same reason that my previous paper is not acceptable. For let G be any set of events earlier than or simultaneous with F. G might be some exceedingly biased selection of individual molecules, such as those that are proceeding south at a thousand miles per hour. Consider the expression $P(E|G) - P(E|G.F)$. Normally this will be positive for some G, say G_1, and negative for some G, say G_2. We now imagine G_1 to be gradually distorted into G_2. The above expression must change sign at some point during this gradual distortion, at which "time" its value will be zero. Hence the second part of Reichenbach's definition seems to be vacuous. In order to patch up the definition it seems to be necessary to take G as the complete state of the universe at the time F started.

In my previous paper, conditions C7 to C10 were vacuous for much the same reason, though it may be possible to patch the thing up, as stated therein (inserted in proof), by insisting that G should be in some sense a "natural" event.

APPENDIX II. HOLMES, MORIARTY, AND WATSON (See Section 2)

Sherlock Holmes is at the foot of a cliff. At the top of the cliff, directly overhead, are Dr Watson, Professor Moriarty, and a loose boulder. Watson, knowing Moriarty's intentions, realizes that the best chance of saving Holmes's life is to push the boulder over the edge of the cliff, doing his best to give it enough horizontal momentum to miss Holmes. If he does not push the boulder, Moriarty will do so in such a way that it will be nearly certain to kill Holmes. Watson then makes the decision (event F) to push the boulder, but his skill fails him and the boulder falls on Holmes and kills him (event E).

This example shows that $Q(E:F)$ and $\chi(E:F)$ cannot be identified, since F had a tendency to prevent E and yet caused it. We say that F was a cause of E because there was a chain of events connecting F to E, each of which was strongly caused by the preceding one.

APPENDIX III. THE MEANING OF "STATE" IN QUANTUM MECHANICS (See Section 4)

The seven relevant interpretations of "state" in quantum mechanics are the first seven on the following list. All seven of these meanings, and perhaps others, should be taken into account in a comprehensive discussion of the place of probabilistic causality in quantum mechanics.

(*i*) The class of all past phenomena, classically describable. (*ii*) The class of phenomena extending only a short way into the past. (*iii*) The wave function of a physical system, under observation by another physical system. (*iv*) The joint wave function of the pair of systems. (*v*) The wave function of one system conditional on an assumed wave function of another system. This is the "relative state" of Hugh Everett III (1957). (*vi*) The wave function of the entire universe if this has any meaning. See Everett, *loc. cit.* (*vii*) The wave function of the entire universe together with all other past phenomena. (*viii*) An ensemble of wave functions. See, for this eighth interpretation, R. C. Tolman (1938, Section 98).

CHAPTER 22

A Simplification in the "Causal Calculus" (#1336)

A quantitative explication was given in **#223B** for $Q(\mathrm{E}:\mathrm{F})$, defined as the degree to which an event F tends to cause a later event E. The argument depended on assigning to a causal network a "resistance" R and a "strength" S. By considering a parallel network, a functional equation was found for S (p. 205). On the other hand, by considering a special causal Markov chain $\mathrm{F} \to \mathrm{E}_1 \to \mathrm{E}_2$, in which $P(\mathrm{F}) = x$, $P(\mathrm{E}_1|\mathrm{F}) = p_1$, $P(\mathrm{E}_1|\overline{\mathrm{F}}) = 0$, $P(\mathrm{E}_2|\mathrm{E}_1) = p_2$, $P(\mathrm{E}_2|\overline{\mathrm{E}}_1) = 0$, and then coalescing E_1 and E_2 into a single event $\mathrm{E} = \mathrm{E}_1 \mathbin{\&} \mathrm{E}_2$, we can obtain the further functional equation

$$R(p_1p_2, 0, p_1p_2x) = R(p_1, 0, p_1p_2x) + R(p_2, 0, p_1p_2x).$$

Here $R(p, q, x_0)$ is supposed to be the resistance of a simple chain $\mathrm{F}_0 \to \mathrm{E}_0$ where $P(\mathrm{F}_0) = x_0$, $P(\mathrm{E}_0|\mathrm{F}_0) = p$, $P(\mathrm{E}_0|\overline{\mathrm{F}}_0) = q$. This functional equation for R was only implicit in **#223B** with the result that the arguments in that paper were somewhat more complicated than necessary. The two functional equations have solutions of the form

$$R(p, 0, r) = -u(r/p)\log p$$

$$S(p, q, r) = v\left\{\frac{r-q}{p-q}\right\}\log\frac{1-q}{1-p} \quad (p>q)$$

where u and v are non-negative functions of their arguments. Since S is assumed to be a function of R it can be deduced that u and v are constants. By an appropriate choice of units it then follows, as in **#223B**, that

$$R = -\log(1-e^{-S}),\ S = -\log(1-e^{-R}).$$

Explicativity: A Mathematical Theory of Explanation with Statistical Applications (#1000)

By *explicativity* is meant the extent to which one proposition or event explains why another one should be believed. Detailed mathematical and philosophical arguments are given for accepting a specific formula for explicativity that was previously proposed by the author with much less complete discussion. Some implications of the formula are discussed, and it is applied to several problems of statistical estimation and significance testing with intuitively appealing results. The work is intended to be a contribution to both philosophy and statistics.

1. INTRODUCTION

By *explicativity* I mean the extent to which one proposition or event F explains why another one E should be believed, when some of the evidence for believing E might be ignored. Both propositions might describe events, hypotheses, theories, or theorems. For convenience I shall not distinguish between an event and the proposition that states the event. In practice usually only putative explanations can be given and this is one reason for writing "should be believed" instead of "is true," but explanation in the latter sense can be regarded as the extreme case where belief is knowledge.

The word "explanatoriness" is not used here because it is defined in the *Oxford English Dictionary* as a quality, where "explicativity" is intended to be quantitative as far as possible. Also it has a more euphonic plural.

The concept of explicativity can be thought of as a "quasiutility," which is a substitute for utility, preferably additive, when ordinary utility is difficult to judge. The condition of additivity for quasiutilities is necessary to justify the maximization of their expected values (#618). The need for at least a rough measure of explicativity arises in pure science more obviously than in commerce where utilities can often be judged in financial terms. But if a measure of explicativity is proposed in general terms it should make sense whatever the field of application. One such field consists of the estimation of statistical parameters since any such estimate can be regarded as a hypothesis that helps to explain observations. Examples of statistical estimation and of significance testing will be given in this paper.

The topic of explicativity belongs to the mathematics of applied philosophy. The present account is based on (#599, **#846**, Good, 1976) and goes much further, though it does not cover everything on the topic in the previous publications.

The advantage of the mathematics of philosophy over classical philosophy is that a formula can be worth many words. The topic is mathematical because it depends on probability. In this respect explicativity resembles some explications for information, weight of evidence, and causal propensity, and it will be convenient to list these explications first, without details of their derivations.

It may be possible sometimes to invert our approach, and to use explicativity inequalities to aid us in our probability judgments.

2. NOTATION

Let A, B, C, E, F, G, H, J, K, sometimes with subscripts or primes, usually denote propositions, or events, or hypotheses, etc. For example, E often denotes an event and *also* the proposition that asserts that the event "obtains." Conjunctions, disjunctions, and negations are denoted by &, v, and a vinculum [macron] respectively. I shall not distinguish between hypotheses, theories, and laws.

Let $P(E|H)$ denote the probability of E given H or assuming H. Similarly let $P(H)$ denote the initial probability of H and let $P(H|E)$ denote its final probability. Often $P(H)/P(H')$ is less difficult to judge than $P(H)$ and $P(H')$ separately. In practice all probabilities are conditional so that $P(E|H)$, $P(H)$ and $P(H|E)$ are abbreviations for $P(E|H \;\&\; G)$, $P(H|G)$, and $P(H|E \;\&\; G)$, where G is some proposition, usually complicated, that is taken for granted. It will sometimes be left to the reader's imagination to decide whether any probability mentioned is physical, logical, or subjective. We shall assume the usual axioms of probability whichever of these interpretations of probability is intended.

The *information concerning a proposition* A *provided by another proposition B, given G throughout,* is denoted by $I(A\!:\!B|G)$ and is defined by

$$I(A\!:\!B|G) = \log \frac{P(B|A \;\&\; G)}{P(B|G)} = \log \frac{P(A|B \;\&\; G)}{P(A|G)}. \qquad (1)$$

(We shall not niggle about zero probabilities.) The base of the logarithms exceeds 1 and determines the unit in terms of which information is measured. For example, if the base is the tenth root of 10, the unit is the deciban, a word suggested by A. M. Turing in 1941 in connection with "weight of evidence." With base 2 the unit is the "bit." When G is taken for granted we write $I(A\!:\!B)$, and a similar abbreviation will be used for other notations. Sometimes $I(\mathrm{A:A})$ is denoted by $I(\mathrm{A})$ and (1) implies

$$I(\mathrm{A}) = -\log P(\mathrm{A}). \tag{2}$$

For a derivation of these formulae see, for example, p. 75 of #13 and #505. Mathematical expectations of (1) occur in Shannon's theory of communication (1948). Information has the additive property

$$I(\mathrm{A:B \;\&\; C}) = I(\mathrm{A:B}) + I(\mathrm{A:C|B}). \tag{3}$$

The *weight of evidence in favour of H_1 as compared with H_2 provided by E given G* is defined by

$$\begin{aligned} W(\mathrm{H_1/H_2:E|G}) &= \log \frac{O(\mathrm{H_1/H_2|E \;\&\; G})}{O(\mathrm{H_1/H_2|G})} \\ &= \log \frac{P(\mathrm{E|H_1 \;\&\; G})}{P(\mathrm{E|H_2 \;\&\; G})} \\ &= I(\mathrm{H_1:E|G}) - I(\mathrm{H_2:E|G}), \end{aligned} \tag{4}$$

where O denotes odds (the ratio of the probabilities of $\mathrm{H_1}$ and $\mathrm{H_2}$). Weight of evidence, which is the logarithm of a Bayes factor, has the additive property

$$W(\mathrm{H_1/H_2:E \;\&\; F}) = W(\mathrm{H_1/H_2:E}) + W(\mathrm{H_1/H_2:F|E}) \tag{5}$$

and of course we can condition on G throughout. If the disjunction $\mathrm{H_1 \vee H_2}$ is also taken for granted, so that $\mathrm{H_2}$ becomes $\overline{\mathrm{H}}_1$, the negation of $\mathrm{H_1}$, then the notation $W(\mathrm{H_1/H_2:E})$ can be abbreviated to $W(\mathrm{H_1:E})$.

For some literature on weight of evidence see Peirce (1878), #13, and numerous papers cited in **#846**.

The *causal support for E provided by F*, or the *propensity of F to cause E*, denoted by $Q(\mathrm{E:F})$, where E and F denote events, is defined (**#223B**) by the equation

$$Q(\mathrm{E:F}) = W(\overline{\mathrm{F}}:\overline{\mathrm{E}}|\mathrm{U \;\&\; L}), \tag{6}$$

the weight of evidence against F if E did not occur, given the state U of the universe just before F occurred, and also given all true laws L of nature. This quantitative explication of causal propensity is basically consistent with the requirements of Suppes (1970) which, however, are only qualitative. The relationship between this monograph and **#223B** is discussed in #754.

The need for mentioning U in (6) is exemplified by the fact that seeing a flash of lightning is not an important cause of hearing loud thunder soon afterwards. Both events were caused by a certain electrical discharge. Equally, the thunder is not *explained* by the visual experience of lightning. On the other hand seeing the lightning does explain why one *believes* that thunder will soon occur; whereas hearing thunder is a good reason for believing that the lightning flash previously occurred. The experiences are thus valid reasons for prediction and retrodiction respectively.

If F occurs after E, it turns out that $Q(\text{E:F}) = 0$. This is because U "screens off" E from F under usual assumptions about the nature of time. This notion of "screening off" is explained in more detail by Reichenbach (1956, pp. 201-205) and herein, p. 216. It is analogous to a Markov chain property.

One potential value of measuring causal tendency quantitatively is for the apportioning of credit and blame, as is done, for example by the British Admiralty if two ships collide, though without using (6), and would be done more generally in the courts of justice if they thoroughly deserved their name.

3. PHILOSOPHICAL ASPECTS

There is a large and interesting literature on the philosophy of explanation (for example, Mill, 1843/1961; Hempel, 1948/65; Braithwaite, 1953; Popper, 1959; Nagel, 1961; Scheffler, 1963; Kim, 1967; Rescher, 1970; Salmon, 1971; and numerous further references in these publications). The present account is succinct but is intended to be full enough for the reader to see how the statistical examples fit into the philosophical background. Also I believe that the philosophical discussion contains some new ideas.

The following terminology is fairly standard: what is to be explained or partially explained is called the *explanandum*, and what explains it the *explanans.*

There are at least three main categories of explanation, with various subcategories. They correspond roughly to the questions "what," "how," and "why."

(1) *Explaining "what," or semantic explanation:* answering the question "What do you mean?"

(1.1) *Dictionary definition.*

(1.2) *Philosophical explication:* extraction of more consistent and precise meaning or meanings by analytic consideration of the usage of words by "good" authors. This definition involves an implicit iterative "calculation" because we should say what is meant by a good author.

(2) *Explaining "how," or descriptive explanation:* answering the question "How is this object constructed?"

(2.1) in Nature;

(2.2) in manufacture.

(3) *Explaining "why," or causal (and probabilistic causal) explanation*

(3.1) The explanandum is an event (or the proposition describing an event).

(3.2) The explanandum is a class of events.

(3.3) The explanandum is a scientific law.

(3.4) Explaining why the explanandum should be believed (*to some extent*), when some knowledge supporting this belief, apart from the explanans itself, might be ignored. (For example, we may "know" E is true and still demand an explanation.) Here the explanans is a (partial) cause of *belief* in E rather than a cause of E itself, though it might be both. (Observing the shadow of an elephant can explain why we believe an elephant is present; whereas observing an elephant can explain *both* why the shadow is there *and* why we believe the shadow should be there.) An explanation of this kind might be a prediction or a retrodiction, or a reasoned argument, or a mixture of two or three of these activities. We might have called this kind of explanation "diction" if this word had not been pre-empted, and anyway a "dictionary" deals with category (1.1). A retrodiction is always a "belief-type" of explanation, rather than a causal type, if it is assumed that causes always precede their effects. I shall make this assumption in this paper though I am not dogmatic about it (see ##882, 1322A).

(3.4.1) The explanandum is a mathematical or logical theorem and the explanans is a proof or heuristic argument. Sometimes an incomplete proof is a better explanation of why a theorem is true than a complete proof. For example, if AOB is a triangle with a right-angle at O, and if a perpendicular is dropped on AB from O, then the three triangles now present all have the same shape so that their areas are proportional to the squares of corresponding linear dimensions. This explains *why* Pythagoras's theorem is true in the sense that the proof is not artificial.

Sometimes "teleological explanation," in which future goals are mentioned, is regarded as forming an additional category, but, unless we allow for precognition, and we shall not do so, this category is not distinct from categories (3.1)-(3.3). This fact is well known. For example, a homing missile, though it acts purposefully, obeys the usual laws of physics. It is its own present *prediction* of the future that affects it, not the future itself.

The present work is an exercise in applied philosophical explication (category 1.2) and its subject matter is category (3). Headings (1.1) and (2) are ignored. The explication of explanation in category (3) often depends on *dynamic* or *evolving* probabilities which can be changed by *reasoning alone* as in a game of chess, and not by new empirical observations. This notion may superficially appear fancy, and is usually overlooked, but I am convinced that it is essential (see especially #**938**). This is obvious when the explanation comes under category (3.4.1), though the above example concerning Pythagoras's theorem shows that the notion of mathematical explanation cannot be fully captured in terms of probabilities alone. We shall soon see that physical explanation also requires something extra.

Dynamic probabilities are also required for the rest of category (3), as shown in ##599, **846**. For example, to give the argument in outline, the apparent motions of the planets (event E), as projected upon the celestial sphere, had their dynamic probabilities enormously increased, in ratio, when it was noticed that the motions are implied by the inverse square law H of gravitation. This was because the inverse square law had, for most scientists, a non-negligible prior probability, owing to its simplicity and to the analogy of light emerging from a point source, and because it explained why objects like apples fall. That apples behave in some respects like planets is an example of what William Whewell called the "consilience of laws": see Kneale (1953, pp. 364-366). Thus $P(\mathrm{E})$, which exceeds $P(\mathrm{E}|\mathrm{H})P(\mathrm{H})$, is much greater than the original value of $P(\mathrm{E})$. This would be true even without bringing in "apples" or the consilience of laws, so that our argument is distinct from Whewell's and Kneale's, and has a somewhat clearer need for the notion of dynamic probability.

To explain why a physical event E occurred is to explain what caused it or tended to cause it, and this requires explicit or implicit reference to a causal chain or causal network that leads to E over some time interval of appreciable duration t. The longer the duration t the fuller the explanation. A causal network cannot be described without at least implicit reference to laws of nature. This shows that probabilities alone, without reference to physical structure, cannot fully capture the notion of physical explanation. Again, if E is itself a law of nature, an explanation of it must be in terms of yet other laws of nature. These will often be more general than E, though explanations by analogy are also possible, and then the explanans might consist of laws no more general than E. Thus, whether the explanandum denotes an event (or set of events) or a law of nature, the explanans will involve laws of nature, and this is a view that has been adopted by many philosophers of science since Mill (1843) or earlier. An immediate consequence of this view is that an event E cannot be regarded as an explanation of itself, since we need $t > 0$, but if you have knowledge that E is true, then this of course fully explains your *belief* in E. Usually in practice our explanations are only putative and only explain beliefs, for real causal networks are enormously complex. Accordingly, the explanation of beliefs will be our main topic.

Sometimes the laws of nature that form part of the explanation of E are taken for granted because of their familiarity. For example, when we say that a window-pane broke because Tom threw a stone at it, we are taking for granted that glass panes are liable to break when hit by fast-moving hard objects that are not too small. Thus a law of physics is here implicit in the explanation. As another example, we might say that it is bad for Ming Vases to leave them unsupported in mid-air.

In deterministic physics a specific event E can sometimes be explained by some boundary conditions B, including initial conditions, combined with differential equations that describe a general law, L. Then B & L explains E, but sometimes, as in the example just given, we call B the explanation when L is taken for granted. The division of an explanation into a *contingent* part and *general laws* is not restricted to physics.

It is difficult to specify sharply whether one law is more general than another. Nagel (1961, pp. 37-42) makes a valiant attempt which he does not regard as fully successful, and I shall here merely point out the relevance of the matter to statistical problems. Suppose that a random scalar or vector x has a probability density function $f(x|\theta)$, where θ is a parameter which is also a scalar or vector. The distribution determined by $f(x|\theta)$ is a "law" in the sense that it says something about a *population* of values of x, and it is often *called* a law (see, for example, Jeffreys, 1939/61). Any proposition of the form $\theta \in \Theta$ (some set of possible values of θ) is a disjunction of laws, and can again reasonably be called a law. Note that θ must be fixed before x can take on a specific value so the time direction is appropriate. If θ itself is regarded as a random scalar or vector containing hyperparameters, as in hierarchical Bayesian techniques (see, for example, ##**26**, 398, **1230**), then a specification of a constrained set of values for these hyperparameters could reasonably be regarded as a law that is more general than $\theta \in \Theta$. For it can be regarded as a proposition about a superpopulation. And similarly for hyper-hyperparameters, etc. A law of the form $\theta \in \Theta$ is a somewhat primitive form of explanation because it does not give detailed information about the structure of the (probabilistic) causal network that leads to an observed value of x, but we cannot usually demand more from statistical estimation procedures. In this example there is no contingent part in the explanans, whereas in regression problems the value of the concomitant ("independent") variable is contingent, when regarded as part of the explanation of a specific value of the dependent variable, whereas the equation of the regression line is lawlike.

There is an intimate relationship between explanation and causation. The broken window was both caused and explained by Tom's naughty behavior. This relation can be formalized to some extent in probabilistic terms: if $P(\mathrm{E}|\mathrm{B} \,\&\, \mathrm{L}) > P(\mathrm{E}|\mathrm{L})$ then B is a probabilistic cause, and a putative partial explanation of E, when the law L is taken for granted. On the other hand, if $P(\mathrm{E}|\mathrm{B} \,\&\, \mathrm{L}) > P(\mathrm{E}|\mathrm{B})$, then L is a putative partial explanation of E, but hardly a probabilistic cause, when B is taken for granted. So causation and explanation are related but are not identical (see also §9).

We shall denote by $\eta(\mathrm{E}:\mathrm{F}|\mathrm{G})$ the explicativity or explanatory power of F with respect to E, given background information G, and shall arrive at a formula for it, based on some desiderata. Here F may or may not include general laws. This notation interchanges the positions of E and F as used in ##599, **846**. The reason for the reversal is that it is more consistent with the notation Q for causal propensity. For grasping the notation we may read $\eta(\mathrm{E}:\mathrm{F}|\mathrm{G})$ from left to right as "the explainedness of E provided by F given G," so that the colon can be pronounced "provided by" whether we are talking about information I, weight of evidence W, causal support Q, or explicativity η. (Having two names "explicativity" and "explainedness" for the same thing is analogous to calling $P(\mathrm{E}|\mathrm{H})$ both a probability of E and a likelihood of H.) By calling G "background information" we mean that it is assumed to be true and that it has already been taken into account for helping to explain E. (See Desideratum (iii) in §4.) There may also be further evidence G′, such as direct evidence that E is true, which is deliberately ignored and is omitted from our notation.

We shall assume that $\eta(\mathrm{E}:\mathrm{F}|\mathrm{G})$ depends only on various probabilities, and we shall not incorporate those requirements that are necessary for regarding F as a partial explanation of E and which do not depend on these probabilities. Thus $\eta(\mathrm{E}:\mathrm{F}|\mathrm{G})$ will denote a putative explicativity when F is a putative explanation of E (given G) and will otherwise denote something more general. In fact it will be a measure of *the degree to which F explains why you should believe E, given G all along, and disregarding evidence for E that is not provided by F and G.* We shall call η "explicativity" in all cases although "dictivity" might be preferred. (See the remark about "diction" under category [3.4].) The name is less important than that η should measure something of interest.

Some philosophers claim, with some justification, that F cannot be a (probabilistic) explanation of E unless F is true. But in practice F can perhaps never be known to be true, even in pure mathematics, so that in this paper we shall regard nearly all explanations as only putative. In practice we talk about "explanations" without saying "putative" each time, and accordingly we sometimes put "putative" in parentheses or omit it.

We regard explanations as good or bad depending in part on whether the probability of the explanans is high or low. Let us then allow the explicativity $\eta(E\!:\!F)$ to depend on $P(F)$. When F is assumed to be known to be true let us use the somewhat hypallagous expression *informed explicativity.* An informed explicativity is of course an extreme case of a (putative) explicativity.

As an example of the distinction between (putative) explicativity and informed explicativity let us again consider the broken window (event E). The hypothesis F that Tom threw a stone at it has more (putative) explicativity than that the Mother Superior did so (hypothesis F_{MS}). For we believe that Tom is naughtier than the Mother Superior as well as being a better shot. On the other hand, if we *saw* the Mother Superior throw the stone vigorously, F_{MS} would have very high *informed* explicativity.

By using the expression "informed explicativity" we do not wish to imply that the whole causal network preceding E is known; we mean only that F becomes known to be true, but is not taken for granted in advance. The informed explicativity of F with respect to E might be high and yet it might turn out that F is not part of the true explanation of E after all.

Both a (putative) explicativity and its extreme case, an informed explicativity, are intended to be measures of the explanatory power of F with respect to E relative to the knowledge that we (or "you") have, and that knowledge will seldom include the certainty of F. We can only hope to measure the extent to which our beliefs about F explain why we should believe E (imagining E to be unobserved). Under this interpretation it is not necessary that F should precede E chronologically; and $\eta(E\!:\!F|G)$ will sometimes measure the predictivity or retrodictivity of F with respect to E, or some mixture. Again, if F is a "law," it need have no position in time, and it might be used for prediction, retrodiction, or putative explanation of E.

Since we regard informed explicativity as an extreme case of (putative) explicativity, we do not need a separate notation for it. It will be merely a matter of putting $P(F|G) = 1$ or $P(F) = 1$ in whatever formula we use for $\eta(E\!:\!F|G)$ or $\eta(E\!:\!F)$.

We conclude this philosophical background with one further property of explanation. Most philosophers believe that an explanation should be based on all relevant evidence apart from the evidence G' that is deliberately ignored such as the direct observation of E. With our notation $\eta(E\!:\!F|G)$ this would mean that F & G must contain all evidence relevant to E, apart from G'. In practice, when we are estimating an explicativity, we must make do with the evidence that appears to us to be sufficiently relevant.

4. THE DESIDERATA AND EXPLICATION FOR EXPLICATIVITY

As a preliminary to proposing some desiderata for explicativity, let us consider a naive approach and an early historical approach to explanation.

Perhaps the most naive suggestion is that E is explained by H if H logically implies E. This is neither a necessary nor a sufficient condition for H to be a good explanation of E. For example, the hypothesis $0 = 1$ logically implies everything and in particular it implies E, but $0 = 1$ is an extremely poor (putative) explanation of anything! Nor does it help to append some irrelevant laws of nature so as to make the explanans lawlike. So we need something less naive. Let us recall a little history.

According to the translation by Charlton (1970, p. 10), Aristotle said " . . . it is better to make your basic things fewer and limited, like Empedocles." In the early fourteenth century the "doctor invincibilis," William of the village of Ockham in Surrey said "plurality is not to be assumed without necessity." This sentiment had been previously emphasized by John of the village of Duns in Scotland who has often been thought, apparently incorrectly, to have been William of Ockham's director of studies (Anon., 1951; Moody, 1967). The saying that "entities should not be multiplied without necessity," though apparently never expressed quite that way by William of Ockham, has come to be known as "Ockham's razor." For a detailed history, but with the Latin untranslated, see Thorburn (1918).

A more modern interpretation of the Duns-Ockham razor is that, of two hypotheses H and H′, both of which explain E, the simpler is to be preferred (see, for example, Margenau, 1949). But the hypothesis $0 = 1$ is simple, at least in the sense of brevity, so we need to sharpen the razor some more. The next improvement is that if H and H′ both imply E, then the hypothesis with the larger initial probability is preferable. In nearly all applications the judgment of whether $P(\mathrm{H}) > P(\mathrm{H}')$ is subjective or personal, although different people often agree about a specific judgment. Note that if $P(\mathrm{H}) > P(\mathrm{H}')$, and H and H′ each imply E, then $P(\mathrm{H}|\mathrm{E}) > P(\mathrm{H}'|\mathrm{E})$, that is, the final probability of H exceeds that of H′. One advantage of this way of interpreting Ockham's razor is that it rules out impossible explanantia such as the hypothesis $0 = 1$.

Whereas the initial probability of a hypothesis has something to do with its simplicity the relationship is not obvious, and if we express all our formalism in terms of probabilities we do not need to refer explicitly to simplicity or complexity. In #599 I defined the complexity of a proposition H as $-\log P(\mathrm{H})$, but I retracted this in #876. There is more than can be and has been said on the relationship between complexity and probability, but to avoid distraction we discuss this matter in appendix A.

What if the two hypotheses H and H′ do not logically imply E but merely increase its probability, so that

$$P(\mathrm{E}|\mathrm{H}) > P(\mathrm{E}) \text{ and } P(\mathrm{E}|\mathrm{H}') > P(\mathrm{E})?$$

Is H a better explanation of E than H′ if $P(\mathrm{E}|\mathrm{H}) > P(\mathrm{E}|\mathrm{H}')$? Not necessarily if $P(\mathrm{H}) < P(\mathrm{H}')$. Some compromise is required, to be discussed later.

Let us assume the following desiderata. (i) The explicativity of H with respect to E, denoted by $\eta(\mathrm{E}:\mathrm{H})$, is a function of at most 52670 variables, namely all probabilities of the form $P(\mathrm{A}|\mathrm{B})$ where A and B run through all the propositions that can be generated from E and H by conjunctions, disjunctions, and negations, and where each of these probabilities is not necessarily equal to 0 or 1. It is not important to check that 52670 is the correct number because an equivalent assumption is that $\eta(\mathrm{E}:\mathrm{H})$ depends at most on $P(\mathrm{E})$, $P(\mathrm{H})$, and $P(\mathrm{E}\ \&\ \mathrm{H})$. (ii) If K and F have nothing to do with H and E then $\eta(\mathrm{E}\ \&\ \mathrm{F}:\mathrm{H}\ \&\ \mathrm{K})$ depends only on $\eta(\mathrm{E}:\mathrm{H})$ and $\eta(\mathrm{F}:\mathrm{K})$. (iii) $\eta(\mathrm{E}:\mathrm{H}|\mathrm{H})$ does not depend on E or H (in fact you can reasonably call it zero). (iv) $\eta(\mathrm{E}:\mathrm{H})$ increases with $P(\mathrm{E}|\mathrm{H})$ if $P(\mathrm{E})$ and $P(\mathrm{H})$ are fixed. (v) $\eta(\mathrm{H}:\mathrm{H}) \geqslant \eta(\mathrm{T}:\mathrm{T})$ where T is a tautology. (vi) $\eta(\mathrm{T}:\mathrm{H}) \leqslant \eta(\mathrm{T}:\mathrm{T})$ (because a tautology needs no explanation).

Then it can be proved [see appendix B] that $\eta(\mathrm{E}:\mathrm{H})$ must be some increasing function of $I(\mathrm{E}:\mathrm{H}) - \gamma I(\mathrm{H})$ where γ does not depend on the probabilities and where $0 < \gamma < 1$ (see appendix B). Since the main purpose is to put explicativities in order we may as well take $\eta(\mathrm{E}:\mathrm{H}) = I(\mathrm{E}:\mathrm{H}) - \gamma I(\mathrm{H})$. Moreover this choice converts (ii) into the strictly additive property

$$\eta(\mathrm{E}\ \&\ \mathrm{F}:\mathrm{H}\ \&\ \mathrm{K}) = \eta(\mathrm{E}:\mathrm{H}) + \eta(\mathrm{F}:\mathrm{K}) \tag{7}$$

(when K and F have nothing to do with H and E), and this justifies us in regarding $\eta(\mathrm{E}:\mathrm{H})$ as a proper quasi-utility. Various forms of $\eta(\mathrm{E}:\mathrm{H})$ are:

$$\eta(\mathrm{E}:\mathrm{H}) = I(\mathrm{H}:\mathrm{E}) - \gamma I(\mathrm{H}) \tag{8}$$

$$= \log P(\mathrm{E}|\mathrm{H}) - \log P(\mathrm{E}) + \gamma \log P(\mathrm{H}) \tag{9}$$

$$= I(\mathrm{E}) - I(\mathrm{E}|\mathrm{H}) - \gamma I(\mathrm{H}). \tag{10}$$

We must adjust equation (9), when dynamic probabilities are relevant, as a formula for "dynamic explicativity," $\eta_{\mathrm{D}}(\mathrm{E}:\mathrm{H})$, namely

$$\eta_{\mathrm{D}}(\mathrm{E}:\mathrm{H}) = \log P_1(\mathrm{E}|\mathrm{H}) - \log P_0(\mathrm{E}) + \gamma \log P(\mathrm{H}). \tag{9 D}$$

Here $P_0(\mathrm{E})$ is the initial probability of E, judged *before* H is brought to your attention, whereas $P_1(\mathrm{E}|\mathrm{H})$ is the conditional probability of E given H *after* H is brought to your attention. When H is a good simple theoretical explanation of E, as in the example of the inverse square law, it can easily happen that $P_1(\mathrm{E}|\mathrm{H})$ $P(\mathrm{H})$, which is equal to $P_1(\mathrm{E}\ \&\ \mathrm{H})$, is much larger than $P_0(\mathrm{E})$. When dynamic probabilities are relevant it is ambiguous to omit the subscripts 0 and 1 from the notations, but sometimes it may not be too misleading to write $\eta(\mathrm{E}:\mathrm{H})$ instead of $\eta_{\mathrm{D}}(\mathrm{E}:\mathrm{H})$. For, in ordinary linguistic usage, the inverse square law is called simply an "explanation" of the planetary motions. It happens to be a dynamic explanation in both senses of "dynamic."

A few exercises, extracted from #846, are:

$$\eta(E{:}0 = 0) = 0, \tag{11}$$

$$\eta(E{:}0 = 1) = -\infty, \tag{12}$$

$$\eta(E \,\&\, F{:}H) = \eta(E{:}H) + \eta(F{:}H|E) + \gamma I(E|H), \tag{13}$$

a modified additivity property. If H and L are mutually exclusive then H v L has less explicativity for E than does H if and only if

$$\frac{P(E|L)}{P(E|H)} < \left[1 + \frac{P(L)}{P(H)}\right]^{1-\gamma} - 1. \tag{14}$$

For example, when $P(H) = P(L)$ and $\gamma = ½$, the right side is 0.414.

5. THE CHOICE BETWEEN HYPOTHESES

More important than assigning an explicativity to a single hypothesis, with respect to E, is deciding which of two hypotheses H and H′ has the greater explicativity and by how much. Then the term $\log P(E)$ in (9) is irrelevant, because it is mathematically independent of the hypotheses. Let us denote H v H′ by J and take it for granted, as is permissible when we are choosing between H and H′. Denote by $\eta(E{:}H/H'|J)$ or $\eta(E{:}H/H')$ the amount by which the explicativity of H exceeds that of H′, or 'the explainedness of E provided by H as against H′ (given J)'. Then

$$\eta(E{:}H/H') = \eta(E{:}H) - \eta(E{:}H') \tag{15}$$

$$= W(H/H'{:}E) + \gamma \log O(H/H') \tag{16}$$

$$= \log O(H/H'|E) - (1-\gamma)\log O(H/H') \tag{17}$$

$$= (1-\gamma)W(H/H'{:}E) + \gamma \log O(H/H'|E). \tag{18}$$

Equation (18) has an interesting interpretation. It exhibits the excess in explicativity of H over its negation as a compromise between two extremes, the weight of evidence on the one hand and the final log-odds on the other. The former of these extremes ($\gamma = 0$) corresponds to the philosophy of "letting the evidence speak for itself" (as advocated by some in the Likelihood Brotherhood), and the latter ($\gamma = 1$) to that of preferring the hypothesis of maximum final probability. Neither of these two philosophies is tenable as we may see clearly by an example, although their implications are reasonably judged to be good enough in some circumstances.

Let E denote the proposition that planets move in ellipses, let H denote the inverse square law of gravitation, and K that there is an elephant on Mars. If we took $\gamma = 0$ we'd find that $\eta(E{:}H \,\&\, K) = \eta(E{:}H)$, in other words that the explicativity of H would be unaffected by cluttering it up with an improbable irrelevant elephant. Thus the size of γ depends on how objectionable we regard it to have clutter, or to "multiply entities without necessity."

The case $\gamma = 0$ of (8), namely the mutual information between E and H, was proposed independently as an explication of explanatory power by Good (1955) and Hamblin (1955), both in relation to Popper's writings. The fact that it did not allow for clutter was pointed out in #599, and this explication was therefore called *weak* explanatory power. In our present terminology it is the "informed (putative) explicativity" of H. Expected amounts of information of this kind, and of the effectively more general notion of weight of evidence . . . , were related to statistical physics by Gibbs (1902, chap. XI) and Jaynes (1957), and to non-physics statistical practice by, for example, Turing in 1941 (see #13), Jeffreys (1946), Shannon (1948), Good (1950/53), Kullback & Leibler (1951), Rothstein (1951), Cronbach (1953), #77, Lindley (1956), Jaynes (1957, 1968), Kullback (1959), ##322, 524, Tribus (1969), #755, over thirty other publications by the present writer, and in several publications by Rothstein and by S. Watanabe.

Next suppose we take $\gamma = 1$, then $\eta(E{:}H)$ would reduce to $\log P(H|E)$ and there would be no better hypothesis than a tautology such as $1 = 1$. This shows, as in Bayesian decision theory, that it is inadequate to choose the hypothesis of maximum final probability as an unqualified principle.

So we must take $0 < \gamma < 1$. There may not be a clearly best value for the "explicativity parameter" γ, but $\gamma = \frac{1}{2}$ seems a reasonable value. It exactly "splits the difference" between the two extreme philosophies just mentioned, and is also the simplest permitted numerical constant.

The *sharpened razor* is the recommendation to choose the hypothesis that maximizes the explicativity with respect to E, or for all known evidence. It differs from a central theme of Popper's philosophy, namely that a useful theory is one that is of low (initial) probability and highly testable. Certainly high checkability is a desirable feature of a theory, and, *if a theory turns out to have a high final probability,* then a low initial probability is desirable because it shows that the theory was informative. But Popper's philosophy does not allow for final probabilities.

It is of interest to consider how much more explicative E itself is than H, relative to E,

$$\begin{aligned}\eta(E{:}E/H) &= \eta(E{:}E) - \eta(E{:}H)\\ &= \gamma \log P(E) - \gamma \log P(H) - \log P(E|H),\end{aligned} \tag{19}$$

or, when dynamic probabilities are used,

$$\eta_D(E{:}E/H) = \gamma \log P_0(E) - \gamma \log P(H) - \log P_1(E|H). \tag{19 D}$$

For $\eta(E{:}E/H)$ we have the following theorem:

When dynamic probabilities are not used, there is no more explicative proposition relative to E than is E itself; in symbols $\eta(E{:}E/H) \geqslant 0$, that is,

$$\eta(E{:}E) \geqslant \eta(E{:}H). \tag{20}$$

Equality occurs only if $P(E|H) = P(H|E) = 1$. *The corresponding result for dynamic explicativities is false.*

Proof. The right side of (19) can be written

$$(1-\gamma)\log P(H) + \gamma \log P(E) - \log P(E \,\&\, H).$$

Since $P(E \,\&\, H)$ exceeds neither $P(E)$ nor $P(H)$, this expression is at least as large as both $(1-\gamma)[\log P(H) - \log P(E)]$ and $\gamma[\log P(E) - \log P(H)]$ and must therefore be non-negative. It vanishes only if $P(H) = P(E) = P(E \,\&\, H)$, that is, only if $P(E|H) = P(H|E) = 1$, which for practical purposes means that E and H are logically equivalent.

That the theorem is false for dynamic explicativities is clear from the example of the planetary motions and the inverse square law. The dynamic explicativity $\eta_1(E{:}H)$ can exceed, equal, or "subceed" $\eta_0(E{:}E)$.

When $\gamma = \tfrac{1}{2}$ we have, when we do not use dynamic probabilities,

$$\eta(E{:}E/H) = \log \frac{[P(E)P(H)]^{1/2}}{P(E \,\&\, H)} \tag{21}$$

which is symmetrical in E and H, just as $I(E{:}H)$ is, though a closer analogue is $I(E{:}E) - I(E{:}H) = I(E|H)$ which is not symmetrical. (Of course it can be forcibly symmetrized by writing $I(E|H) + I(H|E)$.) If we accept the value $\gamma = \tfrac{1}{2}$, (21) could be called the *mutual* explicativity "distance" between E and H, by analogy with the name "mutual information" for $I(E{:}H)$. It equals 0 if H = E and ∞ if $H = \bar{E}$, and resembles $I(E|H)$ in this respect. Symmetry in E and H is an elegant property but it is not a compelling desideratum. The triangle inequality is not satisfied, but it may be of interest that

$$\begin{aligned}\eta(E{:}E/F) + \eta(F{:}F/G) - \eta(G{:}G/E) &= I(E|F) + I(F|G) - I(E|G) \\ &= I(G|F) + I(F|E) - I(G|E),\end{aligned} \tag{22}$$

so that the "triangles" for which the triangle inequality is valid are the same for the functions $(\lambda E)(\lambda F)\eta(E{:}E/F)$ and $(\lambda E)(\lambda F)I(E|F)$(in Alonzo Church's λ notation).

6. REPEATED TRIALS

Sometimes E can be defined as a compound event, or time series, which describes the probabilistic outcome $E_1 \,\&\, E_2 \,\&\, \ldots \,\&\, E_N$ of an experiment performed "independently" N times under essentially similar circumstances. If N is large, the frequencies of the various outcomes settle down, with high probability, to a distribution. A hypothesis H that predicts this distribution has an expected explicativity gain per observation, as compared with another hypothesis H′, and this gain tends in probability to

$$(1-\gamma)\mathcal{E}\,\{W(H/H'{:}E)|H\}, \tag{23}$$

which is proportional to the expected weight of evidence per observation. The second term in (18) gets divided by N and so contributes nothing to the limiting value (23). Thus, for "repeated trials," the application of the notion of explicativity to statistics will lead to the same results as when (expected) weight of evidence is used as a quasiutility, as in numerous publications cited earlier. In particular, if H asserts the true physical probability density $p(x,y)$ of two random variables, whereas hypothesis H′ asserts that the density is $p(x)q(y)$, then $\eta(\text{E}:\text{H}/\text{H}')/N$ tends in probability to

$$(1-\gamma)\int\int p(x,y)\log\frac{p(x,y)}{p(x)q(y)}\,dx\,dy \tag{24}$$

which is $1-\gamma$ times the "rate of transmission of information" concerning x provided by y and can of course be expressed in terms of three entropies. This formula can be used in the choice of an experimental design. The factor $1-\gamma$ is irrelevant for this purpose: see Cronbach (1953), #77, and especially Lindley (1956). Thus, in this application, the value of γ does not matter.

Greeno (1970), unaware of these references, suggested rate of transmission of information as an explication for explanatory power. We see from the above argument how this proposal is deducible from the notion of explicativity, and even from the earlier (Good, 1955; Hamblin, 1955) special case of weak explanatory power (informed explicativity), when E denotes an "infinitely repeated trial."

7. PREDICTIVITY

As we have seen, a probabilistic prediction of the result of an experiment or observation is a special case of a putative explanation, being made before the experimental result occurs. In these circumstances it is natural to measure the *predictivity* of a hypothesis as the mathematical expectation of the putative explicativity, the expectation being taken over the population of possible outcomes. It is appropriate to take expectations of η rather than of some monotonic function of η because of the additive property (7).

The explicativity of H, per trial, with respect to repeated trials, as given by (23), is formally nearly the same as predictivity, owing to the law of large numbers.

For a theory with a wide field of possible applications, the notion of predictivity is necessarily vague; but it might be defined as the expected explicativity over all future observations with discounting of the future at some rate. The concept is important in spite of its vagueness.

For experimental design, predictivities (expected explicativities) are natural quasiutilities. This fact can be regarded as an explication in hindsight why entropies occur in the work of Cronbach (1953) and Lindley (1956). In virtue of these two publications it is not necessary to consider experimental design further here. Instead, we work out in detail only examples of estimating parameters in a distribution law, after observations are taken. In this estimation problem entropies do not occur because expectations are not taken. Hypothesis testing can be regarded as a special case of parameter estimation (and vice versa).

8. "COLLATERAL" INFORMATION VERSUS BACKGROUND INFORMATION

Consider the propositions

E: Jones won the Irish Sweepstake,

H: Jones bought a ticket in this lottery,

and for the sake of simplicity assume that

$$P(\text{H}) = 2^{-8},\ P(\text{E}|\text{H}) = 2^{-20}, \text{ and therefore } P(\text{E}) = P(\text{E} \ \&\ \text{H}) = 2^{-28}.$$

Then, if $\gamma = \frac{1}{2}$, we have $\eta(\text{E:H}) = 8 - 8/2 = 4$ bits. If we knew all along that H was true we would have $\eta(\text{E:H}|\text{H}) = 0$, meaning that H cannot help to explain E if we have already taken H into account. But in another sense, if we discover that H is true we raise the probability of H to 1, and the explicativity of H with respect to E, which is now "informed" explicativity, is $I(\text{E:H}) = 8$ bits. Thus, for the sake of completeness, it is convenient to have a notation for the explicativity of H when its probability is conditional on some *collateral* information K. Let us use a semicolon to mean "given the collateral information." Then we have

$$\eta(\text{E:H; K}|\text{G}) = \log P(\text{E}|\text{H}) \ \&\ \text{G}) - \log P(\text{E}|\text{G}) + \gamma \log P(\text{H}|\text{K} \ \&\ \text{G}) \qquad (25)$$

where we have included G for greater generality. In particular,

$$\eta(\text{E:H; H}) = I(\text{E:H}). \qquad (26)$$

Background information is taken for granted in computing all the probabilities, whereas collateral information affects only the probability of the explanans H and is not taken into account when computing the probability of the explanandum E. Of course $\eta(\text{E:H; H})$ is the informed explicativity of H. No special terminology for $\eta(\text{E:H}|\text{H})$ is proposed because it necessarily vanishes.

The notation $\eta(\text{E:H; K})$ or $\eta(\text{E:H; K}|\text{G})$ helps to formalize the familiar situation in which an explanans H is strengthened by having its *own* probability increased by evidence K. For example, when we discover that Tom was at the scene of the crime, the probability is increased that he threw a stone at the window. Explicativity depends on the explanandum, the explanans, the collateral information, and the background information. We have

$$\eta(\text{E:H; K}) = \eta(\text{E:H} \ \&\ \text{K}) \text{ if and only if } P(\text{E}|\text{H} \ \&\ \text{K}) = P(\text{E}|\text{H}). \qquad (27)$$

9. THE QUANTITATIVE DISTINCTION BETWEEN EXPLICATIVITY AND CAUSAL PROPENSITY

In our lottery example the explicativity of the ticket-purchase, with respect to E, is appreciable (whether the explicativity is "informed" or not), although

$P(E|H)$ is small. There is a distinction between (putative) explicativity and causal propensity: the purchase of the ticket did not do much to *cause* E although it was a necessary condition for it. If Jones had not won the sweepstake, it would have been negligible evidence against his having bought a ticket, so, according to (6), the causal propensity of the purchase is small. Similarly, if Ms Aksed is hit by a small meteorite when out walking, we would not blame her and accuse her of suicidal tendencies. Her decision to go for a walk was a necessary condition for the disaster, but if she had not been hit by a meteorite, it would have been negligible evidence that she was indoors when the meteorite fell. The insurance company would call the incident an Act of God.

10. APPLICATIONS TO STATISTICAL ESTIMATION AND SIGNIFICANCE TESTING

. . . [The eight pages omitted here show that $\eta(E{:}H)$ can be applied to statistics with entirely sensible results. This confirms the reasonableness of η as an explication of explicativity.]

11. FURTHER COMMENTS CONCERNING THE VALUE OF γ

If no other desiderata can be found for fixing γ, the value $\gamma = \frac{1}{2}$ could often reasonably be adopted on grounds of maximum simplicity. This choice can itself be regarded as an application of a form of the Duns-Ockham razor (of higher type so to speak). Moreover there are many scientists who believe that the notion of simplicity is better replaced by that of elegance, or aesthetic appeal. For example, Margenau (1949) says "The physicist is impressed not solely by its [a theory's] far-flung empirical verifications, but above all by the intrinsic beauty of its conception which predisposes the discriminating mind for acceptance even if there were no experimental evidence for the theory at all." Again Dirac (1963) says " . . . it is more important to have beauty in one's equations than to have them fit experiment. . . . That is how quantum mechanics was discovered," and I believe Dirac expressed this view in conversation at least as early as 1940. From this point of view the value $\gamma = \frac{1}{2}$ gains from the elegant symmetry of equation (21). . . . [As a discussion point, I believe that beauty is often a matter of simplicity arising out of complexity arising out of simplicity.]

12. SUMMARY

Philosophical aspects of explanation were discussed in §3 leading up to an informal definition of $\eta(E{:}F|G)$ and to the desiderata and exact explication of η in §4 in terms of probabilities or information. In §5 we showed the relevance of explicativity for a choice between hypotheses. In §6 we saw that if explicativity is used in experimental design it reduces in effect to expected weight of evidence or to rate of transmission of information. In §7 an informal quantification of predicitivity was suggested. In §8 it is pointed out that a distinction between background information and "collateral" information is necessary for formalizing a familiar aspect of explanation, so that η depends on four variables (apart from the evidence G′ that is deliberately ignored: see §3). In §9 it is shown that explicativity and causal propensity can be quantitatively quite different, both in common parlance and in terms of the formalism. In §10 several examples of statistical estimation and significance testing are worked out in terms of explicativity with intuitively appealing results.

APPENDIX A. COMPLEXITY

Although an explication of simplicity or complexity is not required for that of explicativity, the latter depends on the initial probability of a proposition H and this probability surely depends to some extent on the complexity of H. For the complexity of the conjunction H & K of two propositions that are entirely independent is greater than the complexity of either of them separately, in any one's book, and is reasonably assumed to be the sum of the two complexities. If the complexity of H could be defined in terms of P(H) alone then it would have to be $-\log P$(H) as suggested in #599. But the two propositions 0 = 0 and 0 = 1 are about equally simple in my present judgment, though their probabilities are poles apart. So the complexity of H cannot be defined in terms of P(H) alone. Fortunately this error in #599 did not undermine much else in that work. The error was admitted in #876, and on pp. 154-56, where attempts were made to improve the definition. It was proposed that the complexity of a proposition should be defined as the minimum value of $-\log p$ where $p = P$(S) is the probability of some statement S of the proposition *regarded as a linguistic string* and the minimum is taken over all ways of expressing the proposition as a statement. Moreover, the language used must be one that is economical for talking about the topic in question.

A valid objection was raised against this definition by Peter Suzman, as mentioned in Good (1976b). Suzman asked whether the proposition that all caterpillars have chromosomes is more complex than that all dogs have chromosomes. My reply was to concede that these propositions are of (nearly?) *equal* complexity. Nor is it sufficient to modify the proposed definition of complexity, by defining p(S) as the probability of the *syntactic structure* of S, nor by making the definition depend only on the number of dimensionless parameters in a law. For a parameter equal to 5.4603 is more complex than one that is equal to 2. Perhaps one cannot do much better than to define the complexity of a proposition as equal to the *weighted length* of the shortest way of expressing it, measured in words and symbols, where different weights should be assigned to different categories of words such as parts of speech. Perhaps the weights should be minus the logarithms of the frequencies of these *categories* of words (instead of using the frequencies of the individual words and symbols as such). This would reduce the problem to the specification of the categories.

A somewhat different ideal measure of the complexity of a scientific theory is the number of independent axioms in it (see, for example, Margenau, 1949), and I believe this is a useful rule of thumb. But it does not allow for the relative complexities of the axioms.

In practice, the beauty of a theory, rather than its simplicity, might be more important when estimating initial probabilities: see the quotations at the end of the main text. To fall back on beauty as a criterion is presumably to admit that the left hemispheres of the brains of philosophers of science have not yet formalized the intuitive activities of the right hemispheres.

Measurements of complexity or ugliness might help us to judge prior probabilities, but, if the prior probabilities could be adequately judged, the crutches of simplicity and beauty could be discarded. These crutches were not much used in the main text because our aim was to express explicativity in terms of probability.

APPENDIX B. THE FORM OF THE FUNCTION $\eta(E:H)$

. . . [A proof of (8) was given in this appendix.]

References

References

A list of abbreviations for the names of periodicals is provided on p. 251.

Abel, Neils Henrik (1826). "Recherche des fonctions de deux quantités variables indépendantes x et y, telles que f(x, y), qui ont la propriété que f(z, f(x, y)) est une fonction symétrique de z, x, et y," *J. für die reine und angewandte Mathematik, herausgegeben von Crelle, band 1.* Reprinted in *Oevres complètes de Neils Henrik Abel,* Nouvelle édition, Christiania: Grøndhul & Søns, 1881, vol. 1, 61-65; New York: Johnson Reprint Corporation, 1975.

Aczél, J. (1948). "Sur les opérations définies pour nombres réels," *Bull. Soc. Math. France 76,* 59-64.

Aczél, J. (1955). "A solution of some problems of K. Borsuk and L. Janossy," *Acta. Phys. Acad. Sci. Hungar. 4,* 351-362 (*MR 16,* 1128).

Anonymous. (1951). Occam, William of. *Encyclopaedia Brittannica 16,* 678-679.

Anscombe, F. J. (1954). "Fixed sample-size analysis of sequential observations," *Biometrics 10,* 89-100.

Anscombe, F. J. (1968/69). Discussion of a paper by I. J. Good, April 9, 1968, *JASA 64,* 50-51.

Ayer, A. J. (1957). "The conception of probability as a logical relation," in *Observation and Interpretation,* S. Korner, ed., London: Butterworths, 12-30 (with discussion).

Barnard, G. A. (1951). "The theory of information," *JRSS B 13,* 46-64 (with discussion), esp. p. 56.

Barnard, G. A. (1979). "Pivotal inference and the Bayesian controversy," in *Bayesian Statistics* (see #1230 in my bibliography), 295-318 (with discussion).

Bartlett, M. S. (c. 1951). Personal communication.

Bartlett, M. S. (1952). "The statistical significance of odd bits of information," *Biometrika 39,* 228-237.

Bayes, T. (1763/65, 1940/58). An essay toward solving a problem in the doctrine of chances, *Phil. Trans. Roy. Soc. 53,* 370-418; *54,* 296-325. Reprints: the Graduate School, U.S. Department of Agriculture, Washington, D.C. (1940); *Biometrika 45* (1958), 293-315.

Bernoulli, D. (1734). "Recherches physiques et astronomique . . . ," *Recueil des pièces qui ont remportè le prix de l'Academie Royal des Sciences 3,* 93-122.

Bernoulli, D. (1774/78/1961). The most probable choice between several discrepant observations and the formation therefrom of the most likely induction (in Latin), *Acta Academiae Scientiorum Petropolitanae 1* (1777), 3-23. English trans. by C. G. Allen in *Biometrika 48* (1961), 3-13.

Bernstein, S. (1921/22). "Versuch einer axiomatischen Begrundung der Wahrscheinlichkeits-rechnung," *Mitt. Charkow 15,* 209-274. Abstract in *Jahrbuch der Math. 48,* 596-599.

Bishop, Y. M. M., Fienberg, S. E., & Holland, P. W. (1975). *Discrete Multivariate Analysis,* Cambridge, Mass: M.I.T. Press.

Bochner, S. (1955). Personal communication.

Bohm, David (1952). "A suggested interpretation of the quantum theory in terms of 'hidden' variables," *Phys. Rev. 85,* 166-193.

Borel, E. (1920). *Le Hasard,* Paris: Hermann.

Box, G. E. P., & Tiao, G. C. (1973). *Bayesian Inference in Statistical Analysis,* Reading, Mass.: Addison-Wesley.

Braithwaite, R. B. (1951). A lecture at the 1951 weekend conference of the Royal Statistical Society in Cambridge, England.

Braithwaite, R. B. (1953). *Scientific Explanation,* Cambridge: Cambridge University Press.

Brier, G. W. (1950). "Verification of forecasts expressed in terms of probability," *Monthly Weather Review 78,* 1-3.

Bunge, M. (1955). "Strife about complementarity," *BJPS 6,* 1-12 and 141-154.

Burkill, J. C. (1951). *The Lebesgue Integral,* Cambridge: Cambridge University Press.

Carnap, R. (1947). "On the application of inductive logic," *Philosophy and Phenomenological Research 8,* 133-148.

Carnap, R. (1950). *Logical Foundations of Probability,* Chicago: University of Chicago Press.

Carnap, Rudolf, & Jeffrey, Richard C. (1971). *Studies in Inductive Logic and Probability I,* Berkeley & Los Angeles: University of California Press.

Charlton, W. (1970). *Aristotle's Physics,* book I, Oxford: Clarendon Press.

Coolidge, J. L. (1916). *A Treatise on the Circle and the Sphere,* Oxford: Clarendon Press.

Cornfield, J. (1968/70). Contributions to the discussion, in *Bayesian Statistics,* D. L. Meyer & R. O. Collier, Jr., eds., Itasca, Ill.: Peacock, 108.

Cox, R. T. (1946). "Probability, frequency and reasonable expectation," *Amer. J. Physics 14,* 1-13.

Cox, R. T. (1961). *The Algebra of Probable Inference,* Baltimore: Johns Hopkins University Press.

Cronbach, L. J. (1953). "A consideration of information theory and utility theory as tools for psychometric problems," Technical Report, College of Education, University of Illinois, Urbana.

Crook, J. F., & Good, I. J. (1981). "The powers and 'strengths' of tests for multinomials and contingency tables." In preparation.

Daniels, H. E. (1951). "The theory of position finding," *JRSS B 13,* 186-207 (with discussion).

Daniels, H. E. (ca. 1956). Personal communication.

David, F. N. (1949). *Probability Theory for Statistical Methods,* Cambridge: Cambridge University Press.

Davidson, M. (1943). *An Easy Outline of Astronomy,* London: Watts.

Davies, O. L. (1954). *Design and Analysis of Industrial Experiments,* London & Edinburgh: Oliver & Boyd.

De Finetti, B. (1937/64). "Foresight: its logical laws, its subjective sources," trans. from the French of 1937 by H. Kyburg, in *Studies in Subjective Probability,* H. E. Kyburg & H. E. Smokler, eds., New York: Wiley, 95-158. Corrected in 2nd ed., 1980, 55-118.

De Finetti, B. (1951). "Recent suggestions for the reconciliation of theories of probability," *Proc. Second Berkeley Symp. on Math. Stat. and Probability,* Berkeley & Los Angeles: University of California Press, 217-225.

De Finetti, B. (1968/70). "Initial probabilities: A prerequisite for any valid induction," in *Induction, Physics, and Ethics,* P. Weingartner and G. Zecha, eds., Dordrecht, Holland: D. Reidel, 3-17 (with discussion). Also in *Synthese 20* (1969).

De Finetti, B. (1971). "Probabilità di una teoria e probabilità dei fatti," in *Studi di probabilità, statistica e ricerca operativa in onore di Guiseppe Pompili,* Oderisi: Gubbio, 86-101.

De Finetti, B. (1975). *Theory of Probability,* vol. 2, New York: Wiley.

De Morgan, A. (1837/1853). "Theory of probabilities," in *Encyclopedia Metropolitana,* vol. 2, 393-490.

Dempster, A. P., Laird, N. M., & Rubin, D. B. (1977). "Maximum likelihood from incomplete data via the EM algorithm," *JRSS B 39,* 1-38 (with discussion).

Dickey, J. M. (1968). "Estimation of disease probabilities conditioned on symptom variables," *Mathematical Biosciences 3,* 249-265.

Dirac, P. A. M. (1963). "The evolution of the physicist's picture of nature," *Sci. Am. 208* (5) (May), 45-53.

Dodge, H. F., & Romig, H. G. (1929). "A method for sampling inspection," *Bell System Tech. J. 8,* 613-631.

Eddington, Sir Arthur (1933/52). *The Expanding Universe,* Cambridge: Cambridge University Press.

Edgeworth, F. Y. (1910). "Probability," *Encyclopaedia Britannica 22,* 11th ed., 376-403.

Edgeworth, K. E. (1961). *The Earth, the Planets, and the Stars,* New York: Macmillan.

Efron, B. (1971). "Does an observed sequence of numbers follow a simple rule?" *JASA 66,* 552-568 (with discussion).

Eisenhart, C. (1964). "The meaning of 'least' in Least Squares," *J. Washington Acad. Sci. 54,* 24-33.

Everett, H. J., III (1957). " 'Relative state' formulation of quantum mechanics," *Reviews of Modern Physics 29,* 454-462.

Fath, E. A. (1955). *Elements of Astronomy,* 5th ed., New York: McGraw-Hill.

Feibleman, J. K. (1969). *An Introduction to the Philosophy of Charles S. Peirce,* Cambridge, Mass.: M.I.T. Press.

Feller, W. (1940). "Statistical aspects of ESP," *J. Parapsychology 4,* 271-298.

Feller, W. (1950). *An introduction to Probability Theory and Its Applications,* vol. 1, New York: Wiley.

Finney, D. J. (1953). *Statistical Science in Agriculture,* New York: Wiley; Edinburgh: Oliver & Boyd.

Fisher, R. A. (1949). *The Design of Experiments,* 5th ed., Edinburgh & London: Oliver & Boyd.

Fisher, R. A. (1959). *Smoking: The Cancer Controversy,* Edinburgh: Oliver & Boyd.

Frazier, K. (1979). "Schmidt's airing at the APS," *The Skeptical Inquirer: The Zetetic 3* (4), 2-4.

Friedman, K. S (1973). A measure of statistical simplicity, mimeographed, 32 pp.

Freiman, J. A., Chalmers, T. C., Smith, Harry, Jr. and Kuebler, R. R. (1978). "The importance of beta, the type II error and sample size in the design and interpretation of the randomized control trial," *New England J. Medicine 299,* 690-694.

Gauss, K. F. (1798/1809/57). Theory of the motion of heavenly bodies moving about the sun in sections (Latin), in Carl Friedrich Gauss Werke, English trans. by Davis (1857); reprinted by Dover publications.

Gibbs, J. W. (1902). *Elementary Principles in Statistical Mechanics,* New York: Scribner; reprinted by Dover publications.

Godambe, V. P., & Sprott, D. A. (eds.) (1971). *Foundations of Statistical Inference,* Toronto: Holt, Rinehart, & Winston of Canada.

Good, I. J. (1950/53). Contribution to the discussion of papers by E. C. Cherry and by M. S. Bartlett in *Symposium on Information Theory, Report of Proceedings,* London: Ministry of Supply, 167-168, 180-181. Also in *Trans. I. R. E.,* Feb. 1953.

Good, I. J. (1951). Contribution to the discussion of a paper by H. E. Daniels, *JRSS B 13,* 203.

Good, I. J. (1952). Review of G. H. von Wright, "A treatise on induction and probability," *JRSS A 115,* 283-285.

Good, I. J. (1953a). Review of M. H. Quenouille, "Design and analysis of experiment," *Annals of Eugenics 18,* 263-266.

Good, I. J. (1953b). Contribution to the discussion of a paper by D. M. McKay, in *Communication Theory,* Willis Jackson, ed., London: Butterworths, 483-484.

Good, I. J. (1954). Contribution to the discussion of a paper by R. D. Clark, *J. Inst. Actuar. 80,* 9-20.

Good, I. J. (1955). Review of K. R. Popper, "Degrees of confirmation" [*BJPS 5* (1954), 47-48], *MR 16,* 376.

Good, I. J. (1956). Contribution to the discussion of a paper by G. S. Brown in *Information Theory: Third London Symposium 1955,* Colin Cherry, ed., London: Butterworths, 13-14.

Good, I. J. (1958). Review of R. von Mises, *Probability, Statistics, and Truth,* 2nd ed., *JRSS A 121,* 238-240.

Good, I. J. (1960). "Gravitational frequency shift," letter in *New Scientist* (February 25), 483.

Good, I. J. (1962a). Contribution to the discussion in *The Foundations of Statistics,* opened by L. J. Savage, London: Methuen; New York: Wiley, esp. pp. 74 & 78.

Good, I. J. (1962b). "Speculations concerning precognition," in #339, 151-157.

Good, I. J. (1962c). "Physical numerology," in #339, 315-319.

Good, I. J. (1962d). Review of R. T. Cox, *The Algebra of Probable Inference, MR 24,* 107.

Good, I. J. (1964). Contribution to the discussion of A. R. Thatcher, "Relationships between Bayesian and confidence limits for predictions," *JRSS B 26,* 204-205.

Good, I. J. (1965). Letter concerning Tippett's random numbers, *Amer. Stat. 19,* 43.

Good, I. J. (1970). Contribution to the discussion of a paper by H. Vetter, in the Salzburg conference. (See #617.)

Good, I. J. (1971). "The Baker's Oven, XV," *Mensa J. & Bulletin,* No. 150, 10-15.

Good, I. J. (1976). "Explicativity." A lecture in a seminar on Bayesian Inference in Econometrics at Harvard University.

Good, I. J. (1977). "Rationality, evidence, and induction in scientific inference," in *Machine Intelligence 8,* E. W. Elcock and D. Michie, eds., Chichester: Ellis Horwood, Ltd.; New York: Wiley, 171-174.

Greeno, J. G. (1970). "Explanation of statistical hypotheses using information transmitted," *Philosophy of Sci. 37,* 279-294. Reprinted in Salmon (1971).

Greenwood, J. A. (1938). "An empirical investigation of some sampling problems," *J. Parapsychology 2,* 222-230.

Hamblin, C. L. (1955). An unpublished doctoral thesis submitted to the University of London and cited by Popper, 1959, p. 403.

Hamel, G. (1905). "Eine basis aller Zahlen und die unstetigen Lösungen der Funktionalgleichungen $f(x+y) = f(x) + f(y)$," *Math. Annalen 60,* 459-462.

Hardy, G. F. (1889). In correspondence in *Insurance Record,* reprinted in *Trans. Fac. Actuar. 8* (1920), 174-182.

Hardy, G. H., Pólya, G., & Littlewood, J. E. (1934). *Inequalities,* Cambridge: Cambridge University Press.

Hartley, R. V. L. (1928). "Transmission of information," *Bell System Tech. J. 7,* 535-563.

Hempel, C. G. (1948/65). "Studies in the logic of explanation," *Philos. of Sci. 15* (1948), 135-175; reprinted with some changes in *Aspects of Scientific Explanation,* New York: Free Press (1965).

Hempel, C. G. (1967). "The white shoe: No red herring," *BJPS 18,* 239-240.

Hendrickson, A., & Buehler, R. J. (1972). "Elicitation of subjective probabilities by sequential choices," *JASA 67*, 880-883.

Horowitz, I. A., & Mott-Smith, G. (1960). *Point Count Chess*, New York: Simon and Schuster.

Humphreys, P. (1980). "Cutting the causal chain," *Pacific Philosophical Q. 61*, 305-316.

Hurwicz, L. (1951). "Some specification problems and applications to econometric models," *Econometrics 19*, 343-344 (abstract).

Huzurbazar, V. S. (1955). "On the certainty of an inductive inference," *PCPS 51*, 761-762.

Jànossy, L. (1955). Remarks on the foundations of probability calculus, *Acta. Phys. Acad. Sci. Hungar. 4*, 333-349.

Jaynes, E. T. (1957). "Information theory and statistical mechanics," *Phys. Rev. 106*, 620-630.

Jaynes, E. T. (1968). "Prior probabilities," *IEEE Trans. Systems Sc. and Cyb. SSC-4*, 227-241.

Jeffreys, H. (1936). "Further significance tests," *PCPS 32*, 416-445.

Jeffreys, H. (1939/48/61). *Theory of Probability*, Oxford: Clarendon Press.

Jeffreys, H. (1946). "An invariant form for the prior probability in estimation problems," *Proc. Roy. Soc. A 186*, 453-461.

Johnson, William Ernest (1932). Appendix to "Probability: Deductive and inductive problems," *Mind 41*, 421-423. This appendix was edited by R. B. Braithwaite.

Kalbfleisch, J. (1971). *Probability and Statistical Inference* (Lecture Notes for Mathematics, 233; Department of Statistics, University of Western Ontario).

Kalbfleisch, J. G., & Sprott, D. A. (1976). "On tests of significance," in *Foundations of Probability Theory, Statistical Theory, Statistical Inference, and Statistical Theories of Science*, vol. 2, C. A. Hooker and W. Harper, eds., Dordrecht, Holland: D. Reidel, 259-272.

Kemble, E. C. (1941). "The probability concept," *Philosophy of Science 8*, 204-232.

Kemble, Edwin C. (1942). "Is the frequency theory of probability adequate for all scientific purposes?," *Amer. J. Physics 10*, 6-16.

Kemeny, J. G., & Oppenheim, P. (1952). "Degrees of factual support," *Philosophy of Science 19*, 307-324.

Kempthorne, O., & Folks, L. (1971). *Probability, Statistics, and Data Analysis*, Ames, Iowa: Iowa University Press.

Kendall, M. G. (1941). "A theory of randomness," *Biometrika 32*, 1-5.

Kendall, M. G. (1949). "On the reconciliation of theories of probability," *Biometrika 36*, 101-116.

Kendall, M. G., & Stuart, A. (1960). *The Advanced Theory of Statistics*, vol. 2, London: Griffin.

Kerridge, D. F. (1961). "Inaccuracy and inference," *JRSS B 23*, 184-194.

Keynes, J. M. (1921). *A Treatise on Probability*, London & New York: Macmillan; New York: St. Martin's Press, 1952.

Keynes, John Maynard (1933). *Essays in Biography*, London: Rupert Hart Davis.

Kim, J. (1967). "Explanation in science," in *The Encyclopedia of Philosophy 3*, 159-163. New York: Macmillan and Free Press.

Kneale, W. (1953). "Induction, explanation, and transcendental hypotheses," in *Readings in the Philosophy of Science*, H. Feigl and M. Brodbeck, eds., New York: Appleton-Century-Crofts, 353-367.

Kolmogorov, A. N. (1963). "On tables of random numbers," *Sankhyā A 25*, 369-376.

Koopman, B. O. (1940a). "The axioms and algebra of intuitive probability," *Annals of Math. 41*, 269-92.

Koopman, Bernard Osgood (1940b). "The bases of probability," *Bull. Amer. Math Soc. 46*, 763-774.

Koopman, B. O. (1969). "Relaxed motion in irreversible molecular statistics," *Advances in Chemical Physics 15,* 37-63.

Kullback, S. (1959). *Information Theory and Statistics,* New York: Wiley.

Kullback, S., & Leibler, R. A. (1951). "On information and sufficiency," *Ann. Math. Statist. 22,* 79-86.

Laplace, P. S. de (1774). "Mémoire sur la probabilité des causes par les événements," *Mémoires . . . par divers Sarans 6,* 621-656.

Laplace, P. S. de (1878/1912). *Oevres Complètes,* vol. 7, Paris: Gauthiers-Villars.

Lehmann, E. L. (1959). *Testing Statistical Hypotheses,* New York: Wiley.

Lemoine, Émile (1902). *Géométrographie,* Paris: C. Naud.

Leonard, T. (1978). "Density estimation, stochastic processes and prior information," *JRSS B 40,* 113-146 (with discussion).

Levi, Isaac (1973). Inductive logic and the improvement of knowledge, Columbia University, mimeographed.

Levin, B., & Reeds, J. (1977). "Compound multinomial likelihood functions: proof of a conjecture of I. J. Good," *Annals of Statistics 5,* 79-87.

Lidstone, G. J. (1920). "Note on the general case of the Bayes-Laplace formula for inductive or a posteriori probabilities," *Trans. Fac. Actuar. 8,* 182-192.

Lindley, D. V. (1947). "Regression lines and the linear functional relationship," *Suppl. JRSS 9,* 218-244.

Lindley, D. V. (1953). "Statistical inference," *JRSS B 15,* 30-76 (with discussion).

Lindley, D. V. (1956). "On a measure of the information provided by an experiment," *Annals of Math. Statist. 27,* 986-1005.

Lindley, D. V. (1957). "A statistical paradox," *Biometrika 44,* 187-192.

Lindley, D. V. (1965). *Introduction to Probability and Statistics,* part 2, Cambridge: Cambridge University Press.

Lindley, D. V. (1971). "The estimation of many parameters," in *Foundations of Statistical Inference,* V. P. Godambe and D. A. Sprott, eds., Toronto: Holt, Rinehart, and Winston of Canada, 435-455 (with discussion).

Lindley, D. V., & Smith, A. F. M. (1972). "Bayes estimates for the linear model," *JRSS B 34,* 1-41 (with discussion).

Loève, M. (1955). *Probability Theory,* New York: van Nostrand.

McCarthy, John (1956). "Measures of the value of information," *Proc. Nat. Acad. Sci. 42,* 654-655.

Mackie, J. L. (1966). "Miller's so-called paradox of information," *BJPS 17,* 144-147.

MacMahon, P. A. (1892). "The combination of resistances," *Electrician 28,* 601-602.

Mandelbrot, B. (1953). In *Communication Theory,* Willis Jackson, ed., London: Butterworths, 486.

Margenau, H. (1949). "Einstein's conception of reality," in *Albert Einstein: Philosopher-Scientist,* P. A. Schilp, ed., New York: Tudor Publishing Co., 245-268.

Marschak, J. (1951). "Why 'should' statisticians and businessmen maximise 'moral expectation'?" *Proc. 2nd Berkeley Symp. on Math. Stat. and Probability,* Berkeley: University of California Press, 493-506.

Marschak, Jacob (1959). Remarks on the economics of information, in *Contributions to Scientific Research in Management,* Berkeley: University of California Press, 79-98.

Martin-Löf, P. (1969). "The literature on von Mises' Kollektivs revisited," *Theoria 35,* 21-37; *MR 39,* 404.

Mauldon, J. G. (1955). "Pivotal quantities for Wishart's and related distributions, and a paradox in fiducial theory," *JRSS B 17,* 79-85.

Mayr, E. (1961). "Cause and effect in biology," *Science 134,* 1501-1506.

Michie, D. (1977). "A theory of advice," *Machine Intelligence 8,* E. W. Elcock and D. Michie, eds., Chichester: Ellis Horwood, Ltd.; New York: Wiley, 151-168.

Mill, J. S. (1843/1961). *A System of Logic, Ratiocinactive and Inductive: Being a Connected*

View of the Principles of Evidence and the Methods of Scientific Investigation, Book III, Chapter XII, Section I, London: Longmans.
Miller, D. W. (1970). Personal communication.
Minsky, Marvin, & Selfridge, Oliver G. (1961). "Learning in random nets," in *Information Theory*, Colin Cherry, ed., London: Butterworths, 335-347.
Moody, E. A. (1967). "William of Ockham," *Encyclopedia of Philosophy 8*, 306-317.
Mosteller, F., & Wallace, D. L. (1964). *Inference and Disputed Authorship*, Reading, Mass.: Addison-Wesley.
Nagel, E. (1961). *The Structure of Science: Problems in the Logic of Scientific Explanation*, New York: Harcourt, Brace & World.
Newcomb, S. (1910). "Bode," in *Encyclopedia Britannica 4*, 11th ed., 108.
Neyman, J. (1952). *Lectures and Conferences on Mathematical Statistics and Probability*, 2nd ed., Washington, D.C.: Graduate School, U.S. Department of Agriculture.
Neyman, J. (1977). "Frequentist probability and frequentist statistics," *Synthese 36*, 97-131.
Neyman, J., & Pearson, E. S. (1933). "On the problem of the most efficient tests of statistical hypotheses," *Philosophical Transactions of the Roy. Soc. of London A 231*, 289-337.
Nieto, M. M. (1972). *The Titius-Bode Law of Planetary Distances*, Oxford & New York: Pergamon Press.
Orear, J., & Cassel, D. (1971). "Applications of statistical inference to physics," in *Foundations of Statistical Inference*, V. P. Godambe and D. A. Sprott, eds., Toronto: Holt, Rinehart, & Winston of Canada, 280-288 (with discussion).
Orwell, George (1949). *1984*, New York: Harcourt, Brace and World.
Pascal, B. (1670). "Article II. Qu'il est plus advantageux de croire que de ne pas croire ce qu-enseigne la religion chrétienne," *Pensées*, Garnier Frères, Paris, after the ed. of 1670. (Reference given by Marschak [1951].)
Patil, G. P. (1960). "On the evaluation of the negative binomial distributions with samples," *Technometrics 2*, 501-505.
Payne-Gaposchkin, C. (1961). *Introduction to Astronomy*, London: Methuen.
Peirce, Charles Sanders (1878). "The probability of induction," *Popular Science Monthly*; reprinted in *The World of Mathematics 2*, James R. Newman, ed., New York: Simon and Schuster (1956), 1341-1354.
Pelz, W. (1977). Topics on the estimation of small probabilities, doctoral thesis, Virginia Polytechnic Institute and State University, 123 pp.
Perks, W. (1947). "Some observations on inverse probability including a new indifference rule," *J. Inst. Actuar. Students' Soc. 73*, 285-334 (with discussion).
Poisson, S. D. (1837). *Recherches sur la probabilités des jugements en matière criminelle et matière civile, précédées des règles générales du calcul des probabilités*, Paris: Bachelier.
Pólya, G. (1941). "Heuristic reasoning and the theory of probability," *Amer. Math. Monthly 48*, 450-465.
Pólya, G. (1950). "On plausible reasoning," *Proc. Int. Congress Math. 1*, 739-47.
Pólya, G. (1954). *Mathematics and Plausible Reasoning*, 2 vols. Princeton, N.J.: Princeton University Press.
Popper, K. R. (1954). "Degree of confirmation," *BJPS 5*, 143-149.
Popper, K. R. (1957a). *The Poverty of Historicism*, London: Routledge and Kegan Paul; Boston: Beacon Press.
Popper, K. R. (1957b). "Probability magic or knowledge out of ignorance," *Dialectica 11*, 354-357.
Popper, K. R. (1959). *The Logic of Scientific Discovery*, London: Hutchison.
Popper, K. R. (1962). *Conjectures and Refutations*, New York: Basic Books.
Raiffa, H., & Schlaifer, R. (1961). *Applied Statistical Decision Theory*, Boston: Graduate School of Business Administration, Harvard University.

Ramsey, R. P. (1926/31/50/64). "Truth and probability," a 1926 lecture published in *The Foundations of Mathematics and Other Logical Essays,* London: Routledge and Kegan Paul; New York: Humanities Press, 1950. Reprinted in *Studies in Subjective Probability,* H. E. Kyburg and H. E. Smokler, eds., New York: Wiley (1964), 63-92; 2nd ed., Huntington, N.Y.: Kriege (1980), 23-52.

Reichenbach, H. (1949). *The Theory of Probability,* Berkeley & Los Angeles: University of California Press. (Trans. by E. H. Hutten and Maria Reichenbach from the German of 1934.)

Reichenbach, H. (1956). *The Direction of Time,* Berkeley & Los Angeles: University of California Press.

Reichenbach, H. (1958). *The Philosophy of Space and Time,* New York: Dover Publications; from the German of 1928.

Reichenbach, H. (1959). *Modern Philosophy of Science,* London: Routledge & Kegan Paul; New York: Humanities Press.

Rényi, A. (1961). "On measures of entropy and information," *Proc. Fourth Berkeley Symposium Math. Statist. and Prob.,* vol. 1, Berkeley, University of California Press, 547-561.

Rescher, N. (1970). *Scientific Explanation,* New York: Free Press; London: Collier-Macmillan.

Riordan, J. (1958). *An Introduction to Combinatorial Analysis,* New York: Wiley.

Robbins, H. E. (1952). "Some aspects of the sequential design of experiments," *Bull. Amer. Math. Soc. 58,* 527-535.

Robbins, Herbert E. (1968). "Estimating the total probability of the unobserved outcomes of an experiment," *Ann. Math. Statistics 39,* 256-257.

Rogers, John Marvin (1974). Some examples of compromises between Bayesian and non-Bayesian statistical methods, doctoral dissertation, Virginia Polytechnic Institute and State University, 125 pp.

Rosen, D. A. (1978). "Discussion: In defense of a probabilistic theory of causality," *Phil. Sc. 45,* 604-613.

Rothstein, J. (1951). "Information, measurement, and quantum mechanics," *Science, N.Y. 114,* 171-175.

Russell, B. (1948). *Human Knowledge, Its Scope and Limits,* New York: Simon & Schuster, part V.

Salmon, W. C. (1971). *Statistical Explanation and Statistical Relevance* (with contributions by R. C. Jeffrey and J. C. Greeno), Pittsburgh: University of Pittsburgh Press.

Salmon, W. C. (1980). "Probabilistic causality," *Pacific Philosophical Q. 61,* 50-74.

Samuelson, P. A. Measurement of utility reformulated, unpublished paper. (Reference given by Marschak [1951, p. 495], who calls the article "recent.")

Savage, L. J. (1951). Notes on the foundation of statistics, privately circulated paper.

Savage, L. J. (1954). *Foundations of Statistics,* New York: Wiley.

Savage, L. J. (1959). Personal communication, July.

Savage, L. J. (1959/62). "Subjective probability and statistical practice," in *The Foundations of Statistical Inference,* G. A. Barnard and D. R. Cox, eds., London: Methuen; New York: Wiley, 9-35.

Savage, L. J. (1971). "Elicitation of personal probabilities and expectations," *JASA 66,* 783-801.

Sayre, K. M. (1977). "Statistical models of causal relations," *Phil. of Sci. 44,* 203-214.

Scheffler, I. (1963). *The Anatomy of Inquiry,* New York: Alfred A. Knopf, part I, Explanation.

Schrödinger, E. (1947). "The foundation of probability," *Proc. Roy. Irish Acad. 51A,* 51-66 and 141-146.

Scott, Christopher (1958). "A review of a book by G. S. Brown," in *J. of the Soc. of Psychical Res. 39,* 217-234.

Shackle, G. L. S. (1949). *Expectation in Economics,* Cambridge: Cambridge University Press.

Shannon, C. E. (1948). "A mathematical theory of communication," *Bell System Technical J. 27,* 379-423, 623-656. (Reprinted in C. E. Shannon & W. Weaver, *The Mathematical Theory of Communication,* Urbana: University of Illinois Press.)

Simon, Herbert A. (1957). *Models of Man,* New York: Wiley.

Simon, H. A. (1978). "Causation," in *International Encyclopedia of Statistics,* W. H. Kruskal and J. M. Tanur, eds., New York: Free Press; London: Collier-Macmillan, 35-41.

Smith, C. A. B. (1961). "Consistency in statistical inference and decision," *JRSS B 23,* 1-25 (with discussion).

Smith, C. A. B. (1965). "Personal probability and statistical analysis," *JRSS A 128,* 469-499 (with discussion).

Stone, M. H. (1948). Notes on integration, I, II, III, IV, *Proc. Nat. Acad. Sci. 34 & 35.*

Sunnucks, Anne (1970). *The Encyclopedia of Chess,* New York: St. Martin's Press.

Suppes, P. (1970). *A Probabilistic Theory of Causality,* Amsterdam: North-Holland Publishing Co.

Thorburn, W. M. (1918). "The myth of Occam's razor," *Mind 27,* 345-353.

Todhunter, I. (1865). *A History of the Mathematical Theory of Probability.* (Reprint, New York: Chelsea Publishing Co., 1949 and 1965.)

Tolman, R. C. (1938). *The Principles of Statistical Mechanics,* Oxford: Clarendon Press.

Tribus, M. (1969). *Rational Descriptions, Decisions, and Designs,* New York: Pergamon Press.

Tullock, G. (1979). Personal communication.

Turing, A. M. (1937). "On computable numbers, with an application to the Entscheidungsproblem," *Proc. London Math. Soc., ser. 2, 42,* 230-265; *43,* 544.

Turing, A. M. (1941). Personal communication.

Vajda, S. (1959). Personal communication.

Valéry, P. A. (1921). "Eupalinos ou d'architecte-dialogue des morts," *La Nouvelle Revue Francaise 16* (90), 257-285. (Kraus reprint; Nendeln/Leichtenstein, 1968.)

Van de Kamp, Peter (1965). "The Galactocentric revolution, a reminiscent narrative," *Publications of the Astronomical Soc. Pacific 77,* 325-335.

von Mises, R. (1942). "On the correct use of Bayes's formula," *Annals of Math. Statist. 13,* 156-165.

von Mises, R. (1957). *Probability, Statistics and Truth,* 2nd rev. ed., prepared by Hilda Geiringer, London: Allen and Unwin; New York: Macmillan.

Von Neumann, J., & Morgenstern, O. (1947). *Theory of Games and Economic Behavior,* 2nd ed., Princeton, N.J.: Princeton University Press.

Wald, A. (1950). *Statistical Decision Functions,* New York: Wiley.

Weaver, W. (1948). "Probability, rarity, interest and surprise," *Scientific Monthly 67,* 390-392.

Weizel, W. (1953). "Ableitung der Quantentheorie aus klassischem, kausal determiniertem Modell," *Zeit. Phys. 134,* 264-285.

Wiener, N. (1956). "The theory of prediction," in *Modern Mathematics for the Engineer,* E. F. Beckenbach, ed., New York: McGraw-Hill, 165-190.

Wigner, Eugene P. (1962). "Remarks on the mind-body question," in #339, 284-302.

Wilks, S. S. (1963). Personal communication.

Wilson, E. B. (1927). "Probable inference, the law of succession, and statistical inference," *JASA 27,* 209-212.

Wrinch, Dorothy, & Jeffreys, Harold (1921). "On certain fundamental principles of scientific discovery," *Philos. Mag. ser. 6, 42,* 369-390.

Young, Charles (1902). *Manual of Astronomy,* New York: Ginn.

Zellner, A. (ed.) (1980). *Bayesian Analysis in Econometrics and Statistics: Essays in Honor of Harold Jeffreys,* Amsterdam: North Holland Publishing Co.

Bibliography: Main Publications by the Author

Bibliography: Main Publications by the Author

The numbering of these publications agrees with that on my private list and is used to avoid errors. A few selected contributions to discussions have been included. There is an index to this bibliography on pp. 269-332.

Abbreviations

AMS—Annals of Mathematical Statistics
Biom—Biometrika (*Biometrics* is spelled out)
BJPS—British J. Philosophy of Science
CSSC—Communications in Statistics: Simula. Comput.
JASA—J. American Statistical Association
JLMS—J. London Mathematical Society
JRNSS—J. Royal Naval Science Service
JRSS—J. Royal Statistical Society
JSCS—J. Statist. Comput. & Simul.
MR—Mathematical Reviews
MTAC—Mathematical Tables and Other Aids to Computation
PCPS—Proc. Cambridge Philosophical Society
QJM—Quarterly J. Mathematics, Oxford
*—Book
§—A paper whose merit is its brevity
C—Contribution
P—Paper
R—Review
BR—Book review
IP—Informal paper
JP—Joint paper
I—Very informal

P 1. "The approximate local monotony of measurable functions," *PCPS 36* (1940), 9-13.
P 2. "The fractional dimensional theory of continued fractions," *PCPS 37* (1941), 199-228. (Smith's Prize Essay.)
P 3. "Some relations between certain methods of summation," *PCPS 38* (1942), 144-165.

P 4. "Note on the summation of a classical divergent series," *JLMS 16* (1941), 180-182.

P 5. "On the regularity of moment methods of summation," *JLMS 19* (1944), 141-143.

P 6. "On the regularity of a general method of summation," *JLMS 21* (1946), 110-118.

P 7. "Normal recurring decimals," *JLMS 21* (1946), 167-169. (The "teleprinter problem"; solved in 1941 but not then submitted for publication.)

P 8. "A note on positive determinants," *JLMS 22* (1947), 92-95.

JP 8A. IJG and G. E. H. Reuter. "Bounded integral transforms," *QJM 19* (1948), 224-234. (See #18.)

P 12. "The number of individuals in a cascade process," *PCPS 45* (1949), 360-363. (See #55.)

* 13. *Probability and the Weighing of Evidence* (London, Charles Griffin; New York, Hafners; 1950, 119 pp.).

P 14. "A proof of Liapounoff's inequality," *PCPS 46* (1950), 353.

P 17. "On the inversion of circulant matrices," *Biom. 37* (1950), 185-186.

P 18. "Bounded integral transforms, II," *QJM (2), 1* (1950), 185-190. (See #8A.)

P 20. "Random motion on a finite Abelian group," *PCPS 47* (1951), 756-762; *48* (1952), 368.

P **26.** "Rational decisions," *JRSS B 14* (1952), 107-114. (See #43.)

P 29. "A generalization of Dirichlet's multiple integral," *Edin. Math. Notes 38* (1952), 7-8.

P 33 and 33A. "Skin banks," *The Lancet* (August 9, 1952, and February 7, 1953), 289 and 293-294. (Corrects a broadcast by P. B. Medawar.)

P 36. "The serial test for sampling numbers and other tests for randomness," *PCPS 49* (1953), 276-284. (See #123.)

JP 37. F. G. Foster and IJG. "On a generalization of Pólya's random-walk problem," *QJM 4* (1953), 120-126.

P 38. "The population frequencies of species and the estimation of population parameters," *Biom. 40* (1953), 237-264. (See #86.)

P **43.** "The appropriate mathematical tools for describing and measuring uncertainty," Chapter 3 of *Uncertainty and Business Decisions* (Liverpool: Liverpool University Press, 2nd ed., 1957 [based on a symp. in the Economics section of the British Association, 1953]), 20-36. (Similar to #**26** but includes material on "surprise.")

P 50. "On the marking [grading] of chess players," *Math. Gaz. 39* (1955), 292-296.

P 52. "On the substantialization of sign sequences," *Acta. Cryst. 7* (1954), 603. (See #186.)

P 55. "The joint distribution for the sizes of the generations in a cascade process," *PCPS 51* (1955), 240-242. (See ##12, 200, 337, 413.) (Reprinted in *Proc. Roy. Soc. A, C 68* [1962], 256-259.)

P 56. "A new finite series for Legendre polynomials," *PCPS 51* (1955), 385-388. (See #972.)

P 62. "Conjectures concerning the Mersenne numbers," *MTAC 9* (1955), 120-121.

BR 75. Review of L. J. Savage, "The Foundations of Statistics," *JRSS A 118* (1955), 245-246.

P 77. "Some terminology and notation in information theory," *Proc. Institution Elec. Engrs., Part C (3), 103* (1956), 200-204; and Monograph 155R (1955).

P 78. "On the weighted combination of significance tests," *JRSS B 17* (1955), 264-265. (See #174.)

C 80. Contribution to the discussion in a symposium on linear programming, *JRSS B 17* (1955), 194-196.

P 82. "The surprise index for the multivariate normal distribution," *AMS 27* (1956), 1130-1135; *28* (1957), 1055.

P 83. "On the estimation of small frequencies in contingency tables," *JRSS B 18* (1956), 113-124.

P 84. "The likelihood ratio test for Markoff chains," *Biom. 42* (1955), 531-533; *44* (1957), 301.

P **85A.** "Which comes first, probability or statistics?" *J. Inst. Actuaries 82* (1956), 249-255.

JP 86. IJG and G. H. Toulmin. "The number of new species, and the increase of population coverage, when a sample is increased," *Biom. 43* (1956), 45-63. (See #38.)

P 110. "A classification of rules for writing informative English," *Methodos 7* (1955), 193-200. (Reprinted in #339.)

BR 112. Review of M. Allais, "Foundements d'un théorie positive des choix comportant un risque et critique des postulats et axiomes de l'école Américaine," *JRSS A 119* (1956), 213-214.

BR 115. Review of D. Blackwell and M. A. Girschick, "Theory of Games and Statistical Decisions," *JASA 51* (1956), 388-390.

P 123. "On the serial test for random sequences," *AMS 28* (1957), 262-264. (See #36.)

I 125. "Variable-length multiplication," *Computers and Automation 6* (1957), 54.

P 127. "Saddle-point methods for the multinomial distribution," *AMS 28* (1957), 861-881. (See #238.)

IP 130. "Distribution of word frequencies," *Nature 179* (1957), 595.

JP 136. R. B. Dawson and IJG. "Exact Markov probabilities from oriented linear graphs," *AMS 28* (1957), 946-956.

P 140. "Legendre polynomials and trinomial random walks," *PCPS 54* (1958), 39-42.

P 141. "Random motion and analytic continued fractions," *PCPS 54* (1958), 43-47.

P 142. IJG and K. Caj. Doog. "A paradox concerning rate of information," *Information and Control 1* (1958), 113-126. (See ##192 and 210.) (K. Caj. Doog is a pseudonym.)

P 146. "The interaction algorithm and practical Fourier analysis," *JRSS B 20* (1958), 361-372. (See #209.) (A Fast Fourier Transform.)

C 153. Contribution to the discussion of a paper by J. Neyman and Elizabeth L. Scott, "Statistical approach to problems of cosmology," *JRSS B 20* (1958), 35.

BR 156. Review of L. Hogben, "The Relationship of Probability, Credibility, and Error," *Nature 181* (1958), 1687. (Review entitled "Sociology of statistics.")

BR 162. Review of G. S. Brown, "Probability and Scientific Inference," *BJPS 9* (1958), 251-255.

P 169. "How much science can you have at your fingertips?" *IBM J. Res. Dev. 2* (1958), 282-288. (Invited lecture at opening of IBM San Jose Laboratories.)

P 174. "Significance tests in parallel and in series," *JASA 53* (1958), 799-813. (See #78.)

P 180. "A theory of causality," *BJPS 9* (1959), 307-310. (See #**223B**.)

P 181. "Lattice structure of space-time," *BJPS 9* (1959), 317-319.

P **182.** "Kinds of probability," *Science 129* (1959), 443-447. (Italian translation by Fulvia de Finetti in *L'Industria,* 1959.) (Reprinted in *Readings in Applied Statistics,* William S. Peters, ed. [New York, Prentice-Hall, 1969], 28-37.)

P 183. "Could a machine make probability judgments?" *Computers and Automation 8* (1959), 14-16 and 24-26.

P 185. "Speculations on perceptrons and other automata," *IBM Research Report,* RC 115, 2 vi 59, 19 pp.

IP 186. "Randomized and pseudo-randomized substantialization of sign sequences," *Acta. Cryst. 12* (1959), 824-825. (See #52.)

BR 191. Review of K. R. Popper, "The Logic of Scientific Discovery," *MR 21* (1960), 1171-1173.

P 192. "A paradox concerning rate of information: Corrections and additions," *Information and Control 2* (1959), 195-197. (See #142.)

P § 195. "Monte Carlo method," *McGraw-Hill Enc. of Sc. and Tech. 8* (1960), 586-587.

P 196. "A classification of fallacious arguments and interpretations," *Methodos 11* (1959), 147-159. (Reprinted, with minor modifications, in *Technometrics 4* [1962], 125-132.) (Spanish translation by D. J. B. Monistrol in *Cuadernos de Estadistica Apl. e Inv. Op. 2* [1963], 41-51.) (See #520.)

C 198. Contribution to the discussion of a paper by E. M. L. Beale, "Confidence regions in non-linear estimation," *JRSS B 22* (1960), 79-82. (See #622.)

P 199. "The paradox of confirmation," *BJPS 11* (1960), 145-148. (See #245.)

P 200. "Generalizations to several variables of Lagrange's expansion, with applications to stochastic processes," *PCPS 56* (1960), 367-380. (See ##413, 857, & 899.)

C 203. Contribution to the discussion of a paper by W. E. Thomson, "ERNIE—a mathematical and statistical analysis," *JRSS A 122* (1959), 326-328.

C 207. Contribution to the discussion of a paper by W. F. Bodmer, "Discrete stochastic processes in population genetics," *JRSS B 22* (1960), 240-242.

P 209. "The interaction algorithm and practical Fourier analysis: An addendum," *JRSS B 22* (1960), 372-375. (See #146.)

P 210. "Effective sampling rates for signal detection: Or can the Gaussian model be salvaged?" *Information and Control 3* (1960), 116-140. (See ##142 and 192.)

P 211. "Weight of evidence, corroboration, explanatory power, information, and the utility of experiments," *JRSS B 22* (1960), 319-331; *30* (1968), 203.

P 217. "Speculations concerning information retrieval," *Res. Rep. RC-78,* December 10, 1958, IBM Research Center, Yorktown Heights, N.Y., 14 pp.

P 218. "Some numerology concerning the elementary particles or things," *JRNSS 15* (1960), 213. (See #339.)

P 221. "Weight of evidence, causality, and false-alarm probabilities," *Fourth London Symp. on Information Theory* (London, Butterworths, 1961), 125-136.

P 222. "A comparison of some methods of calculating covariance functions on an electronic computer," *Computer J. 3* (1960), 262-265.

P 223B. "A causal calculus," *BJPS 11* (1961), 305-318; *12* (1961), 43-51; *13* (1962), 88. (See ##754 & 1263.)

P § 224. "The real stable characteristic functions and chaotic acceleration," *JRSS B 23* (1961), 180-183.

P 225. "An asymptotic formula for the differences of the powers at zero," *AMS 32* (1961), 249-256. (A corollary of #127.)

C 228. Contribution to the discussion of a paper by C. A. B. Smith, "Consistency in statistical inference and decision," *JRSS B 23* (1961), 28-29. (See #**230**.)

P **230**. "Subjective probability as the measure of a non-measurable set," *Logic, Methodology, and Philosophy of Science: Proc. of the 1960 International Congress* (Stanford, Stanford University Press, 1962), 319-329.

P 235. "The colleague matrix, a Chebyshev analogue of the companion matrix," *QJM 12,* 115-122; *13* (1962), 61-68.

P 236. "Analysis of cumulative sums by multiple contour integration," *QJM 12* (1961), 115-122; *13* (1960), 80.

P 237. "The frequency count of a Markov chain and the transition to continuous time," *AMS 32* (1961), 41-48.

P 238. "The multivariate saddlepoint method and chi-squared for the multinomial distribution," *AMS 32* (1961), 535-548. (See #127.)

P 243. "The mind-body problem or could an android feel pain?" (March, 1960). *Theories of the Mind*, J. Scher, ed. (New York, Glencoe Free Press, 1962), 490-518. (Corrected reprint, 1966.)

P 245. "The paradox of confirmation, II," *BJPS 12* (1961), 63-64. (See ##199 & **518**.)

P § 263. "A short proof of MacMahon's 'master theorem,' " *PCPS 58* (1962), 160.

P § 264. "Proofs of some binomial identities by means of MacMahon's 'master theorem,' " *PCPS 58* (1962), 161-162.

P 290. "How rational should a manager be?" *Management Science 8* (1962), 383-393. (Reprinted with numerous minor imporvements in *Executive Readings in Management Science*, Martin K. Starr, ed. [New York, Macmillan; and London & Toronto, Collier-Macmillan; 1965], 89-98.)

C 293. Contribution to the discussion of a paper by Charles Stein, *JRSS B 24* (1962), 289-291.

BR 294 and 294A. Review of H. Jeffreys, "Theory of Probability" [3rd ed.] *Geophysical J. of the Roy. Astr. Soc. 6* (1961), 555-558; and *JRSS A 125* (1962), 487-489.

P 315. "Measurements of decisions," in *New Perspectives in Organization Research*, W. W. Cooper, H. J. Leavitt, and M. W. Shelly II, eds. (New York, London, Sydney, Wiley, 1964), 391-404.

P 316. "Analogues of Poisson's summation formula," *Amer. Math. Monthly 69* (1962), 259-266.

P 322. "Maximum entropy for hypothesis formulation, especially for multidimensional contingency tables," *AMS 34* (1963), 911-934.

P 323. "Weighted covariance for estimating the direction of a Gaussian source," *Proc. Symp. on Time Series Analysis*, Brown University, June, 1962, Murray Rosenblatt, ed. New York, Wiley, 1963), 447-470.

P 337. "Cascade Theory and the molecular weight averages of the sol fraction," *Proc. Roy. Soc. A 272* (1963), 54-59. (See #55.)

* 339. Edited, with the help of A. J. Mayne (associate editor) and John Maynard Smith (biology editor). *The Scientist Speculates: An Anthology of Partly-Baked Ideas* (London, Heinemann, 1962; New York, Basic Books, 1963; German trans.: Dusseldorf, Econ. Verlag, 1965; French trans.: Paris, Dunod, 1967; paperback: New York, Capricorn Books, 1965).

P 368. "The relevance of semantics to the economical construction of an artificial intelligence," *IEEE Special Publication S142*, 157-168. (For a much expanded version, see #397.)

P 374. "On the independence of quadratic expressions," (with an appendix by L. R. Welch), *JRSS B 25* (1963), 377-382; *28* (1966), 584.

P 375. "Quadratics in Markov-chain frequencies, and the binary chain of order 2," *JRSS B 25* (1963), 383-391.

IP 376. "Information theory: Survey," CRD-IDA Working Paper #83, April, 1963, 33 pp.

P 385. "Quantum mechanics and Yoga," *Res. J. Philosophy and Social Sc. 1* (1963), 84-91.

P 391. "The human preserve," *JRNSS* (1964), 370-373; and *Spaceflight 7* (1965), 167-170 & 180. (See #476.)

P 397. "Speculations concerning the first ultraintelligent machine," *Advances in Computers 6* (1965), 31-88.

* 398. *The Estimation of Probabilities: An Essay on Modern Bayesian Methods* (Cambridge, Mass., M.I.T. Press, 1965), xii & 109. (Paperback, 1968.) (Out of print.) (See ##522 & 547.)

IP 400 "A note on Richard's paradox," October, 1963, *Mind 75* (1966), 431.

IP 409. "The loss of information due to clipping a wave-form," ARL/GAMMA 50/R1, August 1964. 409A: *Information and Control 10* (1967), 220-222.

P 411. "Categorization of classification," *Proc. of a Conference on Mathematics and*

Computer Science in Biology and Medicine, Oxford (London, HMSO, 1965), 115-128.

P 413. "The generalization of Lagrange's expansion, and the enumeration of trees," *PCPS 61* (1965), 499-517; *64* (1968), 489. (See ##12, 55, & 200.) (Contains a conjecture about the frequencies of polymers in nature. The conjecture was correct, according to Manfred Gordon.)

C 416. Contribution to the discussion of J. W. Pratt, "Bayesian interpretation of standard inference statements," *JRSS B 27* (1965), 196-197.

C 418. Seconding of vote of thanks for A. Q. Morton's "The authorship of the Pauline epistles: A scientific approach," *JRSS A 128* (1965), 225-227 & 623.

JP 425. Bernard Meltzer & IJG. "Two forms of the prediction paradox," *BJPS 16* (1965), 50-51.

P 426. "Logic of man and machine," *New Scientist 26* (1965), 182-183; *21* (1965), 606; *27* (1965), 518. (See also #540.)

P § 476. "Life outside the earth," *The Listener 73* (June, 1965), 815-817. (See ##391 & 644.)

IP 486. "Regression of a phenotypic value on the values for the parents and grandparents," *Nature 208* (October, 1965), 203-204.

P 498. "The probability of war," *JRSS A 129* (1966), 268-269.

BR 499. Review of H. A. Bethe, "Intermediate Quantum Mechanics," *Math. Gaz. 50* (1966), 359-360.

P 505. "A derivation of the probabilistic explication of information," *JRSS B 28* (1966), 578-581.

P **508.** "On the principle of total evidence," *BJPS 17* (1967), 319-321. (See #855.)

BR 516. Review of R. von Mises, "Mathematical Theory of Probability and Statistics," *JRSS A 128* (1966), 289-291.

P **518.** "The white shoe is a red herring," *BJPS 17* (1967), 322. (See ##245 & **600.**)

P 520. "Fallacies, statistical," *International Enc. Social Sc. 5* (New York, Macmillan and Free Press, 1968), 292-301. (See ##196 & 928.)

P 521. "A five-year plan for automatic chess," in *Machine Intelligence 2* E. Dale and D. Michie, eds. (London, Oliver and Boyd, 1968), 89-118. (Paraphrased by Baruch Wood, ed., in *Chess 34* [1969], 245-250, who says " . . . a look into the future with, *en passant,* a masterly guide to positional judgment.")

P 522. "How to estimate probabilities," *J. Inst. Maths. Applics. 2* (1966), 364-383. (See ##398 & 547.)

P 523. "The decision-theory approach to the evaluation of information-retrieval systems," *Information Storage and Retrieval 3* (1967), 31-34.

P 524. "Statistics of language," in *Encyclopedia of Linguistics, Information and Control,* A. R. Meetham, ed. (New York, Pergamon Press, 1969), 567-581.

P 525. "The function of speculation in science exemplified by the subassembly theory of mind," *Theoria to Theory 1* (1966), 28-43. (See ##339 & 397.)

JP 526. IJG & T. N. Grover. "The generalized serial test and the binary expansion of $\sqrt{2}$," *JRSS A 130* (1967), 102-107; *13* (1968), 434. (See ##36 & 123.)

IP 533. "Square-root law for solving two-ended problems," *Nature 212* (December, 1966), 1280.

P § 540. "Human and machine logic," *BJPS 18* (1967), 144-147. (See ##426 & 626.)

BR 541A. Review of I. Hacking, "Logic of Statistical Inference," *Nature 213* (1967), 233-234.

P 547. "A Bayesian significance test for multinomial distributions," *JRSS B 29* (1967), 399-431 (with discussion); *36* (1974), 109. (See ##398 & 522.)

C 565. (Analysis of log-likelihood ratios, "ANOΛ.") Contribution to the discussion of a paper on least squares by F. J. Anscombe, *JRSS B 29* (1967), 39-42.

C 570. Contribution to discussion on a paper by S. F. Buck and A. J. Wicken on the risk of mortality from lung cancer and bronchitis, *JRSS C, Applied Statistics 16* (1967), 206-208.

JP 574. IJG & G. H. Toulmin, "Coding theorems and weight of evidence," *J. Inst. Math. Applics. 4* (1968), 94-105.

P 592. "Some statistical methods in machine-intelligence research," *Virginia J. Sc. 19* (1968), 101-110. (A slightly improved version appeared in *Math. Biosc. 6* [1970], 185-208.)

P 598. "Science in the flesh," in *Cybernetics, Art and Ideas,* Jasia Reichart of the Institute of Contemporary Arts, ed. (London, Studio Vistas, 1971), 100-110.

P 599. "Corroboration, explanation, evolving probability, simplicity, and a sharpened razor," *BJPS 19* (1968), 123-143. (See #876.)

P § **600**. "The white shoe qua herring is pink," *BJPS 19* (1968), 156-157. (See #**518**.)

P **603B**. "A subjective evaluation of Bode's Law and an 'objective' test for approximate numerical rationality," *JASA 64* (1969), 23-66 (with discussion). See #705.)

P 604. "The characteristic functions of functions," *Proc. Roy. Soc. (London) A 307* (1968), 317-334. (See also *Nature 218* [1968], 603-604.)

P 606. "A generalization of the Bernouilli-Euler partition formula," *Scripta Math. 28* (1970), 319-320.

P 607. "Some applications of the singular decomposition of a matrix," *Technometrics 11* (1969), 823-831; *12* (1970), 722.

JP 610. IJG and R. F. Churchhouse, "The Riemann hypothesis and pseudorandom features of the Möbius function," *Mathematics of Computation 22* (1968), 857-861.

IP § 612. "The number of possible strategies when writing compilers," *Comm. ACM 11* (1968), 474.

P 615. "Creativity and duality in perception and recall," in *Pattern Recognition I.E.E./N.P.L.* (London, Institution of Electrical Engineers, July 1968), 228-237.

P § 617. Discussion of Bruno de Finetti's paper, "Initial probabilities: A prerequisite for any valid induction," *Synthese 20* (1969), 17-24. (Also in *Induction, Physics, and Ethics: Proceedings and Discussions of the 1968 Salzburg Colloquium in the Philosophy of Science,* P. Weingartner and G. Zecha, eds., [Dordrecht, Holland, Synthese Library, D. Reidel, 1970], 18-25.)

P 618. "What is the use of a distribution?" in *Multivariate Analysis-II,* P. R. Krishnaiah, ed. (New York, Academic Press, 1969), 183-203. (See #622.)

P 621. "Conditions for a quadratic form to have a chi-squared distribution," *Biom. 56* (1969), 215-216; *57* (1970), 225.

IP § 622. "Utility of a distribution," *Nature 219* (1968), 1392. (See #618.)

P § 626. "Gödel's theorem is a red herring," *BJPS 19* (1969), 357-358. (See #540.)

IP 629. "A proposal for an eye-brain experiment," *Nature 220* (1968), 1127.

C 631. Contribution to the discussion of a paper by M. R. Novick, "Multiparameter Bayesian indifference procedures," *JRSS B 31* (1969), 59-61.

JP 636. W. I. Card and IJG. "The estimation of the implicit utilities of medical consultants," *Mathematical Biosciences 6* (1970), 45-54.

IJP§ 637. IJG & R. A. Gaskins. "The centroid method of integration," *Nature 222* (May, 1969), 697-698. (See #696.)

P 643. "How random are random numbers?" *American Statistician 23* (October, 1969), 42-45.

P 644. "The chief entities," *Theoria to Theory 3* (April, 1969), 71-82. (See ##391 & 476.)

P 645. "Polynomial algebra: An application of the fast Fourier transform," *Nature 222* (1969), 1302.

P 646. "The factorization of a sum of matrices and the multivariate cumulants of a set of quadratic expressions," *J. Combinatorial Theory Ser. A 11* (1971), 27-37; *12* (1972), 309.

P **659.** "The probabilistic explication of information, evidence, surprise, causality, explanation, and utility," in *Foundations of Statistical Inference,* V. P. Godambe and D. A. Sprott, eds. (Toronto, Holt, Rinehart, and Winston of Canada, 1971) 108-141 (with appendix, discussion, and replies). (See #**679.**)

JP 660. IJG & R. A. Gaskins. "Some relationships satisfied by additive and multiplicative congruential sequences, with implications for pseudorandom number generation," in *Computers in Number Theory: Proc. Sc. Res. Council Atlas Symp. at Oxford, 18-23 Aug. 1969,* A. O. L. Atkin and B. J. Birch, eds. (New York, Academic Press, 1971) 125-136.

JP 662. John A. Cornell & IJG. "The mixture problem for categorized components," *JASA 65* (1970), 339-355.

JP 665. IJG, T. N. Gover, & G. J. Mitchell, "Exact distributions for X^2 and for the likelihood-ratio statistic for the equiprobable multinomial distribution," *JASA 65* (1970), 267-283; *66* (1971), 229. 665B: Second Corrigenda (by IJG and J. F. Crook), *JASA 73* (1978), 900.

P 666. "Some future social implications of computers," in *Cybernetics, Simulation and Conflict Resolution,* Douglas Knight, ed. (New York, Spartan Books, 1971), 221-249; and in *Intern. J. Envir. Studies 1* (1970), 67-79; *3* (1972), 331.

P § 668. "A short proof of a conjecture by Dyson," *J. Math. Physics 11* (June 1970), 1884.

P § 670. "The interpretation of X-ray shadowgraphs," *Physics Letters A 31A* (3) (February 9, 1970), 155. (Tomography.)

P 672. "A suggested resolution of Miller's paradox," *BJPS 21* (1970), 288-289.

P 673. "The inverse of a centrosymmetric matrix," *Technometrics 12* (1970), 925-928.

P **679.** "Twenty-seven principles of rationality," Appendix to #659; pp. 124-127.

IP 686. "Words, diagrams and numbers in the communication of science," *Times Literary Supplement,* No. 3558 (May 7, 1970), 513.

P 688. "An analogy between sunspots, the planets and satellites," *JRNSS 25* (July, 1970), 211-213.

P 690A. "Information, rewards, and quasi-utilities," in *Science, Decision, and Value,* J. J. Leach, R. Butts, and G. Pearce, eds. (Dordrecht, Holland, D. Reidel, 1973), 115-127. (Based on an invited lecture at the Second World Conference of the Economic Society, Cambridge, England, September, 1970.)

JP 696. IJG & R. A. Gaskins. "The centroid method of numerical integration," *Numerische Mathematik 16* (1971), 343-359. (See #637.)

BR 697. Review of Philip McShane, "Randomness, Statistics, and Emergence" [Gill and Macmillan, 1971, p. 268], *Times Literary Supplement* (September 18, 1971), 1043.

IP 699. "Non-parametric roughness penalty for probability densities," *Nature Physical Science 229* (1971), 29-30. (Owing partly to the British postal strike, this contains 21 misprints.) (See #701.)

JP 700. IJG & W. I. Card, "The application of rationality to medical records," *Math. Biosc. 10* (1971), 157-176.

JP 701. IJG and R. A. Gaskins. "Nonparametric roughness penalties for probability densities," *Biometrika 58* (1971), 255-277. (See ##699 & 810.)

P 705. "The evolving explanation of a numerological 'law,' " an invited "rebuttal" to Bradley Efron's paper "Does an observed sequence of numbers follow a simple rule? (Another look at Bode's law)," *JASA 66* (1971), 559-562. (See ##**603B** & 764.)

P § 707. "Free will and speed of computation," *BJPS 22* (1971), 48-50.

P 708. "The relationship between two fast Fourier transforms," *IEEE Trans. on Com-*

puters, C20 (March 1971), 310-317. (Reprinted in *Number Theory in Digital Signal Processing,* J. H. McClellan and C. M Rader, eds. [Englewood Cliffs, N.J.: Prentice-Hall, 1969], 150-157.)

P 709. "The proton and neutron masses and a conjecture for the gravitational constant," *Physics Letters A 33* (November, 1970), 383-384.

C 719-738. Contribution to the discussion of twenty papers in the Proceedings of the Symposium on the Foundations of Statistical Inference held at the University of Waterloo, Ontario, Canada, 1970. (See #**659**.)

750. "The Bayesian influence," mimeographed notes of lectures at VPI&SU, April 2 to June 4, 1971, 123 pp.

IP 751. "Speculation—how to save democracy," *Futures 3* (1), (March, 1971), 77-79.

P 753. "Statistics and today's problems," *American Statistician* (3), (June, 1972), 11-19.

BR 754. Review of Patrick Suppes, "A Probabilistic Theory of Causality," [Acta Philosophica Fennica, Fasc. XXIV], *JASA 67* (March, 1972), 245-246.

JP 755. IJG & W. I. Card. "The diagnostic process with special reference to errors," *Methods of Information in Medicine 10* (1971), 176-188.

P 758. "The average on a sphere of the exponential of a homogeneous function," *Iranian. J. Sci. and Technology, 1* (1971), 11-20. (A slight revision of an unpublished report of 1963.)

BR 761. Review of J. R. Lucas, "Freedom of the Will" [Oxford, 1970], *BJPS 22* (1971), 382-387.

IP 762. "The optimal size of an organization," *Eureka 34* (October, 1971), 28-30.

IP 764. "Christian Wolff, Titius, Bode, and Fibonacci," letter in *American Statistician 26* (February, 1972), 48-49.

IP **765**. "46656 varieties of Bayesians," letter in *American Statistician 25* (December, 1971), 62-63.

I 771. Letter in *Scientific American 225* (October, 1971), 8, concerning an article, "Eye movements and visual perception," by D. Noton and L. Stark.

P 777. "Human and machine intelligence: Comparisons and contrasts," *Impact of Science on Society 21* (1971), 305-322. (Also in French in *Impact: Science et Société, 21* [1971], 343-362.)

P 788A. "Chinese universes," *Physics Today* (July, 1972), 15. (See #999.)

P 792. "Correlation for power functions," *Biometrics 28* (1972), 1127-1129; *29* (1973), 829. (See #1477.)

IP 793. "Scientific induction and exponential-entropy distributions," *American Statistician 26* (April, 1972), 45.

JP 795. IJG & Lawrence S. Mayer, "On surfaces of constant societal loss in a model of social choice," *J. Mathematical Sociology 2* (1972), 209-219.

P 796. "Food for thought," in *Interdisciplinary Investigation of the Brain,* J. P. Nicholson, ed. (New York, Plenum Press, 1972), 213-228.

JP 798. W. I. Card & IJG, "A logical analysis of medicine," in *A Companion to Medical Studies, 3,* R. Passmore and T. S. Robson, eds. (Oxford, Blackwell's, 1974), Chapter 60.

JP 810. IJG & R. A. Gaskins, "Global nonparametric estimation of probability densities," *Virginia J. of Science 23* (December, 1972), 171-193. (This paper was invited by the editor after #701D was awarded the Horsley Prize. It is a much expanded form of #701.)

P **814**. "Is the size of our galaxy surprising?" *American Statistician 27* (February, 1973), 42-43.

P **815**. "Random thoughts about randcmness," in *PSA 1972* (Boston Studies in the Philosophy of Science; Dordrecht, Holland, D. Reidel, 1974), 117-135.

P 817. Reprinting of #777 in *RUR: Journal of Cybernation and Public Affairs* (Summer,

1972) (issue on intelligence, machine and human), 4-12. (Journal edited by T. D. C. Kuch.)

P 822. "The joint probability generating function for run-lengths in regenerative binary Markov chains, with applications," *Annals of Statistics 1* (1973), 933-939.

IP 827. Continued fractions for the exponential function, *Amer. Math. Monthly 80* (February, 1973), 209; *81* (1975), 532-533.

P 829. "A reciprocal series of Fibonacci numbers," *Fibonacci Quarterly 12* (1974), 346.

I 837. "Bode, von Weizacker, and Fibonacci," *American Statistician 27* (June, 1973), 127.

P 838. "The Bayesian influence, or how to sweep subjectivism under the carpet," *Foundations of Probability Theory, Statistical Inference, and Statistical Theories of Science*, Proc. of a Conference in May, 1973, C. A. Hooker and W. Harper, eds., vol. 2 (Dordrecht, Holland, D. Reidel, 1976), 125-174.

I 839. "Larsen unnecessarily bent," *CHESS 37* (October, 1971), 10. (Analysis of a position, Bent Larsen vs. Bobby Fischer.)

BR 844. Review of Terrence L. Fine, "Theories of Probability: An Examination of Foundations" [Academic Press, 1973], *IEEE Trans. Infn. Th. IT-20* (1974), 298-300.

P **846.** "Explicativity, corroboration, and the relative odds of hypotheses," *Synthese 30* (1975), 39-73. (See #890.)

* 854. D. B. Osteyee and IJG, *Information, Weight of Evidence, the Singularity between Probability Measures and Signal Detection* (Berlin, Heidelberg, New York, Springer Verlag, 1974).

P 855. "A little learning can be dangerous," *BJPS 25* (1974), 340-342. (See #**508**.)

P 857. "The Lagrange distribution and branching processes," *SIAM J. Appl. Math. 28* (1975), 270-275.

P 858. "The number of orderings of n candidates when ties are permitted," *Fibonacci Quarterly 13* (1975), 11-18.

P 860. "The Bayes factor against equiprobability of a multinomial population assuming a symmetric Dirichlet prior," *Annals of Statistics 3* (1975), 246-250.

I 861. Cassette recording, interviewed by Christopher Evans, 1973, in the series "Brain Science Briefings"; Ferranti Ltd., transferred to Ars Magna Ltd., April, 1974.

JP 862. IJG & J. F. Crook, "The Bayes/non-Bayes compromise and the multinomial distribution," *JASA 69* (1974), 711-720.

JP 871. IJG and T. N. Tideman, "From individual to collective ordering through multidimensional attribute space," *Proc. Roy. Soc. London A 347* (1976), 371-385. (See #1014.)

BR 875. Review of Arnold Zellner, "An Introduction to Bayesian Inference in Econometrics" [Wiley, 1971], *Technometrics 17* (1975), 137-138; *18* (1976), 123.

IP 876. "A correction concerning complexity," *BJPS 25* (1974), 289. (See #599.)

P 882. "And Good say that it was God(d)," *Parascience Research J. 1* (February, 1975), 3-13. (See #1322A.) (Reprinted with minor changes, *Parasc. Proc.* [1973/77], 55-56.)

JP 883. IJG and L. S. Mayer, "Estimating the efficacy of a vote," *Behavioral Science 20* (1975), 25-33. (Errata mimeographed in 1975.) (See #901.)

P 890. Reply to the discussion at the Conference on "Methodologies: Bayesian vs. Popperian," *Synthese 30* (1975), 83-93. (See #**846**.)

P 895. "A new formula for cumulants," *Mathematical Proceedings of the Cambridge Philosophical Society 78* (1975), 333-337. (See #871.)

IP 898. "The infinite speed of propagation of gravitation in Newtonian physics," *American Journal of Physics 43* (1975), 640.

P 899. "The relationship of a formula of Carlitz to the generalized Lagrange expansion," *SIAM Journal on Applied Mathematics 30* (1976), 103.

JP 900. P. S. Bruckman and IJG, "A generalization of a series of de Morgan, with applications of Fibonacci type," *Fibonacci Quarterly 14* (1976), 193-196.

P 928. "Statistical fallacies," for *The International Encyclopedia of Statistics,* William Kruskal and Judith Tanur, eds. (New York, Free Press, 1978, [issued January, 1979]), 337-349. (An expanded version of #520.)

P 929. "On the application of symmetric Dirichlet distributions and their mixtures to contingency tables," *Annals of Statistics 4* (1976), 1159-1189. (See #974.)

JP 937. Wolfgang Pelz and IJG, "Approximating the lower tail-areas of the Kolmogorov-Smirnov one-sample statistic," *JRSS B 38* (1976), 152-156.

P **938.** "Dynamic probability, computer chess, and the measurement of knowledge," in *Machine Intelligence 8,* E. W. Elcock and D. Michie, eds. (Chichester: Ellis Horwood Ltd.; New York: Wiley, 1977), 139-150.

JP 939. D. B. Osteyee and IJG, "Regeneration of a binary signal in a uniform transmission line," *IEEE Communications Society Transactions* (1976), 1054-1057.

BR 956. Review of Bruno de Finetti, "Theory of Probability" [Wiley, 1974 & 1975], *Bull. Amer. Math. Soc. 83* (January, 1977), 94-97.

BR 957. Review of Y. M. M. Bishop, S. E. Fienberg, and P. W. Holland, with the collaboration of R. J. Light and F. Mosteller, "Discrete Multivariate Analysis: Theory and Practice" [M.I.T. Press], *MR 52* (1976), 286.

BR 958. Review of Richard Swinburne, "An Introduction to Confirmation Theory" [Methuen, 1973], *BJPS 27* (1976), 289-292.

JP 960. L. V. Holdeman, IJG, and W. E. C. Moore, "Human fecal flora: Variation in bacterial composition within individuals and a possible effect of emotional stress," *Applied & Environmental Microbiology 31* (1976), 359-375.

JP 966. IJG & T. N. Tideman, "Stirling numbers and a geometric structure from voting theory," *J. Combinatorial Theory A 23* (1977), 34-45.

IP 970. Discussion of "On rereading R. A. Fisher" (by L. J. Savage, edited by John W. Pratt), *Annals of Statistics 4* (1976), 492-495.

C 972. Contribution (formula 7.249.2) to Gradshteyn and Ryzhik, *Tables of Integrals, Series, and Products,* reprint of the English trans. of the 4th Russian ed. (inscribed 1965 but issued in 1966), 825. (Acknowledgement on p. 6.) (See #56.)

JP 974. IJG and J. F. Crook, "The enumeration of arrays and a generalization related to contingency tables," *Discrete Mathematics 19* (1977), 23-45. (See #929.) (Imprinted July but issued in December.)

P 980. "The Botryology of Botryology," in *Classification and Clustering,* J. van Ryzin, ed. (New York, Academic Press, 1977), 73-94.

P 981. "A new formula for k-statistics," *Annals of Statistics 5* (1977), 224-228. (See #895.)

I 986-992. "Minicommunications," *CSSC B5* (1976), 81-82.

P 999. "Black and white hole hierarchical universes: A synthesis of the steady state and big bang theories," *Theoria to Theory, 10* (1976), 191-201. (See #788A.)

P **1000.** "Explicativity: A mathematical theory of explanation with statistical applications," *Proc. Roy. Soc. (London) A 354* (1977), 303-330. (See #1161 for a reprinting.)

P 1001. "Justice in voting by demand revelation," *Public Choice XXIX-2* (special supplement to Spring, 1977; July, 1977), 65-70.

JIP 1014. IJG and T. N. Tideman, Letter regarding voting in *Scientific American 235* (October, 1976), 10 & 12.

P 1015. "Early work on computers at Bletchley," Number 1 in a series of special lectures on "The pioneers of computing," National Physical Laboratory Report. Com. Sci. 82 (September, 1976), Department of Industry, 16 pp. (See ##1178, 1218, & 1299.)

I 1016-1027. "Minicommunications," *CSSC B5* (1976), 141-144.

IP 1033. "A simple cure for grade inflation," *Journal of Educational Data Processing 13* (1976), 29-32.

JP 1034. A. I. Khuri and IJG, "The distribution of quadratic forms in nonnormal variables and an application to the variance ratio," *JRSS B 39* (1977), 217-221.

JP 1035. IJG and A. I. Khuri, "Forms for the distribution of a ratio in terms of characteristic functions," Minicommunication M24, in *CSSC B5* (1976), 209-211.

JP 1036. Brian Conolly and IJG, "A table of discrete Fourier transform pairs," *SIAM J. Appl. Math. 32* (1977), 810-822; *33* (1977), 534.

R 1060. Review of Thomas S. Ferguson, "Prior distributions on spaces of probability measures" [*Ann. Statist. 2* (1974), 615-629], *MR 55* (1978), 1546-1547.

IP 1065. "Adenine arabinoside therapy," item C7 in CCC in *JSCS 6* (1978), 314-315.

R 1068. Review of Bruno de Finetti, "La probabilita: Guardarsi dalle contraffazioni!" (with an English translation) [*Scientia* (Milano) *111* (1976), 255-303], *MR 56* (1978), 1302.

BR 1075. Review of Benoit B. Mandelbrot, "Fractals: Form, Chance, and Dimension" (W. H. Freeman, 1977), *JASA 73* (1978), 438.

I 1078. "Are maximum-likelihood estimates invariant?" C10 in CCC in *JSCS 7* (1978), 80-81.

C 1080. Contribution to the discussion of T. Leonard, "Density estimation, stochastic processes and prior information," *JRSS B 40* (1978), 138-139.

JP 1100. IJG and T. N. Tideman, "Integration over a simplex, truncated cubes, and Eulerian numbers," *Numerische Mathematik 30* (1978), 355-367.

P 1111. "The inversion of the discrete Gauss transform," *Applicable Analysis 9* (1979), 205-218.

IP 1137. "Alleged objectivity: A threat to the human spirit?" *International Statistical Institute Review 46* (1978), 65-66.

IP 1146. "Ethical treatments," C16 in CCC in *JSCS 7* (1978), 292-295.

IP 1148. "A fuzzy Bayesian method for estimating probabilities given two related multinomial distributions," C18 in CCC in *JSCS 7* (1978), 296-299.

IP 1157. "Path analysis and correlation for power functions," C22 in CCC in *JSCS 8* (1978), 80.

JIP 1159. IJG and D. Michie, "Improved resolution of a form of Mackie's paradox," *Firbush News 8* (April, 1978), 22-24.

P 1160. "The contributions of Jeffreys to Bayesian statistics," Chapter 3 in *Bayesian Analysis in Econometrics and Statistics: Essays in Honor of Harold Jeffreys*, Arnold Zellner, ed. (Amsterdam, North Holland Publishing Co., 1980), 21-34.

P 1161. Reprint of #1000 in *Bayesian Analysis . . .* (1980), 397-426.

IP 1166. "Monozygotic criminals," C23 in CCC in *JSCS 8* (1978), 161-162.

IP 1173. "A comparison of some statistical estimates for the numbers of contingency tables," C26 in CCC in *JSCS 8* (1979), 312-314.

P 1178. Updated version of #1015 for the *Annals of the History of Computing 1* (1) (1979), 38-48. (See #1218.)

IP 1181. "Proper fees in multiple-choice examinations," C38 in CCC in *JSCS 9* (1979), 73-74.

IP 1186. "On the combination of judgements concerning quantiles of a distribution with potential application to the estimation of mineral resources," C41 in CCC *JSCS 9* (1979), 77-79 & 159.

JP 1199. J. F. Crook and IJG, "On the application of symmetric Dirichlet distributions and their mixtures to contingency tables, Part II," *Annals of Statistics 8* (1980), 1198-1218. (See #929.)

JP 1200. IJG & R. A. Gaskins, "Density estimation and bump-hunting by the penalized likelihood method exemplified by scattering and meteorite data," *JASA 75*

(1980), 42-73 (with discussion). (The invited paper for the Applications Section of the Annual meetings of ASA in August, 1979, in Washington, D.C.)

P 1201. "Turing's statistical work in World War II" [Studies in the history of probability and statistics, XXXVII], *Biom. 66* (1979), 393-396. (See #1361.)

1212. Interview by C. R. Evans, with a photo by Pat Hill, in *OMNI* (January, 1979), 70-73 & 117-121. (My corrections to the transcript, submitted in April, 1978, were omitted in error.) (See also April, 1979, 10; August, 1979, 10 & 28.)

BR 1217. Review of "Logic, Laws, and Life: Some Philosophical Complications," Robert G. Colodny, ed. (University of Pittsburgh Press, 1977), *JASA 74* (1979), 501-502.

P 1218. Version of #1015 for *Cryptologia 3,* (2) (1979), 67-77, and photograph on front cover.

BR 1221. Review of Arthur W. Burks, "Chance, Cause, Reason: An Inquiry into the Nature of Scientific Evidence" (University of Chicago Press, 1977), *JASA 74* (1979), 502-503.

P 1222. "The impossibility of complete mutual observation," *Physics Letters A70* (February, 1970), 81-82.

IP 1227. "The Bayesian meaning of invariant estimation," C43 in CCC in *JSCS 9* (1979), 160-161.

IP 1228. "Bayes's billiard-table argument extended to multinomials," C44 in CCC in *JSCS 9* (1979), 161-163.

P **1230.** "Some history of the hierarchical Bayesian methodology," International Meeting on Bayesian Statistics, May 28-June 2, 1979, Valencia, Spain. In *Bayesian Statistics,* Bernardo, J. M. *et al,* eds. (Valencia, Spain: University of Valencia Press), 489-519, with discussion.

P 1234. "Some logic and history of hypothesis testing," in *Philosophical Foundations of Economics,* Joseph C. Pitt, ed. (University of Western Ontario series on the Philosophy of Science, Dordrecht, Holland, D. Reidel, 1981), 149-174.

BR 1235. Review of P. Hájek and T. Havránek, "Mechanizing Hypothesis Formulation" [Springer-Verlag, 1978], *Bull. Amer. Math. Soc.* (New Series) *1* (1979), 650-654.

C 1238. Contribution to the discussion of D. V. Lindley, A. Tversky, and R. V. Brown, "On the reconciliation of probability assessments," *JRSS A 142* (1979), 173-174.

R 1241. Review of Henry Kyburg, "Subjective probability: Criticisms, reflections, and problems," *J. Philos. Logic 7* (1978), 157-180; *MR 58* (1979), 3642.

IP 1245. "Predictive sample re-use and the estimation of probabilities," C50 in CCC in *JSCS 9* (1979), 238-239.

IP 1246. Demonstration that Levy could have won the historic fourth game versus Chess 4.7 in Toronto, 1978, *Personal Computing* (May, 1979), 50.

IP 1248. "The clustering of random variables," C52 in CCC in *JSCS 9* (1979), 241-243.

IP 1249. "Partial correlation and spherical trigonometry," C53 in CCC in *JSCS 9* (1979), 243-245.

JP 1250. IJG, Byron C. Lewis, Raymond A. Gaskins, & L. W. Howell, Jr., "Population estimation by the removal method assuming proportional trapping," *Biom. 66* (1979), 485-494.

P 1263. "Some comments on probabilistic causality," *Pacific Philosophical Q.* (formerly *The Personalist*) *61* (1980), 301-304.

1266. "Key words for Bayesian publications by I. J. Good," June 28, 1979 (from papers #750 to #1238), mimeographed, 5 pp.

C 1267-1278. Discussion of papers at the Valencia conference. (See #**1230**.)

BR 1290. Review of collected papers of R. A. Fisher (vols. III to V), [University of Adelaide Press, 1974], *JASA 75* (1980), 239.

1297. "Partly-baked ideas, 28 columns edited by IJG, *Mensa J.* (1968-1980), pbi's up to number 767.

P 1298. "The chief entities," a shortening of #644 for *Cosmic Search 2* (Spring, 1980), 13-17.

P 1299. An updated version of #1015 in *A History of Computing in the Twentieth Century,* N. Metropolis, ed. (New York, Academic Press, 1980), 31-45. (See ##1178 & 1218.)

P 1300. "The axioms of probability," *Encyclopedia of Statistical Sciences,* Vol. 1, N. L. Johnson & S. Kotz, eds. (New York, Wiley, 1982), 169-176.

IP 1303. "Another relationship between weight of evidence and errors of the first and second kinds," C67 in CCC in *JSCS 10* (1980), 315-316.

P 1313. "Degrees of belief," *Encyclopedia of Statistical Sciences,* Vol. 2, N. L. Johnson & S. Kotz, eds. (New York, Wiley, 1982, 287-293.

P 1317. "Degrees of causation in regression analysis," C71 in CCC in *JSCS 11* (1980), 153-155.

IP 1320. "The diminishing significance of a P-value as the sample size increases," C73 in CCC in *JSCS 11* (1980), 307-309.

P 1322A. "Scientific speculations on the paranormal and the parasciences," *The Zetetic Scholar,* No. 7 (December, 1980), 9-29.

P 1330. "The philosophy of exploratory datum analysis," ASA annual meetings at Houston, August, 1980, *Proc. Bus. & Econ. Stat. Section* (1981), 1-7.

P 1331. "A further comment on probabilistic causality: Mending the chain," *Pacific Philos. Q. 61* (1980), 452-454.

IP 1332. "On Godambe's paradox," C78 in CCC in *JSCS 12* (1980), 70-72.

IP 1333. "Feynman's path integrals and Sewall Wright's path analysis," C80 in CCC in *JSCS 12* (1980), 74-77.

IP 1334. "Functions of distinct random variables having identical distributions," C77 in CCC in *JSCS 12* (1980), 68-70.

IP 1335. "Vinograde's lemma, singular decompositions, and k-frames," C79 in CCC in *JSCS 12* (1980), 72-74.

IP 1336. "A simplification in the 'causal calculus,'" C81 in CCC in *JSCS 12* (1980), 77-78.

P 1338. "Vigor, variety and vision—The vitality of statistics and probability" (title chosen by Jim Swift), keynote speech for the sessions on the teaching of statistics and probability at the school level in *Proc. Fourth International Congress on Mathematical Education,* Berkeley, California, August 10-26, 1980. In press.

IP 1341. "Roughness penalties, invariant under rotation, for multidimensional probability density estimation," C87 in CCC in *JSCS 12* (1981), 142-144.

IP 1343. "The autocorrelation function $\rho|\tau|^c$," C90 in CCC in *JSCS 12* (1981), 148-152.

IP 1345. A problem concerning regular polygons. Elementary Problem #E2889. *Amer. Math. Monthly 88* (1981), 349.

P 1350. "Ethical machines," *Machine Intelligence 10,* D. Michie, ed. (Chichester: Ellis Horwood, 1982), 555-560.

I 1351. Opening of the after-dinner discussion on October 31, 1981 at the 21st SREB-NSF meeting on Bayesian Inference in Econometrics, University of Chicago. In the Report of the Meeting, edited by A. Zellner (1981), 13-15.

P 1354. "Generalized determinants and generalized generalized variance," C91 in CCC in *JSCS 12* (1981), 311-313.

JIP 1357. IJG & T. N. Tideman, "The relevance of imaginary alternatives," C93 in CCC in *JSCS 12* (1981), 313-315.

IP 1358. "An approximation of value in the Bayesian analysis of contingency tables," C88 in CCC in *JSCS 12* (1981), 145-147.

P 1361. Reprinting of #1201 in *Machine Intelligence and Perception: A Turing Commemorative,* Judith M. S. Prewitt, ed. In process.

P 1365. "When is G positive in the mixed Dirichlet approach to contingency tables?" C94 in CCC in *JSCS 13* (1981), 49-52.

P 1366. "The Monte Carlo computation of Bayes factors for contingency tables," C95 in CCC in *JSCS 13* (1981), 52-56.

P 1367. Reprint of #777 in *Machine Intelligence and Perception: A Turing Commemorative,* Judith M. S. Prewitt, ed. In process.

IP 1369. "The weight of evidence provided by uncertain testimony or from an uncertain event," C96 in CCC in *JSCS 13* (1981), 56-60.

IP 1371. "Generalized determinants and generalized Jacobians," C97 in CCC in *JSCS 13* (1981), 60-62.

IP 1382. "An error by Peirce concerning weight of evidence," C102 in CCC in *JSCS 13* (1981), 155-57.

JP 1383. IJG and Michael L. Deaton, "Recent advances in bump-hunting," in *Computer Science and Statistics: Proceedings of the 13th Symposium on the Interface,* William F. Eddy, ed. (Springer, 1981), 92-104 (with discussion).

P 1386. "Some comments on Rejewski's paper on the Polish decipherment of the Enigma," *Annals of the History of Computing 3* (1981), 232-234.

IP 1389. "The effect of permutations of rows and columns on measures of association," C103 in CCC in *JSCS 13* (1981), 309-312.

IP 1396. Comment on a paper by Glen Shafer. *JASA 77* (1982), 342-344.

P 1397. "Quadratic and logarithmic indexes of diversity and surprise," a discussion note on a paper by Patil and Taillie, *JASA 77* (1982), 561-563.

P 1399. "An analogue of chi-squared that is powerful against bumpy alternatives," C108 in CCC in *JSCS 13* (1981), 319-323.

P 1401. "Randomly connected genetic nets," C110 in *JSCS 13* (1981), 324-327.

JP 1402. IJG & Byron C. Lewis, "Probability estimation for 2 x s contingency tables and predictive criteria," for the meetings of the International Statistical Institute, Buenos Aires, 1981 December.

P 1404. "The cumulants of an analogue of Pearson's chi-squared," *JSCS 15* (1982), 171-181.

P 1408. "The fast calculation of the exact distribution of Pearson's chi-squared and of the number of repeats within the cells of a multinomial by using a Fast Fourier Transform," C119 in CCC in *JSCS 14* (1981), 71-78; addendum C138 in press. (See also #1500.)

P 1414. "Is the Mars effect an artifact?" *Zetetic Scholar #9* (1982), 65-69.

P 1419. A report on a lecture by Tom Flowers on the design of Colossus. *Annals of the History of Computing 4* (1982), 53-59.

P 1420. "The robustness of a hierarchical model for multinomials and contingency tables," in *Scientific Inference, Data Analysis, and Robustness.* G. E. P. Box, T. Leonard and Chien-Fu Wu, eds. (New York: Academic Press, 1983), 191-211.

P 1421. "A good explanation of an event is not necessarily corroborated by the event," *Philosophy of Science 49* (1982), 251-253.

IP 1430. "When is maximum-likelihood estimation O.K. for a Bayesian?", C127 in CCC in *JSCS 15* (1982), 75-77.

P 1432. "An index of separateness of clusters and a permutation test for its statistical significance," C129 in CCC in *JSCS 15* (1982), 81-84.

JP 1434. IJG & D. R. Jensen, "Determining the noncentral Wishart distribution via quadratic expressions," C 131 in CCC in *JSCS 15* (1982), 85-87.

P 1436. "Some applications of Poisson's work," for the Poisson Bicentennial Commemoration at George Washington University, 1982 March 15.

JP 1444. J. F. Crook and IJG, "The powers and 'strengths' of tests for multinomials and contingency tables," *JASA 77* (1982), 793-802.

P 1445. "Succinct speculations," *Speculations in Science & Technology 5* (1982), 363-373.

P 1450. "The irregular shapes of polypeptide chains," C136 in CCC in *JSCS 15* (1982), 243-247.

P 1451. Foreword to *Against the Odds: Mathematical Challenges to the Neo-Darwinian Interpretation of Evolution,* 2nd edn. Santa Barbara: Ross-Erikson. In press.

JP 1453. W. E. C. Moore, L. V. Holdeman, R. M. Smibert, IJG, J. A. Burmeister, K. G. Palcani & R. R. Ranney, "Bacteriology of experimental gingivitis in young adult humans," *Infection and Immunity 38* (1982), 651-667; *39* (1983), 1495.

JP 1457. D. R. Jensen & IJG, "A representation for ellipsoidal distributions," *SIAM J. Appl. Math.* In press.

IP 1458. "The calculation of X^2 for two-rowed contingency tables," *The American Statistician 37* (1983), 94.

IP 1462. "The standard error of the estimated 'coverage' of a sample of species or vocabulary," C139 in *JSCS 15* (1982), 337.

P 1468. IJG & Golde Holtzman, "On s-fold repeats and their additive and multiplicative properties," C141 in *JSCS 16* (1982), 66-69.

P 1469. IJG & Golde Holtzman, "The moments of the number of s-fold repeats," C142 in *JSCS 16* (1982), 69-75.

BR 1471. Review of Gordon Welchman, "The Hut Six Story," *New Scientist 96* (1982 October 7), 42.

I 1472. "Who is a Bayesian?," letter in *The American Statistician 37* (1983), 95.

IP 1474. "Scientific induction: universal and predictive," C143 in *JSCS 16* (1983), 311-312.

IP 1475. "The diminishing significance of a fixed P-value as the sample size increases: a discrete model," C144 in *JSCS 16* (1983), 312-314.

IP 1476. "A measure of adhockery," C145 in *JSCS 16* (1983), 314.

IP 1477. "Correlation between power functions," C146 in *JSCS 16* (1983), 314-316. (See #792.)

IP 1481. "Probability estimation by Maximum Penalized Likelihood for large contingency tables and for other categorical data," C148 in *JSCS 17* (1983), 66-67.

IP 1482. "A simple consequence of Daniel Bernoulli's logarithmic formula for utility," C149 in *JSCS 17* (1983), 67-68.

IP 1484. "Antisurprise," C151 in *JSCS 17* (1983), 69-71.

IP 1485. "On the number of duplicated fingerprints," C152 in *JSCS 17* (1983), 71-73.

P 1492. "The philosophy of exploratory data analysis," *Philosophy of Science 50* (1983), 283-295. (See #1330.)

I 1499. "Where have all the residents gone? Gone to infinity every one," *Amer. Math. Monthly.* In press.

P 1500. "An improved algorithm for the fast calculation of the exact distribution of Pearson's chi-squared," C157 in *JSCS 17.* In press. (See #1408.)

JIP 1503. IJG & Eric Smith, "The early history of the quadratic index of diversity or repeat rate", C155 in *JSCS 17.* In press.

P 1504. "A note on the extraction of coefficients from power series", C158 in *JSCS 17.* In press.

BR 1505. Review of "The Maximum Entropy Formalism", Raphael D. Levine & Myron Tribus, eds. (1979); JASA. In press.

JP 1512. IJG & G. Tullock, "Judicial errors and a proposal for reform," *J. Legal Studies.* In press.

P 1515. "Weight of evidence: a brief survey," Second Valencia International Meeting on Bayesian Statistics, 1983.

P 1517. "A correction concerning my interpretation of Peirce, and the Bayesian interpretation of Neyman-Pearson 'hypothesis determination'," C165 in CCC in *JSCS 18.* In press.

Indexes

Subject Index of the Bibliography

This index covers only those items listed in the author's bibliography that are not reprinted in this book. The references are to publication numbers. Occasionally, there are numbers in parentheses referring to pages within the publications. There are many keywords that appear also in the index of the book itself. Items beyond #1457 are not indexed.

A. P. Dempster, on p. 81 of the 1970 Waterloo conference proceedings (see #659), commented that an information retrieval system for my work would be helpful. The following index, combined with that for the book, should go a long way in that direction, although the bibliography is not comprehensive.

Change of mind: by four statisticians, 719 (302); with respect to a class of acts, 315
Channel capacity, 854
Chaos, acceleration in, 224
Character recognition, 398. *See also* Pattern recognition
Characteristic function: discrete, 13; of functions, 604, 1034; multivariate, 1334; of powers, 604; of products and quotients, 604; of a quadratic form in nonnormal variables, 1034; of quadratic forms, 374, 604; and ratios, 1035; stable, 224, 604
Characteristic functionals, 237, 516
Charity, all or nothing (Dan Herrell), 1445
Cheating, unconscious (wishful thinking), 13, 398
Chebyshev polynomials: as applied to characteristic functions, 1399; and the colleague matrix, 235; and cumulants, 1404
Checkers program, 185, 521
Cheek, bulgy (in a probability density), 1080
Chemical combination in a near vacuum, 1250
Chemical research by intersteller communication, 339 (239-40)
Chess, 13, 397 (34, 50); analysis of, automatic, 397 (49), 592, 1212, 1235; analysis of, effective depth of, 521; analysis of, of the first victory by computer versus master, 1246; analysis of, five-year plan for, 521; on computers, 1212; control of squares in, a hypothesis concerning, 1235; number of possible games and positions in, 521; openings in, statistics of, 38, 86, 398; players of, grading of, based on an economic principle (better than Elo's system), 50; players of at Bletchley, 1015, 1178, 1218; programs for, 183, 185, 796; programs for, as encouraging AI, 521; programs for, history of, 521; programs for, tactics and strategy in, 183; quiescence, turbulence, and agitation in, 521; randomized, 397 (35), 521; relevance of to the philosophy of probability, 844; strategy of, 521; values of pieces and squares in, 521; White's advantage in as half a move, 183
Chief entities. *See* Cosmic Club
Chimpanzees: as all Bayesians, but not conversely, 750 (6); gambling by, 243 (506)
Chinese remainder theorem, 146, 209, 708
"Chinese" universes (universes within universes), 788A, 882, 999, 1323A
Chi-squared: analogue of, for bumpy alternatives, 1399, 1404; analogy of, with confidence intervals, 13; asymptotic distribution of, 127; and contingency tables, 13; as distribution of quadratic form, 621; formula for distribution of, 13; generalization of, 992; interpretation of, 13; and log-factor, relationships between, 398; for multinomials, 127, 238, 665, 1408; Pearson's, exact distribution of obtained by a Fast Fourier Transform, 1408; for rank of population contingency table, 398; supplementation of, 174. *See also* X^2
Chomskyism, an alternative to, 1445
Chondrites, at least three kinds of, 1080, 1200
Chords: detection of, and logarithm of acoustic frequencies, 796; pleasant and unpleasant, 796; and whether their recognition is innate, 796
Christoffel-Darboux formula, 701
Christoffel three-index symbol, 699, 701, 810
Chromocriminology, 1445
Chromosomes. *See* Genetic(s)
Chronon, 181
Church's (Alonzo) lambda calculus, 701
Chymotrypsinogen A, 1450
Cicero versus Cicero, 1338
Circuits, reverberating, 243 (506)
Circulant matrices. *See* Circulices; Circulix inversion
Circularity in definitions, 397 (42, 75); inevitability of, 980
Circularization, 36, 84, 136; in communication theory, 142; in density estimation, 1383; in signal detection, 210; of signals, 323
Circularized multinomial, 1404
Circulices (circulant matrices), 17, 1399; recursively blocked, 708; in signal detection, 210
Circulix inversion, 708
Circumcircle of spherical triangle, statistical application of, 1249
Classification, 397 (71), 398; arboresque (or dendroidal), 397 (53), 398; categorization of, 411; one-way, 398; *see also* Multinomial distribution; recognition that it is improved, as requiring a measure of simplicity, 185; as regeneration, 397 (46); two-way, 398. *See also* Botryology
Clifford matrices, and fundamental particles, 218

formulation of, 890. *See also* Null hypotheses
Hypothetical population, 13

I Ching, 524
"I think I think," 243 (494)
IAS (Institute for Advanced Studies computer), 666
IBP (International Business People), a firm staffed by robots, 169
Idealized problems, 13. *See also* Additivity, complete; Distributions, improper; Proposition, idealized
Ideas, half-baked: journal proposed for, 169; and notebooks of great scientists, 169
Idempotent operator, 854
Identifiability: and Bayesian method, 875; and Nelson Goodman's "gruesome" paradox, 875
Identity: as an illusion(?), 861; sense of, its dependence upon speed of communication, 1212
"Idol of the market place" ("Tyranny of words"), 1313
"Iffff," 1217
I. J. Good, and one person who has read all he has written, 958
Ignorance: explosion of (Lukasiewicz), 666; formal expression of, 398; about our unconscious minds, and praising ourselves for it, 1212; partial, when probability is used, 13
Ignoring of information. *See* Evidence, ignoring of
Ill-conditioned calculation in population estimation, 1250
Illusions, optical, and economy, 796
IM (intelligent machine), 666
Image reconstruction, 1111
Image stabilization, 629
Imaginary alternatives, relevance of, 1357
Imaginary results, device of. *See* Device of Imaginary Results
Imaginary universe or world, 13
Imagination, 397 (34); in dreams, 397 (58); and language, 397 (36)
Immortal consciousness, integrated. *See* Godd
Immortal men, extinction of, by natural selection, 796, 1445, 1451
Immunity, 33
Importance versus urgency, 13
Improper priors, not always justifiable, 1268
Improper theories, 13
Impulse response function, 854
"In principle" as often meaning "impracticable," 1420
"Inaccuracy," 618, 622
Incentives for probability estimation. *See* Fees, fair
Incentives, principle of none, in a grading system for game players, 50
Incidence matrix, 136
Independence, 13; causal, 397 (67, 68); in contingency tables, 13: *see also* Contingency tables; in contingency tables, evidence in the marginal totals, 929, 1199, 1358, 1420; maximization of, 397 (83), 398; of quadratic expressions, 374
Indeterminism. *See* Determinism
Index of a library, size proportional to $n \cdot \log(n)$, 217
Index terms, 397 (43), 398
Indicant space, 700
Indicants in medicine, 636, 700, 755, 798
Indifference, principle of, 13
Induction, scientific, 13, 183, 398, 522, 617, 956, 958, 970, 1160; always used by good mathematicians, 970: *see also* Probability, dynamic; Bayesian arguments for, criticized, 294; as a consistent metaphysical hypothesis, 191; defense of, 191; and exponential-entropy distributions, 793; large expectation of, 191; mechanization of, 183; philosophy of, 1217, 1221; predictive, robustness of, 1420; simplicity of as an argument in its favor, 844; universal, valid only for extremely large samples, 1420; universal and predictive, 1420
Inductive methods, continuum of, history of, 1221
Inductivist, as sometimes less prepared to use induction than his opponents, 890
Inequality: for determinants, 8; Liapounov's, 14
Inference: Bayesian (rational), 398; fiducial, 398: *see also* Fiducial argument
Infinite, "approximately," 13
Infinite Bayes factor, 13
Infinite expectation, 13
Infinite number of hypotheses, 13
Infinite number of parameters, 13
Infinite number of propositions, 13
Infinite population, hypothetical, 13
Infinite probability, 13
Infinite succession of trials, 13
Infinity. *See* Mathematical convenience
Inflexion, points of, and bumps, 13
Information, 854; amount of, concerning

Name Index

Subject Index

M. H. A. Newman once remarked that he reads the index of a book to find out what the book is about. The following index is intended to serve such readers well. The references are to page numbers.

Born in London and educated at Cambridge University, **I. J. Good** worked during World War II on cryptanalysis as a statistician in the British Foreign Office, where he was associated with the prehistory of computers. For many years he held research posts in the British government and in the Universities of Manchester, Princeton, and Oxford, and he is now University Distinguished Professor of Statistics at Virginia Polytechnic Institute and State University. Good has written extensively; his books include *Probability and the Weighing of Evidence* and *The Estimation of Probabilities.* He was the editor of *The Scientist Speculates: An Anthology of Partly Baked Ideas.* Good is also happy to recall that he was runner-up in the West of England Chess Championship in 1958.